APPLIED STATISTICS

APPLIED STATISTICS

Proceedings of the Conference
at Dalhousie University, Halifax,
May 2-4, 1974

Edited by

R. P. GUPTA
Dalhousie University

1975

NORTH-HOLLAND PUBLISHING COMPANY – AMSTERDAM • OXFORD
AMERICAN ELSEVIER PUBLISHING COMPANY, INC. NEW YORK

ISBN North-Holland: 0 7204 2 825 4
ISBN American Elsevier: 0 444 10 772 X

PUBLISHERS:

NORTH-HOLLAND PUBLISHING COMPANY - AMSTERDAM
NORTH-HOLLAND PUBLISHING COMPANY, LTD. - OXFORD

SOLE DISTRIBUTORS FOR THE U.S.A. AND CANADA:

AMERICAN ELSEVIER PUBLISHING COMPANY, INC.
52 VANDERBILT AVENUE
NEW YORK, N.Y. 10017

PRINTED IN THE NETHERLANDS

PREFACE

These proceedings are the outgrowth of a conference on "Applied Statistics" held at Dalhousie University on May 2-4, 1974. The aim of the conference was to bring together statisticians and those who apply statistics in order that each might appreciate and understand the problems of the other. The conference was attended by faculty members and students from various universities of Canada and the United States and also scientists from government and industries. Much emphasis was placed on the applications of statistics, but a session on mathematical statistics was also held. The papers read at the Conference are valuable not only for their content but also for the promise of a future for the dissemination of the results of statistical research. A panel discussion on "The Universities' Response to Increasing Demands for Applied Statisticians from Government and Industries" was also a main part of the conference. The invited talks were given by W.T. Federer (Cornell University), Alvan R. Feinstein (Yale University), A.G. Laurent (Wayne State University), John W. Tukey (Princeton Univ. and Bell Laboratories), and Arnold Zellner (University of Chicago). The proceedings contain the invited addresses, contributed papers and the panel discussion. It is hoped that this publication will stimulate the interaction of statistics with mathematical, physical and social sciences.

R. P. Gupta

ACKNOWLEDGEMENT

I wish to express my appreciation to many persons for their help in organizing the conference. Drs. W.R.S. Sutherland, Chairman of the Mathematics Department and A.J. Tingley, Professor of Mathematics took great personal interest in all phases of organizing the conference. I also take pleasure in thanking my colleagues, especially, Drs. C.A. Field, I.T. Jolliffe, S. Swaminathan and K.L. Weldon for their help and advice.

The Canada Council, Faculty of Graduate Studies, Dalhousie University; Department of Preventive Medicine, Dalhousie University and the Atlantic Provinces Inter-University Committee on the Sciences provided financial support for the conference. They are gratefully acknowledged.

Finally I am grateful to Miss Paula Flemming and Gretchen Smith who typed the manuscript with admirable accuracy.

R. P. Gupta

CONTENTS

CONTENTS

Applied Statistics, R.P. Gupta, (editor)

ALTERNATIVE CALCULATIONS FOR REGRESSION AND ANALYSIS OF VARIANCE PROBLEMS

D. F. Andrews

Abstract

Some inadequacies of least-squares procedures are illustrated by a regression example. Robust procedures for regression and analysis of variance are described. A robust measure for comparing models is described. The motivation for these procedures is to give reasonable answers in the presence of "outliers" and non-Gaussian distributions and to provide clean residuals clearly pointing to inadequacies in the contemplated model.

1. Introduction

Regression and analysis of variance calculations account for a large amount of statistical computing. The common approach to these problems, using least-squares, was used widely by the first half of this century. The application of least-squares in this period was dependent on the existence of efficient computational procedures based on simple sums and sums of squares. In the last 25 years computational potential has increased substantially so that this simplicity is no longer essential. However, at least initially, the procedures developed for desk calculators were implemented on computers. More recently numerical analysts have provided procedures more suitable for computers.

However this development has been on least-squares procedures, procedures used initially because of their computational feasibility. Other rationalizations for their use are based on linearity, unbiasedness, or an assumption about independent Gaussian random variables.

The ease of computation is no longer a factor restricting attention to least-squares. Linearity and unbiasedness are only minor luxuries to be afforded only after small mean square error and other essential properties have been secured. Finally, considerable evidence supports the assumption of not Gaussian but of longer tailed distributions. All of this indicates that some alternatives to least-squares are possible. To see that they are necessary consider the following example.

2. An Example of the Use of Least-Squares.

Early last year Professor Stephen Stigler drew my attention to a paper by Johnson and Raven in the March issue of Science 1973. These authors investigated the number of species on the Galápagos islands. The available data contained observations on the number of species, S , the elevation, E , and the area, A , for each of 23 islands. The question of interest was whether S depended more on E or A .

A decade earlier, Hamilton et al found that E was more important than A . Simple linear (least-squares) regression of S on E yielded $r^2 = 0.5$ while that of S on A yielded $r^2 = 0.4$ suggesting that E was more important. By taking logarithms, Johnson and Raven found that the regression of log S on log A yielded $r^2 = 0.8$ while that of log S on log E yielded only $r^2 = 0.5$ suggesting that A was more important.

A number of difficulties are associated with this problem. A plot of S vs. A is shown in Figure 1.

FIGURE 1

Plot of Number of Species S vs. Area A for 23 Galápagos islands

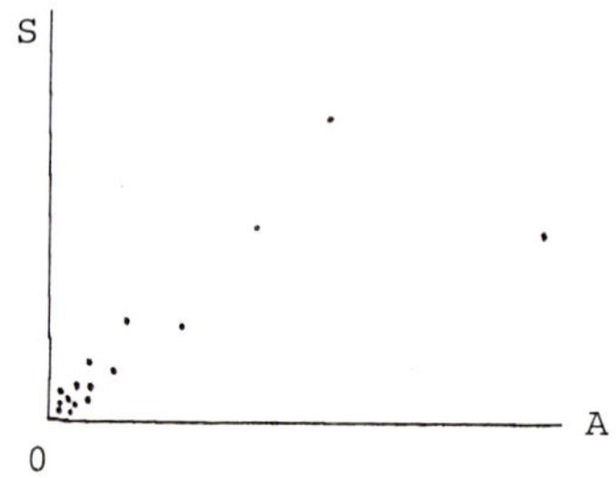

A glance at this plot shows clearly that one large island will dominate the regression producing a fit determined by essentially one island. Certainly transforming with logarithms will help. However the problems are far from solved. Figure 2 is a plot of log E vs. log A . Volcanic action produced islands resembling similar cones. Log E and log A appear linearly related and no regression analysis can separate the effects of E from those of A .

This example illustrates the way regression analysis is used by many research workers. It is not the best usage but it is far from the worst. In this example there is no basis for any Gaussian

FIGURE 2

Plot of log E vs. log A
for 23 Galápagos islands

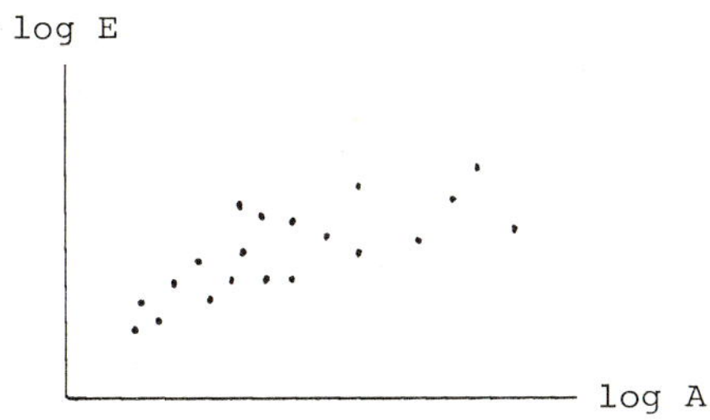

assumption. The interpretation of r^2 was misleading. One large island dominated the first results of Hamilton *et al*.

Alternatives for fitting and measuring the quality of fit are required. These alternatives should be i) not greatly affected by a small number of unusual observations and ii) not greatly affected by departures from the Gaussian distribution. The motivation here is to give good estimates under realistic conditions and, more importantly, to give residuals which can clearly indicate inadequacies in the contemplated model.

One approach has been to test for and delete 'outliers'. This is a very difficult problem. One outlier tends to mask another. Most tests are designed to detect 'outliers' sequentially. Tests for moderate numbers of 'outliers' are only marginally feasible. Another approach has been to test and correct for non-Gaussian distributions. Again this is difficult. Least-squares has poor efficiency under even slight departures from the Gaussian distribution. Yet distribution tests have low power against these alternatives.

The approach used here is to use procedures not greatly affected by either (or both) 'outliers' and non-Gaussian distributions.

3. Robust Regression.

Let the regression model be given by

$$\begin{aligned} y &= x_1\beta_1 + \ldots + x_k\beta_k + \sigma \text{ (error)} \\ &= \underset{\sim}{x}\,\underset{\sim}{\beta} + \sigma \text{ } (\underset{\sim}{\text{error}}) , \end{aligned}$$

and for a set of n observations by

$$\underset{\sim}{y} = \underset{\sim}{X}\,\underset{\sim}{\beta} + e \quad .$$

Any estimate $\underset{\sim}{b}$ yields a set of residuals

$$\underset{\sim}{r} = \underset{\sim}{r}(\underset{\sim}{b}) = \underset{\sim}{y} - \underset{\sim}{X}\underset{\sim}{b} \quad .$$

Maximum likelihood (including least-squares) estimates $\hat{\underset{\sim}{\beta}}$ are solutions of equations of the form

$$\sum_i x_{ij}\phi(r_i(\hat{\beta})/\hat{\sigma}) = 0 \quad , \tag{3.1}$$

for some function ϕ . Least-squares estimates are solutions obtained with $\phi(x) = x$. Huber (1964, 1973) developed the theory for estimates based on

$$\phi(x) = \begin{cases} -k & k < x \\ x & |x| < k \\ k & x < k \end{cases} \quad .$$

A great variety of related location estimates are studied in Andrews, Bickel, Hampel, Huber, Rogers, and Tukey (1972). Andrews (1974) extends one of these to the regression case using the function

$$\phi(x) = \begin{matrix} 0 & |x| > 2.1\,\pi \\ \sin(x/2.1) & |x| \leq 2.1\,\pi \end{matrix} \quad . \tag{3.2}$$

The equations (3.1) may be solved iteratively using weighted least-squares by noting

$$\sum_i x_{ij}\phi(r_i(\underset{\sim}{\hat{\beta}})/\hat{\sigma}) = \sum_i x_{ij}\; w_i\; r_i(\hat{\beta})$$

where

$$w_i = \phi(r_i(\underset{\sim}{\hat{\beta}})/\hat{\sigma})/r_i(\underset{\sim}{\hat{\beta}}) \quad .$$

The efficiency of the estimate does not depend markedly on the choice of the estimate of σ . A robust estimate of scale may be used

$$\hat{\sigma} = \text{median}\{|r_i(\hat{\beta})|\} \quad .$$

A good starting point $\hat{\beta}_0$ may be obtained as the solution of the equations

$$\sum_i \phi(s_{ij}) = 0 \qquad j = 1,\ldots,k$$

where $s_{ij} = x_{ij}r_i(\hat{\beta}_0)$. This 'score' estimate bounds the influence of any one observation. In the example of section 2 this large observation would dominate a Huber type estimate.

4. Robust Analysis of Designed Experiments.

Designed experiments typically have a structure that may be exploited for robust estimation. Tukey, in an unpublished report discusses one procedure based on medians. He discusses a familiar example found in Kempthorne (1962, page 208) and elsewhere. The 32 observations may be grouped into 16 pairs, the elements of each pair differing only in the level of factor A . The effect of A may be estimated by the median of these 16 differences. The effect of A may then be removed (by subtraction) from the data. Similarly the effect of B may be estimated and removed from the reduced data. The remaining main effects may be similarly treated.

The 32 reduced observations may now be divided into 8 groups of 4, the elements of each group differing only in the levels of factors A and B . Each group yields an estimate of the interaction of A and B which may be better estimated by the median of the 8 quantities just calculated. This estimated interaction may be removed from the data and the remaining 2-factor interactions treated in the same way. The 3-factor interactions may be obtained from 4 groups of 8 and so on. At each stage the residuals from the fit may be examined for unusual observations.

This procedure may be applied to any factorial experiment. A single unusual observation may appear as a high order interaction in the standard analysis. It will be difficult to detect from an examination of the residuals from the standard fit. Such an unusual observation will affect the median fit only slightly and will be clearly visible from an examination of residuals.

5. Assessing the Fit.

The example of section 2 illustrates the difficulties associated with comparing fits to data under various transformations. Indeed if y is strictly increasing as a function of x , values of r^2 arbitrarily close to 1 may be obtained by repeated exponentiation of y and x . In such cases it is clear that the value of r^2 says nothing useful about the appropriateness of the model.

Some of the difficulty with r^2 arises because it compares the fit of the contemplated model with the fit of the trivial constant model. Under transformation this latter fit changes.

Consider two sets of data y , z representing the same phenomenon though not necessarily in the same units. Let N be the number of observations on each of y and z . Let $\hat{y}$, $\hat{z}$ be the corresponding fitted values. The first step is to find a step function to fit each of y and z .

Partition the sorted values of $\hat{y}$ into approximately $\sqrt{N}/2$ groups with approximately the same number of values in each. Partition the sorted values of $\hat{z}$ in the same way. Each observation is associated with a cell consisting of all observations with the same group as $\hat{y}$ and $\hat{z}$ respectively. Let m and n denote the medians of the values of y and z in the same cell. These medians define a step function whose complexity depends on the complexity of both models. Each observational unit is now associated with a pair of responses (y,z) , a pair of fitted values $(\hat{y},\hat{z})$, and a pair of values of the step functions (m,n) . The quality of fit may be assessed by comparing the relative closeness of $\hat{y}$ to y and $\hat{z}$ to z . This may be done by assigning ranks.

Let $R(x)$ denote the rank of x among the set of y's , and let $S(x)$ denote the rank of x among the z's . Then define

$$Q(\hat{y}) = \Sigma\{|R(m) - R(y)| - |R(\hat{y}) - R(y)|\}$$

$$Q(\hat{z}) = \Sigma\{|S(n) - S(z)| - |S(\hat{z}) - S(z)|\} \quad .$$

$Q(\hat{y})$ is the number of observations by which $\hat{y}$ is closer to y than the step function m . A good fit should be better than a step function and hence Q should be greater than N . The difference $Q(\hat{y}) - Q(\hat{z})$ measures the relative quality of the two fits.

If y and z are two fits to the same data, i.e. $y = z$ and if $|z - \hat{z}| < |y - \hat{y}|$ then $|R(y) - R(\hat{y})| > |S(z) - S(\hat{z})|$ and $Q(\hat{y}) < Q(\hat{z})$. Thus the Q's measure closeness. The grouping, median and ranking operations are invariant under monotone transformations. Thus if z and $\hat{z}$ are the same monotone transformation of y and $\hat{y}$ respectively $Q(\hat{y}) = Q(\hat{z})$ and the measure is invariant. Calibration of Q under a null model would serve no useful purpose.

The measure Q resembles the rank correlation of y and $\hat{y}$ adjusted by the rank correlation of y and the median fit m . This adjustment depends on the quality of the competing model z . Exact invariance under monotone transformations is rarely required. Related measure, more sensitive to the data, may be developed.

REFERENCES

[1] Andrews, D. F., Bickel, P. J., Hampel, F. R., Huber, P. J., Rogers, W. H. and Tukey, J. W. (1972). Robust Estimates of Location: Survey and Advances. Princeton University Press.

[2] Andrews, D. F. (1974). A robust method for multiple linear regression. Technometrics, to appear.

[3] Hamilton, T. H., Rubinoff, I., Barth, R. H. and Bush, G. L. (1963), Science 142, 1575.

[4] Huber, P. J. (1964). Robust estimation of a location parameter. Ann. Math. Statist. 35, 73-101.

[5] Huber, P. J. (1973). Robust regression: asymptotics conjectures and Monte Carlo. Ann. Statist. 1, 799-821.

[6] Johnson, M. P. and Raven, P. H. (1973). Species number and endemism: the Galapagos Archipelago revisited. Science, 179, 893-5.

[7] Kempthorne, O. (1962). Design and Analysis of Experiments. Wiley, New York.

D. F. Andrew
Department of Mathematics
University of Toronto
Toronto, Ontario M5S 1A1
Canada

Applied Statistics, R.P. Gupta, (editor)

THE MISUNDERSTOOD SPLIT SPOT*

W. T. Federer

Abstract

As exemplified in statistical textbooks and other publications, there is often a misunderstanding of the relationship of statistical design and the role of confounding on the resulting statistical analyses and statistical inferences. Since the split plot design and its multitude of variations is a commonly occurring experiment design, it was selected as a basis to discuss appropriate statistical procedures and analyses. In order to have a common ground for discussion a number of items, including the textbook split plot and split block experiment designs, are defined. Standard textbook analysis of variance (ANOVA) procedures are described. Alternate ANOVA's, alternate experiment designs for whole plots, and alternate experiment designs for split plots are then discussed. Analogies, differences, and variations of the standard split plot and split block experiment designs are discussed. The dependence of split plot and whole plot analyses of variance is considered. In the final section of the paper eight rules to follow in analyzing data from complexly designed investigations are presented, and three algorithms are given for keying-out the degrees of freedom in an ANOVA, for computing sums of squares in an ANOVA, and for determining appropriate error variances.

1. Introduction.

During the course of statistical consulting, of serving as a book reviews editor, in talking with statisticians, in reading the statistical literature, and in participating in panels and discussions on statistical applications, it has become quite apparent that many statisticians do not understand the effect of confounding and of statistical design on statistical analyses and the subsequent statistical inferences. The main source of confusion appears to stem from whether a valid error mean square must be ascertained or whether it should be defined. The more mathematically-oriented individuals tend to define the error variance (and linear model) whereas the more experimentally-minded individuals tend to ascertain which

---

*Partially supported by Public Health Research Grant 5-R01-GM-05900, National Institutes of Health.

experimental components give rise to a valid estimate of error variance. The tendency to define an error variance is predominant in statistical pedagogy.

The misunderstanding of the role of confounding on the subsequent statistical procedures is readily apparent in discussions concerning the split plot and split block designs. The purpose of this paper is to first define a number of quantities in order to be certain that the reader is using the same terminology as the author; then a discussion of the classical or textbook version of the split plot and split-split plot designs is given. Since a single form of the analysis of variance (ANOVA) is universally presented, alternate ANOVA's are discussed in the third section. Also, since only one experiment design is usually presented for whole plot treatments and for split plot treatments, alternate experiment designs are presented in the fourth and fifth sections. The analogies and differences between split plot and split block designs are discussed in section six. In section seven, eight variations of standard split plot and split block designs are presented. The dependence of split plot analyses on whole plot analyses is discussed in section eight, while rules and algorithms for obtaining ANOVA's from complex experiments are given in the last section. Error variances for the various contrasts are indicated in each of the above cases.

Before proceeding further a number of definitions are required. First, a <u>treatment</u> is a single entity of interest to an experimenter. The selection of the set of treatments to be studied or compared in a comparative experiment is the <u>treatment design</u>. The arrangement of the treatments in the experiment is denoted as the <u>experiment design</u>. The smallest unit to which one treatment is applied is called an <u>experimental unit</u>. An <u>observational unit</u> is the smallest unit on which an observation is made; often the observational unit and the experimental unit represent the same unit, but in repeated measurements situations and in cases wherein the experimental unit is composed of several separate entities, the observational unit is smaller than the experimental unit. Confusion between these units can, and has, lead to difficulties in statistical analysis of data.

In split plot designs at least two different experimental units are utilized. Also, in this design, a two-factor factorial is usually involved. Let one of the factors (or set of factors) be denoted by a with p different levels, $a_1, a_2, a_3, \ldots, a_p$ and let the second factor (or set of factors) be denoted by b with q

different levels, $b_1, b_2, b_3, \ldots, b_q$. Denote the whole plot treatments to be the a_i and the split plot treatments to be the b_j, $j=1,2,\ldots,q$. A whole plot treatment experimental unit is the smallest unit to which one whole plot treatment, a_i , is applied. Each whole plot is subdivided into split plot treatment experimental units; this unit is the smallest unit to which one split plot treatment, b_j , is applied. A whole plot treatment has sometimes been called a one-way whole plot treatment, and the split plot treatments have sometimes been denoted as sub-plot treatments. If the split plot treatment is further subdivided and if additional treatments, say c_h, $h=1,2,\ldots,k$, are applied to the subunits of the split plot treatment experimental unit, the c_h are denoted as the split-split (or sub-sub) plot treatments and the smallest unit to which a c_h is applied is defined to be the split-split plot treatment experimental unit. There can be additional splits to extend the definitions and concepts (see Federer [1955], page 294, e.g.). A valid estimate of the error variance for a treatment contrast has been defined by Fisher [1966] as one which contains all sources of variation inherent in the variation among treatment effects except that portion of the variance due specifically to the treatments themselves.

A literature coverage is not envisioned here. If such is desired, the reader is referred to Federer and Balaam [1972], classification E12, for literature on this subject. A discussion of these designs may be found in Cochran and Cox [1957] chapter 7, Federer [1955] chapter X and sections XII-3 to XII-5, XIII-4, XVI-5, and XVI-6, and Kempthorne [1952] chapter 19 and section 24.5. Furthermore, specific literature citations of misuse will not be given as the author wishes to stress the positive aspects of appropriate statistical analyses for experimental data. Nothing is to be gained from pointing out published examples of misuse.

2. The Textbook Split Plot and Split-Split Plot Designs.

The almost universally illustrated example of a split plot design appearing in textbooks is the type illustrated by Yates [1937] wherein the whole plots are arranged in a randomized complete block design and the split plot treatments are randomly allotted to the experimental units within each whole plot. Example 2.1 is an illustration of such a layout for $r=4$ complete blocks, $p=3$ whole plot treatments, and $q=4$ split plot treatments.

Example 2.1.

Block I			Block II			Block III			Block IV		
b_1	b_2	b_3	b_1	b_3	b_0	b_3	b_0	b_1	b_2	b_1	b_0
b_0	b_0	b_1	b_3	b_1	b_1	b_2	b_3	b_0	b_0	b_3	b_3
b_2	b_3	b_0	b_2	b_2	b_3	b_0	b_2	b_3	b_1	b_2	b_1
b_3	b_1	b_2	b_0	b_0	b_2	b_1	b_1	b_2	b_3	b_0	b_2
a_1	a_0	a_2	a_1	a_2	a_0	a_2	a_1	a_0	a_0	a_1	a_2

There are $r=4$ randomizations for the a_i treatments and $rp=4(3)=12$ randomizations for the b_j treatments.

The standard textbook analysis of variance presented has the following form almost without exception:

Source of variation	Degrees of freedom		F-test
	General	Example 2.1	
Total	rpq	48	
Correction for mean	1	1	
<u>Whole plot analysis</u>			
Complete blocks = R	r-1	3	
Among whole plot treatments = A	p-1	2	
Error (a) = R × A	(r-1)(p-1)	6	
<u>Split plot analysis</u>			
Among split plot treatments = B	q-1	3	
A × B	(p-1)(q-1)	6	
R × B : A = error (b)*	p(r-1)(q-1)	27	

Occasionally, the A × B interaction mean square is used as the denominator for an F-test involving the B mean square. This is for the situation wherein the a_i represent a random sample from a population of a_i and inferences are being made to the entire population. For the fixed effects case, some discussion is presented concerning the error variances for comparing two levels of b_j for one a_i level, two a_i levels for one b_j level, and two b_j levels from different a_i levels (see Cochran and Cox [1957], section 7.16,

* See following page footnote.

and Federer [1955], example X-1). We shall be considering the treatments as fixed effects and the blocks as random effects throughout the ensuing discussion.

Another type of textbook example is the one wherein ℓ locations (or laboratories) are involved and a designed experiment, usually a randomized complete block design, is conducted at each location. Here the whole plots are locations (or laboratories), and there is only <u>one</u> replicate of whole plots. For this case the analysis of variance (ANOVA) for r blocks of a randomized complete block design with t treatments is given as:

Source of variation	Degrees of freedom	F-test
Total	$r\ell v$	
Correction for mean	1	
Locations (laboratories) = L	$\ell-1$	
Blocks within locations = R : L	$\ell(r-1)$	
Treatments = T	$(t-1)$	
Treatments by locations = T × L	$(\ell-1)(t-1)$	or
Treatments by blocks within locations = R×T:L	$\ell(r-1)(t-1)$	

The above design is not usually included in discussions of split plot design analyses, but is sometimes treated under repetitions of a design. Also, note that the R : L* sum of squares is not partitioned into an R with $(r-1)$ degress of freedom and an R × L sum of squares with $(r-1)(\ell-1)$ degrees of freedom. This would not be correct since the numbering of replications (blocks) at each location (laboratory) is purely arbitrary, that is replication 1 at location 1 has nothing in common with replication 1 at location 2. Any replication at a location could have been designated as replication 1. Hence, it is meaningless to partition the R : L sum of squares into an R and an R × L sum of squares. This is <u>not</u> a three-factor factorial but is a two-factor factorial (blocks and treatments) nested within a third factor, locations.

If the split plot experimental unit is further subdivided and if an additional set of treatments, say $c_1, c_2, \ldots, c_k$ are applied to

*The notation R : L is used to indicate that block contrasts are nested within locations. The symbol to the left of the colon is nested within those to the right of the colon.

the subdivisions of the split plot treatments, then with proper randomization, a split-split plot experiment design results. The c_h are called the split-split plot treatments. Such a design is illustrated in example 2.2. Obviously the splitting can be continued as long as it is feasible and desirable.

Example 2.2. For the split plot design in example 2.1, suppose that it was possible to further subdivide the split plot experimental unit into two split-split plot experimental units for the two split-split plot treatments c_0 and c_1 as follows:

Block I			Block II			Block III			Block IV		
b_1 c_1/c_0	b_2 c_1/c_0	b_3 c_0/c_1	b_1 c_1/c_0	b_3 c_0/c_1	b_0 c_1/c_0	b_3 c_1/c_0	b_0 c_1/c_0	b_1 c_0/c_1	b_2 c_1/c_0	b_1 c_1/c_0	b_0 c_0/c_1
b_0 c_0/c_1	b_0 c_1/c_0	b_1 c_1/c_0	b_3 c_1/c_0	b_1 c_0/c_1	b_1 c_0/c_1	b_2 c_1/c_0	b_3 c_0/c_1	b_0 c_1/c_0	b_0 c_1/c_0	b_3 c_0/c_1	b_3 c_1/c_0
b_2 c_0/c_1	b_3 c_0/c_1	b_3 c_1/c_0	b_2 c_0/c_1	b_2 c_0/c_1	b_3 c_0/c_1	b_0 c_0/c_1	b_2 c_1/c_0	b_3 c_0/c_1	b_1 c_1/c_0	b_2 c_0/c_1	b_1 c_0/c_1
b_3 c_1/c_0	b_1 c_1/c_0	b_1 c_0/c_1	b_0 c_0/c_1	b_0 c_0/c_1	b_2 c_1/c_0	b_1 c_1/c_0	b_1 c_0/c_1	b_2 c_1/c_0	b_3 c_1/c_0	b_0 c_1/c_0	b_2 c_0/c_1
a_1	a_0	a_2	a_1	a_2	a_0	a_2	a_1	a_0	a_0	a_1	a_2

There are r=4 randomizations for the a_i, rp=12 randomizations on the b_j, and rpq=48 randomizations on the c_h treatments. One a_i treatment is applied to a set of eight split-split plot experimental units and one b_j treatment is applied to two split-split plot experimental units. The textbook ANOVA for the above experiment design is:

Source of variation	Degrees of freedom General	Example	F-test
Total	rpqk	96	
Correction for mean	1	1	
Whole plot analysis			
Complete blocks = R	r-1	3	
Among a_i = A	p-1	2	
R × A = error (a)	(r-1)(p-1)	6	
Split plot analysis			
Among b_j = B	q-1	3	
A × B	(p-1)(q-1)	6	
R × B : A = error (b)	p(r-1)(q-1)	27	
Split-split plot analysis			
Among c_h = C	k-1	1	
A × C	(p-1)(k-1)	2	
B × C	(q-1)(k-1)	3	
A × B × C	(p-1)(q-1)(k-1)	6	
R × C : A and B = error (c)	pq(r-1)(k-1)	36	

For situations wherein the pq treatments in a split plot design do not form a factorial arrangement, the first set of analyses in the next section may be appropriate. Some discussion of this situation is given by Federer [1955], section XIII-4, and literature citations on the general topic are presented.

3. Alternate ANOVA for Textbook Split Plot Design

The analysis of variance involving locations suggests an alternate ANOVA for the textbook split plot design. That is, one may consider the a_i whole plot treatments as locations and consider the randomized complete block design for the b_j treatments within each whole plot as follows:

Source of variation	df	a_1 ss	a_2 ss	a_3 ss	...	a_p ss	Sum
Total	rq	T_1	T_2	T_3	...	T_p	ΣT_i
Correction for mean	1	C_1	C_2	C_3	...	C_p	ΣC_i
Blocks	(r-1)	R_1	R_2	R_3	...	R_p	ΣR_1
B treatments	(q-1)	B_1	B_2	B_3	...	B_p	ΣB_i
R × B (blocks × treatments)	(r-1)(q-1)	E_1	E_2	E_3	...	E_p	ΣE_i

Combining these single ANOVA's, we obtain the textbook split plot sums of squares as follows:

Source of variation	df	Sum of squares
Total	rpq	ΣT_i
Correction for mean	1	compute = CT
Whole plot analyses		
Blocks within a_i = R : A	p(r-1)	ΣR_i
Blocks = R	r-1	compute = R
Blocks × A = error (a)	(r-1)(p-1)	ΣR_i - R
A	(p-1)	ΣC_i - CT
Split plot analyses		
B within A = B : A	p(q-1)	ΣB_i
B	q-1	compute = B
A × B	(p-1)(q-1)	ΣB_i - B
R × B : A = error (b)	p(r-1)(q-1)	ΣE_i

The above form of an analysis of variance for split plot experiments is particularly useful for computing single degree of freedom contrasts for the B treatment sums of squares. For example, suppose that the sum of squares for B is partitioned into k sets of contrasts such that each set is associated with n_g degrees of freedom such that $\sum_{g=1}^{k} n_g = q-1$ and with a sum of squares designated as B_{ig} . Then, the $\sum_{i=1}^{p} B_{ig}$ sum of squares is associated with pn_g degrees of freedom; the $B_{.g}$ contrast over all levels of a is subtracted from $\sum_{i=1}^{p} B_{ig}$ to obtain the interaction sum of squares with $(p-1)n_g$ degrees of freedom.

The above form is also useful for partitioning the E_i sum of squares into various parts. One such partitioning would be to compute Tukey's one degree of freedom for nonadditivity sum of squares for each whole plot, say N_i, and to sum these over all whole plots $\sum_{i=1}^{p} N_i$, to obtain a sum of squares with p degrees of freedom. If $N_i = (\sum_h e_{ih}\, \hat{e}_{ih})^2 / \sum_h \hat{e}_{ih}^2$, then $(\sum\sum_{ih} e_{ih}\, \hat{e}_{ih})^2 / \sum\sum_{ih} \hat{e}_{ih}^2 = N$ would be the nonadditivity sum of squares over all whole plots a_i and $\Sigma N_i - N$ would be the corresponding interaction sum of squares with p-1 degrees of freedom. Similarly other single degree of freedom or a set of degree of freedom contrasts could be obtained and the residuals could be summed for possible use as an error variance.

Although the above partitioning and combining may appear obvious after once being pointed out, one wonders why textbook writers do not do this. On the other hand, some authors feel inclined to partition the error (b) sum of squares into $R \times B$ and $R \times A \times B$ sums of squares. Although this computation can always be performed, it may be meaningless and incorrect. These sums of squares are confounded as may be shown from the second ANOVA described in section 2. Note that the replication numbering in each a_i is purely arbitrary and hence the $R \times B$ sum of squares is confounded with the $R \times A \times B$ sum of squares resulting in an $R \times B : A$ sum of squares for the error (b) sum of squares.

The question arises as to when it might be meaningful to compute an $R \times B$ and an $R \times A \times B$ sum of squares for a split plot experiment. If the blocks are a random sample from a population of blocks and if the whole plots are a random sample from whole plots, then it would not be meaningful to partition. However, if the blocks represent large overriding effects (e.g. locations or laboratories) which interact with levels of the factor B, then the $R \times B$ interaction should be partitioned out of the error (b) sum of squares. Hence, it can be seen that the population and sampling structure are the important items to be considered in the analysis of experimental data. The statistical analyst should not be misguided by the numbering system or by an "apparent" analogy to a three-factor factorial. Over-emphasis on the computing aspects of sums of squares has led students and practitioners of statistical methodology into the trap described above.

4. Alternate Experiment Designs for the Whole Plots

The whole plot treatments may be arranged in <u>any appropriate</u> experiment design necessary to control the heterogeneity in the experiment. For example, a completely randomized design, a latin square design, a Youden design, an incomplete block design, a lattice square design, or any one of a number of experiment designs may be utilized. Example 4.1 illustrates the partitioning of the degrees of freedom in the ANOVA for k^2 whole plots arranged in a balanced lattice square design with the split plot treatments being randomly allocated to the split plot experimental units within each whole plot.

Example 4.1. The following is the ANOVA for $k^2 = p$ whole plot treatments arranged in a balanced lattice square design with $k+1=r$ complete blocks. The q split plot treatments are randomly allocated to the experimental units within each whole plot.

Source of variation	Degrees of freedom	
	k	k=3
Total	$q(k^3+k^2)$	36q
Correction for the mean	1	1
Whole plot analysis		
Blocks = R	k	3
A (ignoring rows and columns)	k^2-1	8
Rows (eliminating treatments; ignoring columns)	k^2-1	8
Columns (eliminating treatments and rows)	k^2-1	8
Intrarow-column error	$(k^2-1)(k-2)$	8
Split plot analysis		
B	q-1	q-1
A × B	$(k^2-1)(q-1)$	8(q-1)
R × B : A	$k^3(q-1)$	27(q-1)

Note that the split plot analysis is unchanged by the experiment design utilized for whole plots, and that this analysis is simply a partitioning of the within whole plot variation which is orthogonal to the among whole plot variation.

5. Alternate Experiment Designs for the Split Plots

Any appropriate experiment design may be utilized for the split plot treatments. Two types of design need to be considered. First consider experiment designs for the split plot treatments within each whole plot as described in example 5.1.

Example 5.1. Suppose that the $p=3$ whole plot treatments (a_1,a_2,a_3) are in a randomized complete block design composed of $r=5$ replicates. Furthermore, suppose that the $q=4$ split plot treatments $(b_1,b_2,b_3,$ and $b_4)$ are arranged in a 4-row (order) by 5-column Youden design. One such arrangement, where the order within each whole plot is taken into account is:

Block I			Block II			Block III			Block IV			Block V		
a_2	a_1	a_3	a_1	a_3	a_2	a_2	a_3	a_1	a_2	a_1	a_3	a_3	a_2	a_1
b_2	b_3	b_1	b_4	b_1	b_3	b_4	b_4	b_1	b_1	b_1	b_2	b_3	b_1	b_2
b_4	b_4	b_3	b_1	b_2	b_2	b_1	b_2	b_2	b_4	b_4	b_4	b_1	b_3	b_3
b_3	b_2	b_4	b_3	b_3	b_1	b_2	b_1	b_4	b_3	b_3	b_3	b_2	b_4	b_1
b_1	b_1	b_2	b_2	b_4	b_4	b_3	b_3	b_3	b_2	b_2	b_1	b_4	b_2	b_4

The ANOVA for each whole plot a_i is:

Source	df	a_1 ss	a_2 ss	a_3 ss	Sum of ss
Total	20	T_1	T_2	T_3	ΣT_i
Correction for mean	1	A_1	A_2	A_3	ΣA_i
Columns	4	C_1	C_2	C_3	ΣC_i
Orders (ignoring b_j)	3	O_1	O_2	O_3	ΣO_i
Treatments (eliminating orders)	3	B_1	B_2	B_3	ΣB_i
Residual	9	E_1	E_2	E_3	ΣE_i

Note that a 4-row (order) by 5-column Youden design was constructed for each a_i, and hence the above ANOVA's.

A combined ANOVA for the above is:

Source	df		ss
Total	60		ΣT_i
Correction for mean	1		compute = CT
<u>Whole plot analysis</u>			
Whole plot treatments = A	2		ΣA_i - CT
Columns: A	12		ΣC_i
Complete blocks = R		4	compute = C
R × A		8	ΣC_i - C
<u>Split plot analysis</u>			
Treatments (eliminating orders):A	9		ΣB_i
B		3	compute = B
A × B		6	ΣB_i - B
Orders (ignoring b_j): A	9		ΣO_i
Residual: A	27		ΣE_i

The second type of design for split plot treatments completes the design within each complete block for the whole plot treatments. Such a design is illustrated in example 5.2. The orthogonality aspects are lost in this type of design. Sometimes the statistical analysis becomes cumbersome. The solution of a set of simultaneous equations was avoided for the analysis given in example 5.2.

Example 5.2. The four whole plot treatments A, B, C, and D were laid out in a 4 × 4 latin square design. The four split plot treatments a, b, c, and d were laid out in a 4 × 4 latin square design within each column of the whole plot latin square design. The plan follows:

A cbda	C abcd	B dacb	D abdc
B bcad	A dabc	D bcad	C bdca
D adbc	B cdab	C cbda	A dcab
C dacb	D bcda	A adbc	B cabd

2^4 factorial notation $(e_r f_s g_u h_v)$

Aa = 0000	Ca = 0100
Ab = 0010	Cb = 0110
Ac = 0001	Cc = 0101
Ad = 0011	Cd = 0111
Ba = 1000	Da = 1100
Bb = 1010	Db = 1110
Bc = 1001	Dc = 1101
Bd = 1011	Dd = 1111

Note that for the 16 columns the following confounding scheme results:

1	2	3	4	5	6	7	8
$(EH)_1$ $(EFG)_0$ $(FGH)_1$	$(EH)_0$ $(EFG)_1$ $(FGH)_1$	$(EH)_1$ $(EFG)_1$ $(FGH)_0$	$(EH)_0$ $(EFG)_0$ $(FGH)_0$	$(FH)_1$ $(EFG)_1$ $(EGH)_0$	$(EH)_0$ $(EFG)_0$ $(FGH)_0$	$(FH)_0$ $(EFG)_1$ $(EGH)_1$	$(EH)_1$ $(EFG)_0$ $(FGH)_1$

9	10	11	12	13	14	15	16
$(EG)_0$ $(EFH)_0$ $(FGH)_0$	$(EG)_1$ $(EFH)_1$ $(FGH)_0$	$(EG)_1$ $(EFH)_0$ $(FGH)_1$	$(EG)_0$ $(EFH)_1$ $(FGH)_1$	$(EG)_1$ $(FH)_1$ $(EFGH)_0$	$(EH)_1$ $(FG)_0$ $(EFGH)_1$	$(EG)_0$ $(FH)_0$ $(EFGH)_0$	$(EH)_0$ $(FG)_1$ $(EFGH)_1$

The ANOVA is:

Source	df		SS
Total	64		
Correction for mean	1		
Rows	3		
Columns	3		
Whole plot treatments	3		Sum of following three ss
E = - A - C + B + D		1	
F = - A - B + C + D		1	
EF = A + D - B - C		1	
Error for whole plot treatments	6		
Split plot analysis			
Orders within columns (ignoring interactions)	12		
Split plot treatments	3		Sum of following three ss
G = b + d - a - c		1	From all 16 columns
H = c + d - a - b		1	" " " "
GH = a + d - b - c		1	" " " "
W.P. × S.P. (eliminating orders within columns)	9		Sum of following nine ss
EG' (5/8 information)		1	Compute from columns 1-8,14,16
EH' (1/2 ")		1	" " " 5,7,9-12,13,15
EGH' (7/8 ")		1	" " " 1-4,6,8-16
FG' (7/8 ")		1	" " " 1-13,15
FH' (3/4 ")		1	" " " 1-4,6,8-12,14,16
FGH' (3/8 ")		1	" " " 5,7,13-16
EFG' (1/2 ")		1	" " " 9-16
EFH' (3/4 ")		1	" " " 1-8,13-16
EFGH' (3/4 ")		1	" " " 1-12
Error for split plot treatments	24		Error for above 12 df

By combining knowledge of analyses for factorial treatment designs and confounding concepts, it is possible to obtain the complete analysis of variance and solutions for parameters without solving a set of simultaneous equations.

Note that if the 4×4 latin squares had been within <u>each plot treatment</u> rather than within <u>each column</u>, there would have been no confounding of interaction components with orders within columns and full information would have been obtained on each of the interactions contrasts instead of only partial information. Also, note that there is some interaction information in the orders within columns (ignoring interaction contrasts) sum of squares which could be recovered if desired. These contrasts would involve the comparison of the levels of effects in the columns in which the effect is confounded. Three degrees of freedom would remain for an error sum of squares for these contrasts. The two estimates of the effects could be combined in the usual manner for combining estimates with different variances. Also, one could recover interblock (intercolumn) information in the usual manner for a pseudo-factorial (see, e.g., Federer [1955] and Kempthorne [1952]). The contrast of a level of an effect for columns in which the effect is confounded with the same level in the columns in which the effect is unconfounded, would produce a sum of squares with nine degrees of freedom. These nine together with the three described above would produce a sum of squares with 12 degrees of freedom, which would be free of treatment effects. The resulting mean square could be used to obtain an estimate of the interblock variance $\sigma_\varepsilon^2 + 4\sigma_\gamma^2$. The expectation of the above mean square would be $\sigma_\varepsilon^2 + 3\sigma_\gamma^2$ and is obtained directly from the theory of pseudo-factorial analyses. Then, the usual estimate of treatments means with recovery of interblock information, may be obtained.

6. Split Plot-Split Block Analogies and Differences

In most split block experiment designs discussed in the literature, two-way whole plots are present in each block and a two-factor factorial represents the treatment design. If a randomized complete block design is constructed for the factor a , if one constructs a second set of whole plots for factor b such that any level of factor b contains all levels of factor a , and if there are r randomizations for levels of both factors a and b , a standard textbook example of a split block experiment design results. One such design is presented in example 6.1. Note that either factor could be a factorial treatment design. Also, note that r complete blocks are required for the standard split block design.

Example 6.1. Suppose that there are three levels of factor a, a_1, a_2, and a_3, and that factor b is a 2×2 factorial consisting of the four treatments $b_1 = c_0d_0$, $b_2 = c_1d_0$, $b_3 = c_0d_1$, and $b_4 = c_1d_1$; further suppose that $r=3$ complete blocks are utilized. The nonrandomized plan is:

Block I

	a_0	a_1	a_2
b_1			
b_2			
b_3			
b_4			

Block II

	a_0	a_1	a_2
b_1			
b_2			
b_3			
b_4			

Block III

	a_0	a_1	a_2
b_1			
b_2			
b_3			
b_4			

In order to obtain the split block experiment design randomly allot the three a_i levels to the "columns" in each block and then randomly allot the four $b_j = c_g d_h$ levels to the "rows" in each block. There are three randomizations for the a_i levels and three for the b_j levels. If the experimenter had desired to do so, the column orders within complete blocks could have been used to form a 3×3 latin square for the a_i levels; also, if variation due to row order needs to be controlled, a 4×3 Youden design could have been used. Such a nonrandomized design follows:

Block I

	a_0	a_1	a_2
b_0			
b_1			
b_2			
b_3			

Block II

	a_1	a_2	a_0
b_3			
b_0			
b_1			
b_2			

Block III

	a_2	a_0	a_1
b_2			
b_3			
b_0			
b_1			

This experiment design is not of the standard textbook type but it is a split block or two-way whole plot experiment design. One of the twelve 3×3 latin square arrangements is randomly selected for the a_i treatments and then the b_j are randomly allotted to the "rows" in block I, then to block II except that no b_j is allowed to appear twice in the same row, and finally to block III except that no b_j is allowed to appear more than once in a row.

The ANOVA for the first design above is:

Source	df	df (general)	F-test
Total	36	rpq	
Correction for mean	1	1	
Blocks = R	2	r-1	
Factor a = A	2	p-1	
A × R = error (a)	4	(p-1)(r-1)	
Factor b = B	3	q-1	
B × R = error (b)	6	(q-1)(r-1)	
A × B	6	(p-1)(q-1)	
A × B × R = error (ab)	12	(p-1)(q-1)(r-1)	

There are three error terms introduced because of the confounding scheme utilized. In most experimental situations it would not be appropriate to consider the A × R, the B × R, and the A × B × R mean squares to be estimates of the same error variance σ^2. The reason for this is that the experimental unit size differs and hence can generally be expected to cause unequal variances. This fact is sometimes ignored by statisticians and experimenters alike.

The ANOVA for the second experiment treatment design given above follows:

Source	df	df (general)	F-test
Total	36	qp^2	
Correction for mean	1	1	
Blocks = R	2	p-1	
A	2	p-1	
Column order in blocks	2	p-1	
Error (a)	2	(p-1)(p-2)	
B (eliminating row order)	3	q-1	
Row order (ignoring b_j)	3	q-1	
Error (b)	3	(q-1)(p-2)	
A × B	6	(p-1)(q-1)	
A × B × R = error (ab)	12	$(p-1)^2(q-1)$	

In the above example, it may be noted that the particular analysis of variance required by the experiment design for factor A does not affect the ANOVA for factor B and for the A × B interaction. Also, it should be noted that different error terms are required for factor a, for factor b, and for the two-factor interaction. The number of randomizations on levels of a and on levels of b are the same as the number of complete blocks in the standard textbook split block design.

In actual experimentation involving biological organisms, a common mistake is to consider the split block design as a split plot design. For some experimental situations involving animals or humans, it is sometimes quite difficult to ascertain if there are whole plots and split plots or two-way whole plots. With plant experiments, the difference is usually apparent and the number of randomizations is easily determined. If the experimental units for the a_i and b_j differ in size and/or if the number of randomizations for the a_i is r and for the b_j is rp , then a split plot design is indicated. The experimental unit size can be different and the number of randomizations can differ but the design may not be a split plot design (see example 7.3). Considerable study of the method of experimentation may be involved before one can determine an appropriate ANOVA for a given experiment. Perhaps the most common mistake is to confuse the split block, split plot, and three-factor factorial designs in the resulting ANOVA for an experiment. Several examples in textbooks can be readily cited, when this is the case.

7. Variations and Deviations of Standard Split Plot and Split Block Designs

Eight different deviations or variations from standard textbook split plot and split block designs are discussed in examples 7.1 to 7.8. These examples have been encountered in practice.

Example 7.1. A frequent deviation from a split plot design and one which frequently appears to escape the attention of the statistician and experimenter alike is the experiment wherein only one replicate of the standard split plot design is used. The following experiment was recently encountered where the numbers in the table refer to number of plants used for each treatment:

	Intensity of light		
Plant Types	a_1	a_2	a_3
b_1 = tomato = T	8	8	8
b_2 = pigweed = P	8	8	8
b_3 = T + P	8	8	8

The plant types were grown in three different growth chambers with the light intensity in each growth chamber being one of the a_i levels Furthermore, the eight plants of one plant type, b_2, were grown together in one greenhouse flat. Thus, for each a_i, <u>one</u> replicate of a randomized complete block for the b_j treatments was used. This latter situation is a frequently encountered experiment design where the correct ANOVA should be:

Source	df	df (general)	F-test
Total	24	kq	
Correction for mean	1	1	
Blocks = R	0	0	
Factor b = B	2	q-1	
R × B	0	0	
Plants: R and B	21	q(k-1)	

The last mean square above is often utilized as the error mean square for the b_j levels. There are many situations for which this is an erroneous procedure, one being when competition between items in each experimental unit occurs. In some instances, this mean square could be used to test for an R × B interaction. For the single replicate of a split plot design, the ANOVA for this example is:

Source	df	df (general)	F-test
Total	72	pqk	
Correction for mean	1	1	
Light intensities = A	2	p-1	
Repetitions of a_i = R	0	0	
R × A = error (a)	0	0	
Plant types = B	2	q-1	
A × B	4	(p-1)(q-1)	
R × B : A = error (b)	0	0	
Plants : B and A	63	pq(k-1)	

The experimenter had planned to use, and he probably did so despite advice to the contrary, the plants : B : A mean square as the error mean square in F-tests for A, B, and A × B mean squares. This would be an inappropriate procedure for this experiment.

Example 7.2. A not uncommon deviation of the split plot design is to use a randomized complete block design for the a_i levels, then to lay out the b_j levels as a second way whole plot or split block but to use the same systematic layout in each block, and then to compute an ANOVA and F-tests as if the b_j were split plot treatments within the a_i whole plots. Even though more information was desired on the b_j levels, no randomization was performed because it was simpler experimentally to perform the experiment, because the experimenter forgot to randomize, or because he did not realize the need for randomization. One such example is given by Federer [1955], page 296, and will not be repeated here.

Example 7.3. Suppose that the following schematic plan was the basis for an experiment design involving $p=3$ levels of factor a and $q=4$ levels of factor b:

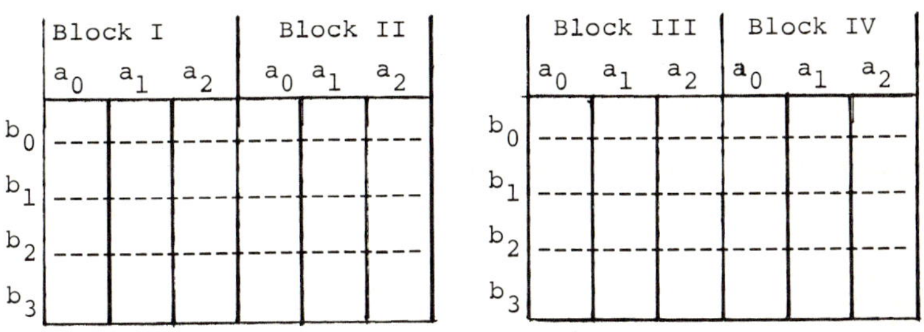

Four randomizations are used for levels of factor a but only two randomizations are used for levels of factor b. Thus the a_i levels are in a randomized complete block design with $r=4$ replicates and the b_j levels are in a randomized complete block design with $r=2$ replicates. If the b_j levels had been randomized within each block, a standard split block experiment design would have resulted. Note this fact in the following analysis. The ANOVA and F-tests for the above design are given below:

Source	df	F-tests
Total	48	
Correction for mean	1	
Blocks for A = RA	3	
Blocks for B = RB	1	
Block I vs. Block II = C	1	
Block III vs. Block IV = D	1	
A	2	⌉
RA × A	6	⌋
B	3	⌉
RB × B	3	⌋
B × C	3	not usable
B × D	3	not usable
A × B	6	⌉
A × B × RA	18	⌋

The experimental unit size for the a_i is one-third of a block whereas the experimental unit size for the b_j is one-fourth of two blocks.

Example 7.4. Another variation of the split block and split plot designs is given on the following page. Four randomizations of treatments I, II, and III were used to form a randomized complete block design for the p_i levels. Likewise, four randomizations for stages 1, 2, and 3 were used to form a randomized complete block design for the s_j levels. Thus, the p_i and s_j levels are arranged in a standard split block design. Then, the t_g levels were designed as split plot treatments within the s_j whole plots and the levels of t_g were randomly allotted to the seven experimental units within each s_j whole plot. Therefore, the s_i and t_g levels are in a standard split plot design but the s_j and t_g levels are split block treatments to the p_i levels. The produce from each combination was subdivided into three parts -- grass, legume, and weeds. Consequently, the analysis of variance and the associated F-tests are:

Source of variation	df	F-test
Total	3×3×7×4×3=756	
Correction for mean	1	
Complete blocks = R	3	
Preconditioning = P	2	
R × P = error (p)	6	
Stages = S	2	
R × S = error (s)	6	
P × S	4	
R × P × S = error (ps)	12	
Treatments = T	6	
T × S	12	
R × T : S = error (t)	54	
T × P	12	
T × P × R = error (tp)	36	
T × P × S	24	
T × P × S × R = error (tps)	72	
Parts (grass, alfalfa, weeds) = C	2	
C × P	4	
C × S	4	
C × P × S	8	
C × T	12	
C × P × T	24	
C × S × T	24	
C × P × T × S	48	
C × R within all others = error (c)	63(2)(3) = 378	

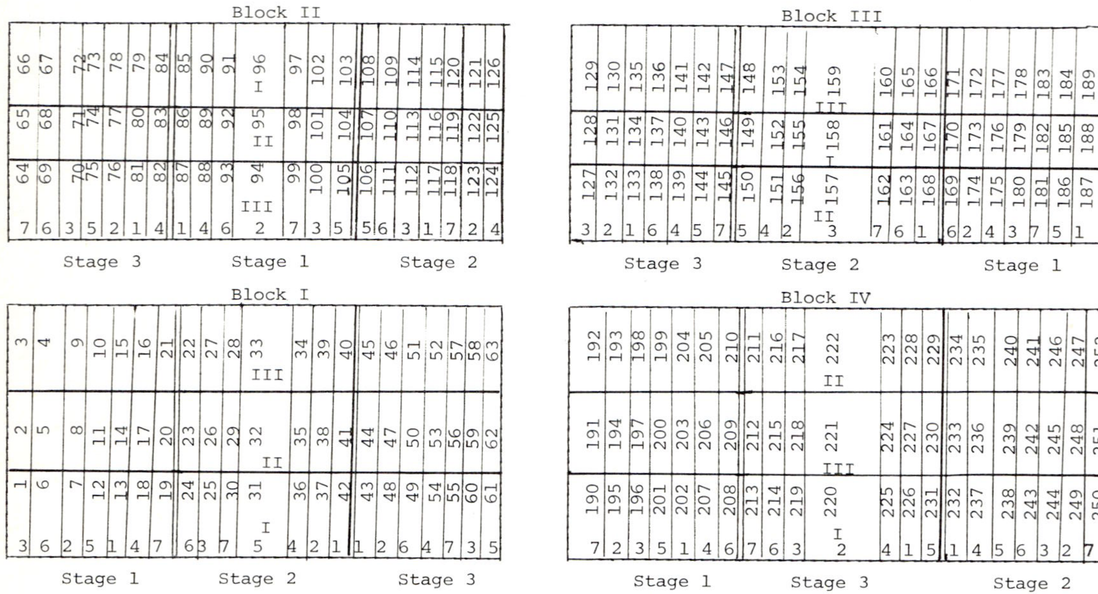

<u>Preconditioning of Soil</u>

p_1 = I = 1/16 lb./A.2-4-D

p_2 = II = 2/16 lb./A.2-4-D

p_3 = III = None

<u>Stage of growth</u>

s_1 = stage 1

s_2 = stage 2

s_3 = stage 3

<u>Treatment of young plants</u>

1 = 2-4-D, 1/4 lb./A. = t_1

2 = 2-4-D, 1/2 lb./A. = t_2

3 = 2-4-D, 3/4 lb./A. = t_3

4 = MCP, 1/4 lb./A. = t_4

5 = MCP, 1/2 lb./A. = t_5

6 = MCP, 3/4 lb./A. = t_6

7 = Check = t_7

Crop: 1st year hay (Alfalfa)

There are seven different error mean squares in the above ANOVA. There is some validity for pooling the $T \times P \times R$ and $T \times P \times S \times R$ mean squares for F-tests of the $T \times P$ and $T \times P \times S$ mean squares. Even with this pooling six different error mean squares result.

The parts of the hay (grass, weeds, alfalfa) could have been designated as X_1, X_2, and X_3 and a multivariate ANOVA computed on the lines in the ANOVA above. Since the interaction terms with C are of importance to agronomists, the above univariate ANOVA is considered satisfactory. Also, before conducting a multivariate ANOVA on data of this nature the reader is referred to Finney [1956].

Example 7.5. As was pointed out in section 5, many experiment designs for the split plot treatments are possible. Some of these have been considered by Kempthorne [1952], Chapter 24, Federer [1955], sections XII-3 to XII-5, and Raktoe [1967]. For example, it would be possible to arrange the whole plots in a randomized complete block design (or some other) and then to arrange the split plot treatments in a lattice square or lattice rectangle design (see Yates [1940], Na Nagara [1957], Federer and Raktoe [1965, 1966], and the above references) within each whole plot treatment a_i . The conditions of the experiment and the need to control the heterogeneity determine the appropriate experiment design for any given situation.

Example 7.6. Another variation of the split plot design is the following example:

Columns

Rows	1	2	3	4	5	6
1	a_0 (b_1,b_2)	a_0 (b_1,b_3)	a_1 (b_3,b_4)	a_1 (b_2,b_4)	a_2 (b_1,b_4)	a_2 (b_2,b_3)
2	a_1 (b_1,b_4)	a_1 (b_2,b_3)	a_2 (b_1,b_2)	a_2 (b_1,b_3)	a_0 (b_2,b_4)	a_0 (b_3,b_4)
3	a_2 (b_3,b_4)	a_2 (b_2,b_4)	a_0 (b_1,b_4)	a_0 (b_2,b_3)	a_1 (b_1,b_3)	a_1 (b_1,b_2)

Within each a_i the $b_j=4$ treatments are arranged in a balanced incomplete block design in blocks of size $k=2$. The a_i treatments are orthogonal to rows and columns, the b_j treatments are orthogonal to the a_i treatments, the b_j treatments are orthogonal to rows but not to columns, and the $A \times B$ interaction is not orthogonal to columns. The ANOVA for each a_i is:

Source of variation	df a_0	df a_1	df a_2
Total	12	12	12
Correction for mean	1	1	1
Blocks (ignoring b_j treatments)	5	5	5
b_j treatments in a_i	3	3	3
Intrablock error	3	3	3

The combined ANOVA would have the following form:

Source of variation	df		F-test
Total	36		
Correction for mean : A	3		
Correction for overall mean		1	
A		2	
Blocks (ignoring b_j) : A	15		
Rows		2	
Columns (ignoring B and $A \times B$)		5	
Remainder (ignoring B and $A \times B$)		8	
B : A (eliminating columns)	9		
B (eliminating columns)		3	
$A \times B$ (eliminating columns)		6	
Intrablock error : A	9		
Blocks (eliminating b_j) : A	15		
Rows		2	
Columns (eliminating B and $A \times B$)		5	
Remainder (eliminating B and $A \times B$)		8	

In the above example, only a portion of the split plot treatments were included in each whole plot treatment experimental unit. This might be called an incomplete split block design as opposed to those considered to this point. These latter could be called complete split plot designs if a name is needed.

Example 7.7. The lattice square designs and the lattice rectangle designs mentioned in example 7.5 could in themselves be called a type of split block design where the rows and the columns represent the two-way whole plots. Even though this analogy could be made and note should be made of it, this is probably carrying the idea further than warranted and pedagogically desirable. If the k^2 treatments in the lattice square (or lattice rectangle) represent a $k \times k$ factorial, then the analogy is not so far-fetched.

Example 7.8. Suppose that one is interested in comparing six diets d_1, d_2, d_3, d_4, d_5, and d_6 and that five animals are to be used for each diet. Thirty animals are required and five are randomly allotted to each treatment. The animals are treated alike except for diet differences. Hence, the variation among animals within diets is considered to be an estimate of the error variance for comparing diets. Now suppose that the weights are obtained at seven different times. One form of analysis for these data is:

	df for weights at time t_i						
Source of variation	t_1	t_2	t_3	t_4	t_5	t_6	t_7
Total	30	30	30	30	30	30	30
Correction for mean	1	1	1	1	1	1	1
Diets	5	5	5	5	5	5	5
Animals within diets	24	24	24	24	24	24	24

Instead of actual weights one might consider weights adjusted for initial weights by covariance and/or might use weight gains in each time period in the above ANOVA's. Also, a similar ANOVA on total weight gain could be constructed.

In pooling the above ANOVA's, experimenters and statisticians sometimes consider the times to be split plot treatments, probably because the data are recorded in a manner for split plot treatments. This would be an erroneous procedure. Note that time periods are unreplicated and that there is only one arrangement of the time periods. Apparently the same kind of reasoning prevails here as for example 7.1. One particular pooled ANOVA for the above experiment would be:

Source of variation	df		F-test
Total	210		
Correction for mean	1		
Among t_j = T	6		
Among d_i : T = D : T	35		
Among d_i = D		5	
D × T		30	
Animals : D and T	7(24)		
Animals : D = A : D		24	
A : D × T		6(24)	

Alternatively, a multivariate ANOVA could have been constructed with the seven weights as multivariates. Other analyses for correlated measurements are possible.

In experiments involving time periods it is important to think clearly and rigorously about the entire experimental procedure. It is necessary to distinguish between <u>calendar time</u> (e.g. May 15) and <u>biological time</u> (number of days to first fruit set, first fruit ripe, etc.). The reader is referred to an experiment described by Snedecor and Cochran [1967], section 12.12, where cutting dates are replicated and to an experiment described by Federer [1955], section X-5.3, where the cutting dates are not replicated. At first glance these experiments appear to be similar but further study indicates the latter experiment is similar to the diets experiment above.

The use of judges, methods, and other similar factors used in repeated measurements studies cause similar difficulties in the statistical analysis of data from experiments. Care must be taken to ascertain the nature of the confounding, the nature of the replication, and the nature of the randomizations utilized before an appropriate ANOVA for the data can be obtained.

8. Dependence of Split Plot and Whole Plot ANOVA's

From the preceding discussion and examples, we can now state the following:

(i) If the q split plot treatments are randomly allotted to the q experimental units within each whole plot, the experiment design for whole plots <u>does not</u> affect the split plot analysis.

(ii) If a complete block experiment design for split plot treatments is used within <u>each</u> whole plot treatment, the form of the split plot analysis is unaffected by the statistical design for whole plot treatments.

(iii) In a standard split block or two-way whole plot experiment design, the analysis of variance for one factor and for the two-factor interaction is unaffected by the experiment design utilized for the second factor.

These facts become apparent from a study of the results to date and from the partitioning of the sums of squares in the ANOVA. If the split plot treatments are nested within the factor a and the blocking, then the within whole plot sum of squares is independent of the among whole plot sum of squares and the partitioning in each part leaves the other unaffected. When nesting does not occur in one of the categories (see, e.g., example 5.2), then the split plot and whole plot analyses are not independent.

9. Rules and Algorithms for Obtaining an ANOVA for a Complex Experiment Design

Eight rules and three algorithms are presented for obtaining an appropriate partitioning of degrees of freedom in the ANOVA and appropriate error variances for interval estimation and F-tests.

In the course of statistical consulting it becomes apparent that the experimenter usually does not know what type of design he has nor what type of confounding of effects is present in the experiment. Perhaps the only type of consulting the author receives is for completely and partially confounded experiments and surveys, but in nearly every case there is no simple textbook answer. Each investigation and its related statistical design must be approached as a unique situation and not one that appears on page X of textbook Y. It *may* be like that one on page X but more often than not there is something in the experiment for which no close analogy to a textbook example can be made. This leads to Rule I.

Rule I: Make no assumptions about the form of the statistical design; always determine the exact experimental procedure, not the stated one.

Quite often the investigator states that his statistical design was D when in fact it was X . One should always have the consultee describe the investigation in minute detail, and then one may come to a conclusion as to the statistical design.

Once one thinks he knows the statistical design, it is then possible to key-out the degrees of freedom for the ANOVA. However in doing this Rules II, III, IV, and V have been found essential.

Rule II: Determine the experimental unit for levels of each category (factor, block, etc.); then determine any common experimental units for combinations of all possible pairs of categories, then for all possible triplets of categories, etc.

Rule III: Count the number of randomizations for each category (factor, block, etc.) in the experiment; then count randomizations for combinations of levels for all possible pairs of categories, then for all possible triplets, etc.

Rule IV: Determine which category levels are nested within another category level and determine which are cross classified.

Rule V: Ignore complexity of design in first key-out of degrees of freedom; relate key-out to nearest known design.

Application of rules I through V should enable one to key-out the degrees of freedom in an ANOVA as described in algorithm I. Note that these rules were applied to the examples throughout the paper.

Algorithm I: Keying-out degrees of freedom in the ANOVA

1. At every step perform simplest key-out of degrees of freedom that is possible.
2. First determine total degrees of freedom and partition into one for the correction term and the remainder for the sum of squares corrected for the mean.
3. Key-out degrees of freedom for category or categories offering the least difficulty.
4. Key-out degrees of freedom for ANOVA's for all possible pairs of categories, then all possible triplets, etc., excluding any pairs, triplets, etc. not needed.
5. Isolate all sets of degrees of freedom in the ANOVA for which the partitioning is not understood.
6. Defer the partitioning of sets of degrees of freedom that are not completely understood.
7. Approach the partitioning in step 6 from different directions in order to reduce steps 5 and 6 to the null set. Note that partitioning may be impossible until more information becomes available.

In using the algorithm always approach the key-out of degrees of freedom from the direction which is simplest and easiest to understand. Keep picking away at the remainder degrees of freedom until one reaches the desired stage, which could of course be single degree of freedom contrasts for the total degrees of freedom. When one knows the total number of observations N, one knows the total degrees of freedom which is N. Then one can always partition these N degrees of freedom into one for the correction for the mean and N-1 for the remainder. Then, if there are r blocks, one can always partition the N-1 into a set of r-1 and N-r degrees of freedom. This procedure is continued until step 7 in the algorithm is reached.

Investigators and statisticians often start computing sums of squares prior to using algorithm I. This practice can result in misspent effort and hence Rule VI.

Rule VI: Do NO computing of sums of squares until the correctness of the degree of freedom key-out in the ANOVA has been ascertained and the appropriate error variances have been designated.

Before computing any sums of squares, it is well to recognize the difficulty encountered in keying-out degrees of freedom in certain types of experiments. It is wise to consider the following two rules whenever human or animal experiments are involved.

Rule VII: With almost probability one, experiments and surveys involving humans and animals will have effects completely or partially confounded and one will need to follow Rules I through V in order to ascertain this.

Rule VIII: Be prepared to spend considerable time and effort unravelling the confounding schemes in any human or animal experiment as planned by the researcher (and perhaps even by a statistician).

When one is satisfied with the key-out of degrees of freedom for an investigation, then and only then should one consider computing totals, solutions for effects, and sums of squares. In connection with the last item algorithm II has been found useful.

Algorithm II: Computing sums of squares in the ANOVA

1. At every step compute the simplest ANOVA sums of squares, that is, sums of squares assuming nesting even though there was no nesting.
2. Compute sums of squares for degree of freedom key-outs in steps 2, 3, and 4 of algorithm I. For many investigations, this is a desk calculator job.
3. For partially confounded effects, it may be necessary to solve a set of normal equations prior to computing the sums of squares. (An exception is given in example 5.2.)
4. If steps 5 and 6 of algorithm I have not been reduced to the null set, nothing should be done about further partitioning of the sums of squares.

All too often computing specialists become imbued with a program or package for high speed computing and do not pay sufficient attention to simplifications. Example 5.2 is a case in point. The method of computing described indicates exactly what is being done whereas the use of a high speed nonorthogonal n-way classification program or a multiple regression program would not indicate the nature of quantities being computed. Likewise, rounding errors from high speed computer programs have always plagued this author, with complete nonsense resulting in several cases.

Once one computes the appropriate sums of squares in the ANOVA, then appropriate error variances need to be determined. Algorithm III is presented in this light.

<u>Algorithm III</u>: <u>Determining appropriate error variances for F-tests</u>

1. Factors with the same type of experimental unit *may* have the same error variance.
2. Factors with different experimental units almost always have different error variances.
3. In order to check the validity of an error variance, determine the appropriate error variance assuming other effects are absent from the experiment for single factors, for pairs of factors, etc.
4. Check to determine if partially confounded effects may be estimated from two sources and with two different error variances as in example 5.2.
5. Check your decisions with known situations.

References

[1] Cochran, W.G. and G.M. Cox (1957) Experimental Designs, 2nd edition (1st edition in 1950). John Wiley and Sons, Inc., New York.

[2] Federer, W.T. (1955) Experiment Design--Theory and Application. The MacMillan Co., New York.

[3] Federer, W.T. and L.N. Balaam (1972) Bibliography on Experiment and Treatment Design Pre-1968. Published for the International Statistical Institute by Oliver and Boyd, Edinburgh.

[4] Federer, W.T. and B.L. Raktoe (1966) Generalized lattice square designs. Jour. Amer. Statist. Assoc. 61: 821-832.

[5] Federer, W.T. and B.L. Raktoe (1965) General theory of prime-power lattice designs. Lattice rectangles for $v=s^m$ treatments in s^r rows and s^c columns for $r + c = m$, $r \neq c$, and $v < 1000$. Jour. Amer. Statist. Assoc. 60: 891-904.

[6] Finney, D.J. (1956) Multivariate analysis and agricultural experiments. Biometrics 12: 67-71.

[7] Fisher, R.A. (1966) The Design of Experiments, 8th edition (1st edition in 1935). Oliver and Boyd, Edinburgh.

[8] Kempthorne, O. (1952) The Design and Analysis of Experiments. John Wiley and Sons, Inc., New York, London, Sydney.

[9] Na Nagara, P. (1957) Lattice rectangle designs. Ph.D. Thesis, Cornell University.

[10] Raktoe, B.L. (1967) Application of cyclic collineations to the construction of balanced ℓ-restrictional prime powered lattice designs. Ann. Math. Statist. 38: 1127-1141.

[11] Snedecor, G.W. and W.G. Cochran (1967) Statistical Methods, 6th edition. The Iowa State University Press, Ames, Iowa, U.S.A.

[12] Yates, F. (1940) Lattice squares. Jour. Agric. Sci. 30: 672-687.

[13] Yates, F. (1937) The design and analysis of factorial experiments. Imp. Bur. Soil Sci. Tech. Comm. 35: 1-95.

W. T. Federer
Biometric Unit
Cornell University
Ithaca, New York
U.S.A.

Applied Statistics, R.P. Gupta, (editor)

STATISTICAL MODELS AND MEDICAL REALITY

Alvan R. Feinstein, M.S., M.D.

Medical research has benefited enormously from the quantitative revolution that followed such epochal publications as R.A. Fisher's The Design of Experiments[1]. The anecdotal experiences that used to fill the medical literature have been largely replaced by enumerations and measurements. Controlled comparisons have become commonplace; numerical differences are regularly checked with tests of statistical significance; and a randomized allocation of therapy is now accepted as a *sine qua non* of plans for the large-scale human experiments that are called clinical trials.

As a result of these changes, a new role has developed for the statistician who works medically as a consultant biostatistician. He was formerly regarded as a mathematical technician, called upon to do the number crunching used for calculations of a t test, a chi square, or a regression co-efficient. He has now often become a scientific arbiter or high priest of research. His authorative approval of research design is sought by investigators, required by editors, and demanded by granting agencies.

My purpose in this discussion is to consider some major shortcomings of the educational background that biostatisticians receive in preparation for their work not as mathematical technicians, but as research consultants. The educational background seldom contains any direct "clinical" experience or "internship" in the actual performance of research. Very few institutions provide a "preceptorship" in which the trainee has a first-hand opportunity to observe the activities of an excellent "practitioner" consultant. More importantly, the intellectual strategies that are currently maintained as the paradigmatic models and concepts of statistical theory are often inadequate or misleading for the realities encountered in clinical research. The flaws of these intellectual strategies will be the focus of attention in the rest of this talk.

A. The Mensuration of Data

The first main topic I want to consider is *data*. The activities of statistics began, and get their name, from efforts to provide a state (or nation) with data. The role of the statist was to assemble descriptive information about commerce and population. Later on, as the chiefs or princes of state wanted help in their recreational gambling, the analysis of probability became a major intellectual challenge, and there ultimately evolved the array of ideas that we now call statistical theory. Statistics thus became a bipartite domain, consisting of the collected information called descriptive statistics and the analytical theories called inferential statistics. This two-part division is also generally true today for the people who are called statisticians: the academic world tends to contain abstract thinkers who advance our ideas about analytic inference; the world outside academia tends to contain pragmatic workers who collect and ponder the descriptive data.

Most statisticians today, however, do not work in academia. In a recent booklet entitled "Careers in Statistics", the Committee of Presidents of Statistical Societies[2] described abundant opportunities in actuarial science, agriculture and fisheries, business and economics engineering, biomedical activities, law, and many other occupations. Employment as a university faculty member was mentioned only in passing. The kind of work described in the various non-academic jobs also dealt mainly with the assembly and contemplation of data. In those activities, a statistician might make use of analytic inferences and tactics, but he would have to start with the real data that were collected and available.

Because of the pre-eminence held by data in the work of practicing statisticians (and certainly in the work of biostatistical consultants), one would think that the principles developed, created, and taught as statistical discipline would contain a major focus on the way that data are produced. After all, data do not exist in nature; data are man-made artefacts. What exists in the reality of nature are phenomena or entities that must be observed, described, and categorized to create data. The name mensuration is sometimes used for this process of observation, description, and categorization, and the term value is often used for the particular item of data that emerges in each class of observation. The value may be a dimensional measurement, such as a serum cholesterol of 230 mgm per 100 ml.; an ordinal rating, such as moderately severe dyspnea; or a nominal designation, such as male gender.

The process of mensuration could therefore be expected to be a major concern when statistics is taught and when creative statisticians choose topics for research. Unfortunately, however, the expectation is wrong. In the current activities of most statistical educators and researchers, almost no attention is given to the acquisition of data. The usual assumption is that the data already exist, and that the statistician's work begins afterward. The predominant emphasis is then directed at the way that theories of statistical inference might be used for analysing and interpreting the data. Those of you who heard the address delivered by John Tukey[3] at a meeting in Ft. Collins several years ago may recall his warning that statistical theory has given excessive emphasis to analytic inference and not enough attention to mensuration. This warning is particularly cogent for the world of medical reality, where most of the unresolved major problems lie in the data themselves, not in the methods used for interpretive manipulation of the results. Let me cite three kinds of examples to illustrate this point.

1. Ingredients of Observer Variability

The first example deals with observer variability. For some of the most important information used in medical research, the apparatus that provides the data is a human being -- a radiologist concluding that an X-ray film does or does not show a duodenal ulcer; a pathologist deciding that a piece of tissue is or is not cancer; a cardiologist determing whether a heart murmur is normal or abnormal; a patient reporting his subjective sensations to a doctor who records them; a family expressing their contentment or dissatisfactions with health care. What are the standards that we use for evaluating whether this information is reliable enough, reproducible enough, and accurate enough to warrant the dignity of statistical tabulation and analysis?

An item of data, of course, can arise in several ways. It can be produced wholly by an inanimate device, or by a person using the scale recorded on an inanimate device, or wholly by a person. As statisticians, we certainly know that the data produced as measurement by machines can be inconsistent and we have developed appropriate theories to deal with the inconsistencies. An entire mathematical discipline, called quality control, has evolved to deal with those problems. We also know that a person using an observational machine (such as a ruler for measuring height) can be inconsistent. This knowledge is the basis for the Gaussian theory of errors, from which arose the parametric concepts of statistics.

But statistical theory has not been well developed for the situation in which a person acts alone -- making the observations and reporting them in verbal units that arise from the contents of that person's mind. There have been a few statistical papers devoted to mathematical analysis of the kinds of discrepancies that arise in either intra-observer or inter-observer variability. But the statistical procedures have been scientifically inadequate. Every thoughtful physician knows that variability exists, and recognizes that an arithmetic can be developed to quantify it. The scientific goal, however, is not to demonstrate and quantify variability; the goal is to get rid of it.

We can obviously get rid of it by building better machines or by replacing human observers by machines; but that tactic will not solve the problem for the many crucial forms of data for which a person is, and will remain, the only observational apparatus. One of the main reasons why statistical theory provides no help for this problem is that our analysis of the variability has not distinguished observational variability from interpretive variability. You and I may both look at a woman and agree observationally that she has blue eyes. We may still disagree, however, in our further interpretation of that and other observations. You may say she looks Scandinavian and I may say she looks Irish. If called upon to specify color of eyes, we would be in agreement. If called upon to cite ethnic appearance, we would disagree. The statistical scoring systems developed for quantifying observer variability usually depend only on the interpretive result, and do not dissect the result into its observational and interpretational components. Without that dissection, it is impossible to establish reasons for disagreements and to develop methods of removing the disagreements.

2. Criteria for Interpretation

Once we have recognized that observer variability contains ingredients from two different processes -- observation and interpretation -- rather than from a single process, we can turn to the second main problem, which deals with criteria for interpretation. If we want to establish consistent, reproducible values in our data, we must have well-defined criteria for each act of observation, description, and categorization. What do I mean by *blue eyes*? What do you mean when you say *she looks Scandinavian*? What is the operational algorithm for looking at a chest X-ray and determining that the heart is *enlarged*? What is the algorithm for categorizing a person's shortness of breath as *moderately severe dyspnea*?

If we are going to establish operational criteria for these designations and other interpretations, what are the criteria for criteria? How do we decide whether a particular set of criteria or algorithms are good or bad for doing their job of identifying and demarcating? For example, I once spent several days working on a committeee containing some prominent biostatistical consultants. Our assignment was to consider the design of systems for the surveillance of adverse reactions to pharmaceutical products. There was lots of discussion about the multivariate analyses and computer processing of data that would be needed to monitor huge numbers of patients and drugs. No one seemed concerned, however, with the fact that the first step in our work was missing. There did not exist any operational algorithms or criteria for determining or identifying whether a particular event that occurred after pharmaceutical treatment was in fact an adverse drug reaction.

3. Dehumanized "Hard Data" and Pertinent "Soft Data"

A third problem in medical data arises as an error of commission rather than omission. We learn our basic statistical theory by studying continuous variables, which are the data of dimensional scalar measurements. Our most cherished analytic tools -- correlation co-efficients, linear regression, the analysis of variance, and the matrix wonderland of multivariate analysis -- are all based on dimensional scalar measurements that can be used to calculate means, variances, and covariances. Since that is the kind of information on which we developed our basic statistical ideas, we like to get that kind of information in our research data.

Unfortunately, however, some of the most important information used in clinical research (particularly in health services research) is not expressible in dimensional scalar values. At best, we may be able to express the data only in ordinal ranks and more commonly, the values are nominal statements. All of the data, in fact, that distinguish people from rats, from agricultural crops, or from molecules, are expressed in verbal terms that must be converted to nominal or ordinal scales before analysis is possible. Pain, distress, discomfort, convenience, satisfaction, love, hate, anguish, happiness -- all of these entities require discrete categorization to be analyzed. Since we have not developed the strategies for creating scalar systems and criteria for the categorizations, however, we tend to ignore such data, or we confine the research analysis to what is available in dimensional units: age, height, serum cholesterol, white blood count.

Consequently, statistical consultants often urge their medical clients to work with "hard data" rather than "soft data". Hard data come from a calibrated, standardized mensurational technique or consist of something like death, sex, and occupation, about which there is relatively little dispute. Soft data consist of all the other variables for which there are no observational scales, established criteria, and demonstrated reproducibility of results. Loathing unreliable information, we urge our clients to avoid making any major decisions based on soft data. This advice is perfectly sensible from the standpoint of statistical reliability, but it is often quite bad for medical science. The research that needed to be done may be transmogrified into data that are unsatisfactory substitutions or total irrelevancies. The question we wanted to ask may become distorted or unanswerable by the choice of data that we analysed.

For example, for almost a generation therapeutic trials have been conducted to test the clinical value of chemical agents that alledly destroy cancer cells. It is no secret that these agents often make a patient miserable with persistent nausea or vomiting; that they predispose to infections which may kill the patient before the cancer has carried out its own lethal potential; and that many other unpleasant human side effects can occur. Nevertheless, in one statistically designed study after another, all of the pertinent "soft data" about the patients' quality of life have been deliberately omitted. From the quantitative results, which are presented in terms of such hard data as survival time, white blood count, and size of the tumor on X-ray, no one can tell whether the patient was comfortable or convulsing, vibrant or vegetating. The critical information about how a person feels is regarded as too soft to warrant statistical attention. Thus, in the name of science and statistics, we may deliberately dehumanize our research[4]. The soft data that would keep it human by making us pay attention to what a patient has to say about his symptoms are eliminated from analysis, and we rely instead on the "hard data" obtained from the X-ray, the microscopic findings, and the lab tests.

You may argue, quite properly, that all of the issues I have just cited are the responsibility of the medical investigator, not the statistician. After all, it is the investigator's job, not the statistician's, to see to it that crucially important variables are not overlooked or omitted merely because they depend on soft data. It is the investigator's job, not the statistician's, to identify those variables, to set up better methods of observation, better scales for reporting, and better criteria for using the scales. I agree with the argument, but the argument itself is a "cop-out". If the statistician is regarded as an authority on the design of the research -- a role that he increasingly receives and accepts -- he becomes responsible for the design of the research, not just for the mathematical technology used in analysis of the data. The medical clinician comes to the statistical "doctor" as a patient seeking advice, and that "patient" will often do what he is told. If statistical consultants do not insist that the client harden his soft data, he won't do it. And if we divert him away from the soft data, so that we analyze only what is already hard, the ultimate responsibility becomes ours, not his.

Let us consider a prominent recent example of the kind of problem I am citing. You are probably familiar with the controversy[5] that has been raging over the validity of a massive clinical trial of treatment for patients with diabetes mellitus, recently conducted in the United States[6]. An international committee of the Biometric Society has been looking into the problems and has not yet reported its conclusions. I do not want to discuss the many biostatistical controversies of that study, but I do want to describe one striking aspect of the collection of data.

A major objective of that trial was to study the development of a vascular complication that is called _diabetic neuropathy._ It occurs in the legs and affects the ability of the peripheral nerves to perceive three important sensations: touch, pain, and vibration. The perception of those sensations is customarily determined by a doctor examining the patient and testing for the patient's ability to distinguish the touch of a piece of cotton wool; the pain of a pin prick; and the vibration of a tuning fork. In the diabetics trial, the clinical investigators and their statistical consultants decided not to use this subjective clinical method of examination. Instead, a patient's diabetic neuropathy was tested with a gadget called a biothesiometer, which is a type of automated tuning fork.

The biothesiometer would presumably provide hard data, but its use required some major scientific clinical sacrifices, because it measures only vibration; it does not measure pain or touch.

Furthermore, unlike a tuning fork, which is applied to the ankle bone, the biothesiometer is applied to the flesh pulp of the big toe or the index finger -- so that the vibration data, which pertain to flesh pulp, are not immediately referable to the usual examination, which deals with vibration in bone. The investigators also had some bad luck. It turned out that the biothesiometer gadgets were not available for use in all patients at the start of the trial. For about half the cohort under study, there were no baseline biothesiometric measurements. Finally, there came the most unfortunate event of all. It turned out that the biothesiometric readings in the big toe were useless. Two consecutive measurements did not give reproducible results. On the other hand, the readings in the index finger were reproducible.

And so, in the most expensive, elaborate, statistically designed study ever conducted for diabetes, the investigators assessed the progress of diabetic neuropathy -- a condition that affects touch, pain, and vibration in the legs -- by getting hard data about vibration in the index finger of the hand. No one, including many of the clinicians involved in the research, now believes that those hard data about diabetic neuropathy have any value whatsoever, except perhaps for their dramatic demonstration of the pitfalls of a statistical infatuation with hard data.

Before closing this part of the discussion, I shall point out that mensuration has not been so badly neglected in other branches of the soft sciences. It was the psychologist S.S. Stevens[7] who helped focus attention on the different nominal, ordinal and other rating scales employed for categorizing data, and who provoked creative analysis of the statistical tests appropriate for the different types of data. Psychologists and sociologists have developed the disciplines of psychometrics and sociometry for constructing the multivariate tests, indexes, and scales used for assessing such phenomena as intelligence and social class. Considerable impetus in the development of contemporary non-parametric statistical techniques also arose from the world of psychometrics and sociometry, and for many years, the most popular textbook in this field was by the psychologist, Sidney Siegel[8].

In modern biometry, however, the process of mensuration has received little or no statistical attention. Consequently, the world of medicine and clinical care is in great need of a clinimetrics concerned not with inference but with evidence -- a clinimetrics that will confront the many mensurational challenges that have been abandoned or never enthusiastically grasped in biometrics. These challenges include the development of improved methods for all of the activities -- in choosing variables, establishing scales, creating interpretive criteria, preparing operational algorithms, developing multi-dimensional indexes, and advancing other methods of observational consistency -- that will harden soft data, produce reliable data, and restore a sense of medical meaning rather than mathematical manipulation to the numbers that emerge from the data of clinical biostatistics.

B. The Design of Research

For my second main topic, I should like to turn from the challenges of getting data to the challenges of planning the research itself. R.A. Fisher called this topic *The Design of Experiments* -- a name that is still used for the courses in which the subject is generally taught to statisticians today. The basic statistical concepts are concerned with the diverse blocks, factorial nests, Latin squares, and analysis of variance arrangements for analyzing the results in the groups who are randomly allocated to receive active treatments or to be "controls". Despite the magnificent artistry and creativity that can often be found in those courses, they have little or nothing to do with the real world of clinical research. As Marvin Zelen[9] has said, they are "so far removed from reality that a heavy dose may be too toxic with regard to future applications". I would like to comment here on three of the most prominent flaws in the theoretical principles of design. These flaws occur in the choice of controls, the model for experiments, and the analysis of factors that produce bias.

1. The Definition and Choice of Control

To plan a statistical design, we begin with the assumption that a suitable *control* has been or will be chosen. The word *control*, however, can be used for many different research ideas[10]. Some of these meanings refer to the concept of regulation. We can talk about *quality control* for the performance of a process, *control* of blood sugar in reference to maintenance of a preferred range of therapeutic values, and *control* of compliance in reference to the patients' fidelity in taking a prescribed medication. Other usages of the word control refer to time. The control period is an interval of time that can be used, before treatment begins, to assess the qualifications of the patients or the stabilization of their test values. A control period can also be used in a crossover trial as a type of "washout" interval between treatments.

The most frequent usage of control refers to an act of comparison. There are several different kinds of comparisons, however: the control treatment; the partition of patient groups into control strata; the baseline value that serves as the "control" of a before and after reading; and the peculiar form of retrospective cause-effect research in which the occurrence or non-occurrence of the effect is used for choosing the cases and control group in an epidemiologic case-control (or trohoc) study[11].

With all these different scientific usages of the word control, it is not surprising that statistical books on experimental design usually contain little or no discussion of the complex issues involved in delineating "control". The books seldom include attention to even the most common problem, which is the choice of a suitable comparison treatment. This choice involves attention to at least five prominent scientific problems whose concepts are far more important in research validity than any of the mathematical ideas about design. The principles are potency, relativity, internal accompaniment, external accompaniment, and concurrency.

Potency refers to dosage or other procedures for administering treatment. If a drug must be given in a flexible dosage that is adjusted to each patient, it is scientific folly to plan an experiment that gives the drug in fixed dosage to everyone. Relativity refers to the type of efficacy we might want to demonstrate. Is the treatment better than nothing or better than some established treatment? Internal accompaniment refers to the ancillary ingredients of a pharmaceutical agent or to the anesthesia and other events that occur during surgery. External accompaniment refers to the ancillary effects of such phenomena as solicitous medical personnel or postoperative recovery rooms after surgery. Concurrency refers to the timing with which the compared treatments precede, parallel, or follow the principal treatment under investigation.

If several hours were available for this talk, I could cite many examples in which each of these elements of control was sadly mismanaged in medical research that was statistically approved and sometimes statistically designed. In the interests of time, however,

I shall mention only a few. In the previously noted clinical trial of diabetes, one group of patients received a fixed dose of insulin throughout the entire investigation, which has lasted more than 10 years. No thoughtful physician would regard such an invariant dosage as satisfactory medical raactice. In other forms of therapeutic research, we regularly assess the efficacy of a drug by comparing it against placebo, rather than against no treatment. We thereby obscure the total effect of the therapeutic environment, which includes the communication, personality, and other human attributes of the physician. Many so-called adverse drug reactions in clinical trials are tabulated without regard to the additional therapies or superimposed illnesses that may have occurred. In highly controversial, statistically designed studies of anticoagulant therapy[12] to prevent blood clots in patients with heart attacks, the therapeutic comparisons did not include attention to certain external physical maneuvers that can also prevent blood clots: such procedures as early ambulation, wearing elastic stockings, and sitting in a chair, rather than resting in bed.

2. The Model of Research

The worst problem of all is revealed in the very words, Design of Experiments, that are used to designate statistical concepts about research. According to these concepts, the investigator takes a relatively homogeneous substance -- such as an agricultural field or a vat of beer -- and imposes his experimental maneuver (or treatment) upon it. The basic idea of an experiment is conveyed in the model[13] of:

$$\text{BASELINE STATE} \xrightarrow{\text{MANEUVER}} \text{SUBSEQUENT STATE}$$

and the basic process is to assign the compared maneuvers randomly to the subjects whose baseline and subsequent states are observed. There are three fundamental flaws in this reasoning.

a. Non-Randomized Maneuvers in Surveys

The first flaw is that most medical research -- certainly research involving the study of people -- is conducted as surveys, not as experiments. The maneuvers under investigation were not chosen and assigned by the investigator. In the case of such agents as cigarette smoking or dietary intake, the maneuvers were self-selected by the patients. In the case of many therapeutic agents, the maneuvers were chosen in a judgmental ad hoc manner by diverse practicing doctors. Because there was nothing random in the process of assigning the maneuver, the surveys are fraught with diverse opportunities for bias that can create major distortions in the results. Yet our statistical teaching focuses on the arithmetic of analyzing a randomized allocation of treatment in an experiment, not on the scientific necessity for contemplating, tracking down, and suitably adjusting the sources of bias in a survey.

b. Non-Random Samples in Surveys or Trials

The second prominent flaw, even when the therapeutic maneuver is randomly assigned, is that none of the many randomized therapeutic trials performed during the past three decades can be generalized[14]. The trouble is that none of the patient groups who entered the trials were selected as a random sample of the disease or condition under treatment. The act of random allocation of treatment means that the compared treatments can be validly contrasted in the patients admitted to the trial -- but we do not know whether the results can be generalized to other patients elsewhere. This same stricture applies to all of the contemporary epidemiologic research dealing with alleged cases of disease. Not a single one of the prominent studies was based on a random sampling of the general population or of people with the disease.

c. Double vs. Single Maneuvers

The third prominent flaw, and probably the most fundamental of all, is that statisticians have been using the wrong intellectual model for thinking about human experiments. According to that model, we study the effects of a single imposed maneuveral intervention. In most acts of clinical research, however, what is under study is a double maneuver. A maneuver of man is imposed upon a maneuver of nature[13,15]. In our single-maneuver thinking, we study causes of disease with the model shown in Fig. 1a. This model is erroneous. What actually happens takes place as two maneuvers not one. The first maneuver is shown in Fig. 1b. This is the underlying maneuver of nature. The second maneuver is shown in Fig. 1c. This is the maneuver of man.

Analogously, in studies of treatment for established disease, the conventional single maneuver model is shown in Fig. 2a. This model is also wrong. The correct model takes place in two maneuvers. The underlying maneuver of nature is shown in Fig. 2b.

The next maneuver is shown in Fig. 2c. This is the superimposed maneuver of man.

3. Factors in Baseline or Detection Bias

Because of this double-maneuver or maneuver-upon-maneuver phenomenon, our single maneuver model has many major deficiencies. By failing to account for a person's natural susceptibility when we study a cause of disease, or for a person's natural prognosis when we study treatment, we allow important sources of bias to enter the results. We cannot simply dismiss these distinctions as being covariates that will somehow be equalized by the randomization, or as confounding variables that one somehow hopes will go away. Our scientific job is to search out and identify those distinctions and to adjust for them accordingly by means of stratification, multivariate analysis, or whatever procedures are necessary.

For example, suppose people who are psychically quite tense are particularly susceptible to develop coronary artery disease, but suppose they are also particularly likely to take up cigarette smoking. We will then find an excellent correlation between cigarette smoking and the development of coronary artery disease. The correlation, of course, would really arise from psychic tension as the underlying source of both the smoking and the coronary disease -- but we would miss this aspect of susceptibility if we examine only the imposed maneuver of cigarette smoking.

Similar problems arise when the treatment of cancer is classified only according to the prognostic distinctions denoted by the anatomic spread of the cancer, while we ignore other prognostic distinctions manifested by weight loss and associated comorbid diseases. By omitting these additional prognostic factors, we use an inefficient staging system that may confuse or distort the results of treatment, and that leaves us unable to generalize the results. In other analyses of treatment, where patients from one doctor or hospital are compared with those of another, we must contemplate the different patterns of referral to different medical settings, and the effects that those patterns might have on the severity of illness and prognosis.

As an example of the problems that can arise when we ignore prognostic distinctions, even in randomized allocations, suppose three groups of patients in a randomized clinical trial have been treated by three different modes of therapy, with the results shown in Table 1. In this trial, we might conclude that there is no difference between the three therapeutic agents. Suppose on the other hand, however, that we had been much more careful about dividing these patients into subgroups with different prognostic outcomes and suppose, in fact, that we had identified a good risk group and a poor risk group. As shown in Table 2, the natural survival rate is 70% in the good risk group and 30% in the poor risk group. With the patients stratified according to prognostic distinctions, let us now look at what the data really showed. The treatments were not at all the same, and were in fact antipodally different. Treatment A gave poor results in the good risk group and good results in the poor risk group, whereas Treatment B had just the reverse effect. In the net result, however, the two effects cancelled one another, leading to our original erroneous conclusion. This type of confusion may be occurring in both epidemiologic studies of cause of disease and in clinical studies of prevention or treatment. We may be making many wrong decisions about etiology and therapy -- and even when the decisions are right, we may be unable to generalize the results because the baseline prognostic characteristics of the patients were so inadequately demarcated.

One of the main reasons for the inadequate demarcations of baseline characteristics is that the factors that delineate natural subceptibility and natural prognosis are often a part of the soft data called "clinical judgement"[16]. Instead of demanding that the clinician specify and articulate the ingredients of that judgment, so that we can define the soft data and harden it, the statistical consultant may admire the judgment and then ignore it. In studying therapeutic outcome in patients with cancer, we know that the clinician's judgment is often much more prognostically accurate than the customary anatomic staging system that has been used for most statistical analyses. Only in recent years, however, has attention been given to dissecting clinical judgment into the factorial components that constitute such variables as symptom severity, auxometry, chronometry, and co-morbidity[17]. In the instance that I cited earlier about psychic tension as a possible cause of coronary artery disease, a consistent mensuration scheme had not yet been developed for classifying patients according to their state of psychic tension. Consequently, psychic factors have been omitted from large-scale epidemiologic studies and no suitable data have been collected to prove or disprove the psychic hypothesis.

Since one of the prime principles of scientific investigation is the testing and ruling out of counter hypotheses, the absence of such tests deprives our biostatistical data of scientific credibility.

Before concluding this topic, let me give you an example of research in which an important baseline factor was neglected by a British working party[18] containing a prominent group of epidemiologists and biostatisticians. In a survey of the treatment of childhood leukemia, the working party compared the survival of patients referred to specialists at large academic centers against the results obtained in smaller urban or rural areas. On finding better survival rates for the patients treated at university centers, the members of the working party knew they had better look for sources of bias because the decision to treat a patient at home or at the university center had not been randomized. In searching for bias, the working party stratified the patients according to age and according to the particular cellular type of leukemia. With those adjustments, there was still a distinct superiority in the survival of patients treated at the university centers, and so the investigators concluded "that children with acute leukemia should be referred...to a centre specializing in the treatment of the disease".

At first glance, the investigation appears quite satisfactory. The working party considered the possibility of bias and apparently adjusted for it appropriately in the analysis of the data. For doctors who work at university centers, the results are appealing, since we seem to do well and we can always use additional customers. Unfortunately, however, the working party failed to consider the way in which their results might have been influenced by a factor that has a much more profound effect than either age or cellular type on the prognosis of acute leukemia. This factor is identified in doctors' clinical judgement[16] by the word, sick. It refers to the severity of the patient's illness, as determined from such elements as appetite, weight loss, fatigue, and other aspects of functional capacity in daily life. Two patients with the same age and the same cell type of acute leukemia may not have the same degree of sickness. One of the patients may clinically feel relatively well, whereas the other may be bedridden and moribund.

Now suppose that general practitioners in the smaller communities of the U.K. make selective decisions about referring children with leukemia to large urban medical centers. Suppose the referred patients are those who are less sick -- able to endure the travel and able to endure the hazards of the chemotherapy offered at those centers. The sicker patients, who seem less able to endure the travel and the treatment are kept at home. The survival rates will then seem much worse for the group treated in the home regions; and would be much better for the healthier referred patients, even if they were wholly untreated at the university centers.

This kind of pseudo-adjustment for bias, while ignoring the really major sources of bias, is a problem that permeates the biostatistics of contemporary epidemiology and clinical therapy. We get into this difficulty by designing survey research to study the effects of a single imposed maneuver -- in this case, the locale of treatment -- without adequate consideration of the antecedent natural maneuvers that may alter the population who receive the subsequent imposed maneuver. Unless we give appropriate attention to the differences in susceptibility or prognosis that are created by the underlying maneuver, we shall continue to overlook important sources of distorting bias that produce erroneous conclusions in the research.

The difficulty I have just discussed creates bias in characterizing the baseline state of the people studies in surveys. I should now like to mention another important but frequently overlooked source of bias, which I shall call _detection bias_. It occurs in identifying the outcome or the subsequent events observed after the imposition of the principal maneuver. The neglect of this form of bias is manifested in the naivete of the many people who believe that the occurrence rate of different diseases is indicated by statistics about the occurrence rate of those diseases. What the statistics indicate, of course, is the frequency with which the diseases have been diagnosed, not the frequency with which they have actually occurred in nature. For example, according to the data of vital statistics for mortality rates, dropsy was one of the great killer diseases of the 19th century. Since almost no one dies of dropsy according to vital statistics today, we might conclude that the eradication of that old killer disease was one of the great triumphs of modern medicine. The truth, of course, is that dropsy still occurs today and is probably just as frequent as it was a century ago. The difference is that we now call it something else. It is diagnosed on death certificates as congestive heart failure and sometimes as hepatic failure or renal failure.

In recent years, a rising death rate has been noted for cancer in black people in the United States. Speculations immediately began about nutritional, social, or even racist causes for this rise. No attention seems to have been given, however, to the detection factors that any thoughtful clinician would cite as a probable cause of the change. In order for someone to die of cancer, the cancer must be diagnosed; and the diagnosis requires that the person be subjected to the necessary technologic tests. In recent years, as blacks have increasingly moved from rural to urban regions, they have been increasingly exposed to the diagnostic technology with which cancers could be detected. Accordingly, instead of dying in some rural area of "old age" or "natural causes" or some other non-specific diagnosis, black people now have their cancers detected and entered on death certificates. The subsequent rise in the death rate due to cancer would be caused by diagnostic detection, and not necessarily by any true change in what is happening in nature.

The problem of detection bias can also affect the analytic statistical surveys that are done to study etiologic associations between a disease and its alleged cause. Let us consider the retrospective epidemiologic surveys[11] that have found an increased frequency of thrombophlebitis (or blood clotting) in association with the use of oral contraceptive pills. It so happens that an ordinary episode of thrombophlebitis can be a relatively simple clinical event -- not much worse than a common cold, and often shorter and milder. The episode may be manifested only by a slight pain in the calf of the leg, lasting for several days and disappearing without any further difficulty. In ordinary circumstances, a patient with mild thrombophlebitis may mention it to no one and do nothing about it. Now let us consider two women who both have an episode of thrombophlebitis. One of the women uses a diaphragm for contraception and has not been to see a doctor for many years. The other woman uses the pill, and has had to be in relatively frequent contact with her doctor to get her prescriptions renewed. Which of these two women do you think is more likely to have her thrombophlebitis detected when it occurs? Obviously, the woman taking the pill is much more likely to be subjected to the medical scrutiny that will enable the thrombophlebitis to be diagnosed. Despite the obvious distortions that can be created by this type of detection bias, it has not been thoroughly considered or analyzed in the many statistical studies associating thrombophlebitis and oral contraceptive pills. The crucial variable to be studied and stratified or adjusted is the intensity of diagnostic surveillance. This variable has been either ignored or dismissed in the existing studies.

C. The Strategy of Significance Tests

Everything I have said so far has had nothing to do with what an academic statistician might regard as statistics. My comments have all dealt with the scientific acquisition and interpretation of data, not with any of the mathematical principles of statistical analysis. Nevertheless, if statistical consultants are to be both authoritative and responsible, these are the major problems with which we must be concerned. As E.B. Wilson[19] once said, "Fifty pages of higher mathematics will not salvage an experiment with a hidden bias".

For my last set of comments, however, I should like to turn briefly to a topic that is intrinsic to statistics itself -- the way we now perform the common tests of so-called "significance". What an investigator usually wants to examine is the difference noted between two groups of people who have been exposed to the maneuvers under comparison. What we usually offer him is a t-test if the data are expressed in means and variances, and a chi square test if the data are expressed in proportions, rates, or percentages. Both of these tests are based on the ideas of estimating appropriate parameters for the population from which the samples were drawn; calculating t or chi-square as a so-called test statistic; noting probabilities associated with the distribution curve of the test statistics; and determining the probability sector that corresponds to the observed value of the test statistic.

The point I should like to make here is that none of this reasoning or calculation corresponds to the question that the investigator really wants answered[20]. He has no hypothetical population in mind, and he is not interested in estimating any parameters for that population. He has no interest in pooled varinace, expected frequencies, or degrees of freedom. He has no intention of repeating his experiment by drawing two random samples from a hypothetical population an infinite number of times. He is probably not even going to repeat his experiment once. He has done his research and has his data. They are the only data that he has or intends to get. There is a numerical difference in the results of the two groups. He is impressed by the clinical or biologic magnitude of that difference, but he is worried about the magnitude of the numbers of people in whom the difference occurred. Is the number of people large enough so that the difference is unlikely to arise from chance alone?

By doing a permutation test, as described by Fisher for attribute data and by Welch for dimensional data, we can tell the investigator exactly what he wants to know. We can rearrange his observed results in all of the appropriate combinations and permutations. We can note the difference that would occur with each such rearrangement and we can determine the frequency of the possible differences. We can then tell the investigator the exact probability, in the data he actually collected, that the difference he actually observed might have occurred by chance. There would be no recourse to ideas about a hypothetical population that is randomly sampled; no ideas about infinite repetitions of the experiment; no parametric estimations, pooled variances or degrees of freedom. He has asked us a straightforward question based on his observed data and we would give him a straightforward answer based on his observed data. It would be an answer that could be readily understood by any intelligent person and it requires no mathematical background.

Why then do we persist in using all the inferential mumbo-jumbo of the estimation theory that is involved for doing either a t test or a chi-square test? Why don't we use permutation tests instead? The only tenable explanation is that the values of t and chi-square are relatively easy to calculate, whereas the exact probabilities determined in a permutation test require some formidable computations. Besides, when the sample sizes are large enough -- which is when the computations are particularly awesome -- the probability values that we get from the t and chi-square procedures are very close to what we would get from the exact probabilities provided by the permutation tests. We use t and chi-square, therefore, not because of any intellectual, scientific, or statistical virtues, but because they are easy to calculate. An exception occurs when the sample sizes are quite small. In these circumstances, the t test -- despite its original focus on small samples -- and the chi-square test can be inaccurate, and so we shift to the permutation procedures. But in most other circumstances, we use t and chi-square to avoid the nuisance of calculating the results of a permutation test.

We achieve this ease in calculation by sacrificing our attention to the design of the research itself. The investigator and the statistician know that the results of a permutation test cannot be generalized. It provides no estimates of parameters for any hypothetical population to which the results might apply. If any generalization is to occur, the basic properties of the groups under study must be analyzed separately to determine whether they represent anything other than themselves. In using t and chi-square, however, we deliberately ignore all the realities and all the possible sources of bias in the research. We assume that our two samples were selected randomly from the hypothetical population and that the compared maneuvers were allocated randomly. In our heart of hearts, we may realize that neither of these assumptions is true, but as our mind of minds gets immersed in all of the theoretical constructions and parametric estimations, we may forget about the bias.

In the days before computers were readily available, it was a horror to calculate a permutation test, and our preference for t and chi-square could readily be justified computationally if not intellectually. But the major purpose of tests of significance in modern medical research is for small sample sizes, where we want the statistical P values to be especially accurate. If the sample sizes are large enough to make the difference in results convincing according to what Donald Mainland once called the traumatic intraocular test (the distinction hits you between the eyes), there is no need for doing statistical tests. The occasions when we want the tests are when sample sizes are small. For these purposes, with the increasing availability and decreasing costs of modern computation, there is no need for us to persist in using suboptimal strategies merely because the tests are easy to calculate. In the era of the computer, all tests are easy to calculate. If we can find the "statistical significance" of a difference between two means or between two proportions by pushing a button and getting the results of a permutation test just as easily as we used to compute t or chi-square on a desk calculator, we can adapt our strategies accordingly.

We can give our respect and homage to all of the magnificent contributions that were made by the t test and chi-square test in biostatistical analysis. On the other hand, just as doctors, in an era of chemical technology, no longer test for diabetes by maintaining the tradition of tasting the sweetness of the patient's urine, there is no reason for statisticians, in an era of computer technology, to use strategies that become traditions because they were easy to carry out with a desk calculator. If a clinical investigator can readily understand what a permutation test does, and if a statistical consultant is freed from having to make inappropriate assumptions about the origin of the numbers, both the clinician and the statistician will be liberated to devote their intellectual energies to concentrating on the architecture of the research.

The architecture of the research[13] is a region far more important than any mathematical theories of inference. It is the region that is greatly in need of creative work today by both clinical investigators and statisticians who can help educate one another into planning better models and modelling better plans. And it is the region where statisticians and clinical investigators can fruitfully collaborate today to advance both the wisdom of science and the sagacity of statistics.

Table 1

Results of a randomized clinical trial for Treatments A, B, and a placebo "control" group (Hypothetical Population)

	Number Improved per Number Treated		
	Treatment A	Treatment B	"Controls"
TOTAL	50/100 (50%)	50/100 (50%)	50/100 (50%)

Table 2

Data of Table 1, with patients stratified prognostically into "good risks" and "bad risks"

	Number Improved per Number Treated		
	Treatment A	Treatment B	"Controls"
"Good Risks"	25/50 (50%)	45/50 (90%)	35/50 (70%)
"Bad Risks"	25/50 (50%)	5/50 (10%)	15/50 (30%)
TOTAL	50/100 (50%)	50/100 (50%)	50/100 (50%)

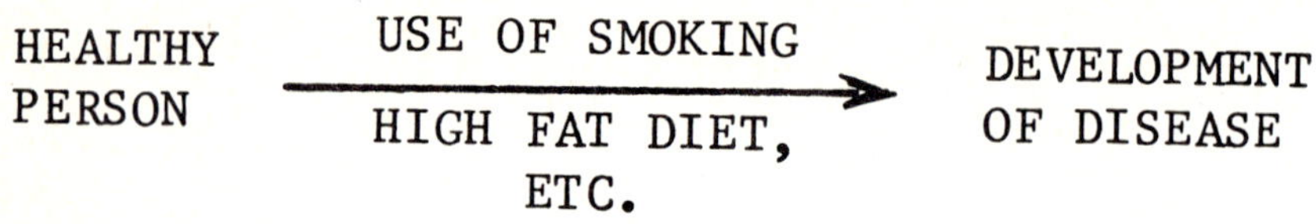

Figure 1a. The usual conceptual model of the "maneuver" for cause of a disease.

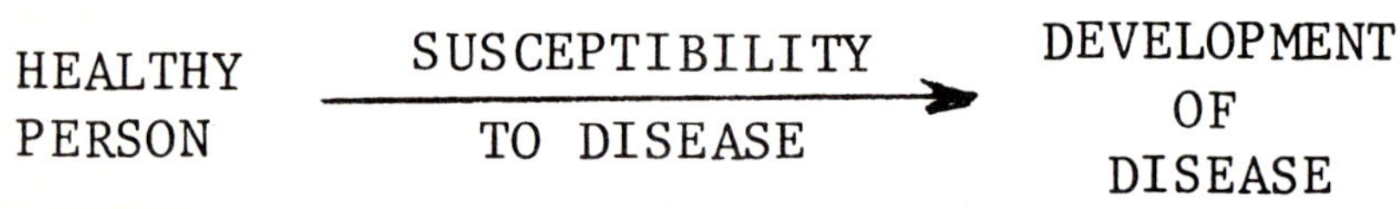

Figure 1b. The natural "maneuver" of susceptibility to a disease.

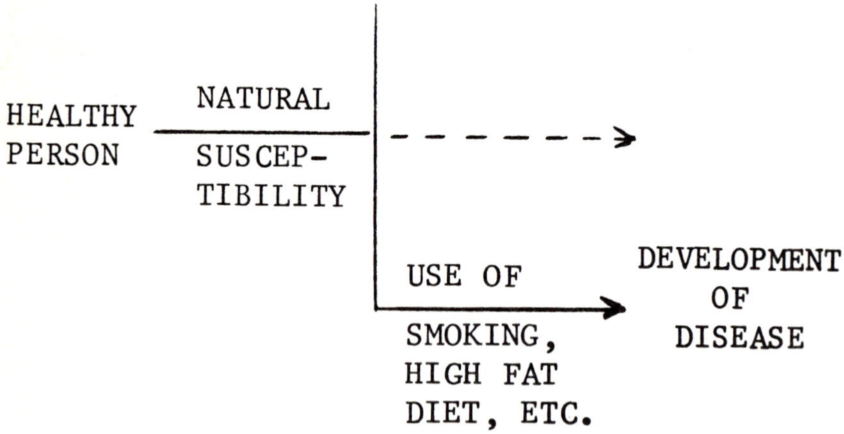

Figure 1c. The imposed maneuver of a possible cause of disease.

Figure 2a. The usual conceptual model of treatment for a disease.

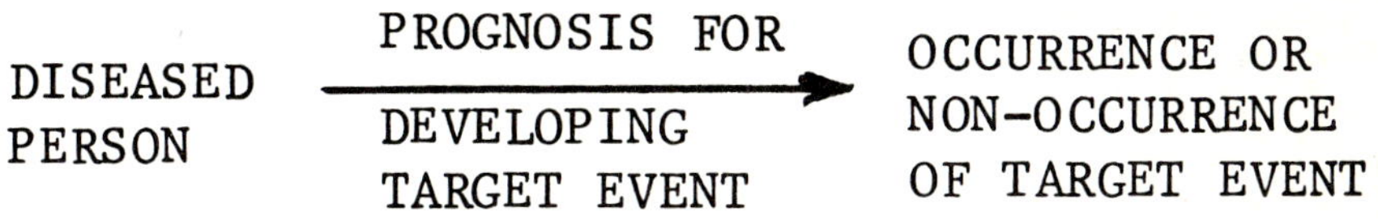

Figure 2b. The natural "maneuver" of prognosis for a target event.

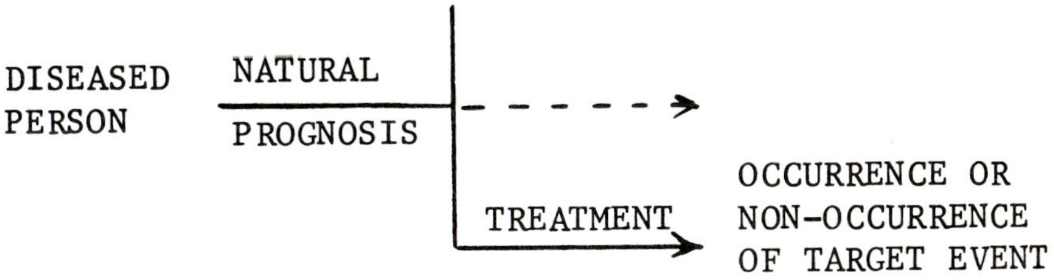

Figure 2c. The imposed maneuver of treatment.

References

[1] Fisher, R.A. (1935), The Design of Experiments. 1st Edn. Oliver and Boyd, Ltd. London.

[2] Committee of Presidents of Statistical Societies (1974). Careers in Statistics. American Statistical Association. Washington.

[3] Tukey, J.W. The future of statisticians. (Reference to be cited later.)

[4] Feinstein, A.R. The need for humanised science in evaluating medication. Lancet 2: 421-423 (August 26, 1972).

[5] Feinstein, A.R. (1971). Clinical biostatistics: VIII. An analytic appraisal of the University Group Diabetes Program (UGDP) study. Clin. Pharmacol. Ther. 12: 167-191.

[6] University Group Diabetes Program. (1970). A study of the effects of hypoglycemic agents on vascular complications in patients with adult-onset diabetes. Part I: Design, methods, and baseline characteristics. Part II: Mortality results. Diabetes 19: (Suppl. 2) 747-830.

[7] Stevens, S.S. (1946). On the theory of scales of measurement. Science 103: 677-680.

[8] Siegel, S. (1956) Nonparametric Statistics for the Behavioral Sciences. McGraw-Hill Book Co., Inc. New York.

[9] Zelen, M. (1969) The education of biometricians. Amer. Statis. 23: 14-15.

[10] Feinstein, A.R. (1973) Clinical biostatistics: XIX. Ambiguity and abuse in the twelve different concepts of 'control'. Clin. Pharmacol. Ther. 14: 112-122 (January-February).

[11] Feinstein, A.R. (1973). Clinical biostatistics: XX. The epidemiologic trohoc, the ablative risk ratio, and 'retrospective' research. Clin. Pharmacol. Ther. 14: 291-307.

[12] Gifford, R.H. and Feinstein, A.R. (February 13), 1969. A critique of methodology in studies of anticoagulant therapy for acute myocardial infarction. New Engl. J. Med. 280: 351-357.

[13] Feinstein, A.R. (1970) Clinical biostatistics: III-V. The architecture of clinical research. Clin. Pharmacol. Ther. 11: 432-441, 595-610, and 755-771.

[14] Feinstein, A.R. (1973). Clinical biostatistics: XXIII. The role of randomization in sampling testing, allocation, and credulous idolatry (Part 2). Clin. Pharmacol. Ther. 14: 898-915.

[15] Feinstein, A.R. (1970). Clinical Biostatistics: II. Statistics vs. science in the design of experiments. Clin. Pharmacol. Ther. 11: 282-292.

[16] Feinstein, A.R. (1967) Clinical Judgement. Williams & Walkins Co. Baltimore.

[17] Charlson, M.E. and Feinstein, A.R. (April 8), 1974. The auxometric dimension. A new method for using rate of growth in prognostic staging of breast cancer. JAMA 228: 180-185.

[18] Report to the Medical Research Council from the Committee on Leukaemia and the Working Party on Leukaemia in Childhood. Duration of survival of children with acute leukaemia. Brit. Med. J. 4: 7-9, 1971.

[19] Wilson, E.B. (1952) An introduction to Scientific Research. McGraw-Hill Book Co., Inc. New York.

[20] Feinstein, A.R. (1973). Clinical Biostatistics: XXIV. The role of randomization in sampling, testing, allocation, and credulous idolatry (Conclusion). Clin. Pharmacol. Ther. 14: 1035-1051.

Alvan R. Feinstein, M.D.
School of Medicine
Yale University
New Haven, Connecticut
U.S.A.

Applied Statistics, R.P. Gupta, (editor)

ATTRIBUTE REGRESSION ON ATTRIBUTES

by J. B. Garner

Summary

The problem is formulated in terms of the logistic models used by Cox (1966, 1970). A linear solution, applicable when all possible combinations of considered attributes are observed, is given based on Berkson's minimum logit chi-square techniques. An alternative linear solution is given based on the methods of Cox (1966) which is appropriate when many observed frequencies are zero.

Suppose there are k regressor attributes defined in the population. Let t be the number of distinct combinations of these k attributes observed in the sample, which is of size N. Then $t \leq N$. This article describes a method of finding the maximum likelihood 'cell' frequencies by iterative multiplication; and concurrently finding the maximum likelihood estimates of the logistic parameters as cumulative sums of the logarithms of the cell multipliers. This method requires computer storage of $2t(\leq 2n)$ cell values as compared with the storage of 2^{k+1} cell values required for Bishop's (1969) technique.

1. Preamble

Suppose that on each member of a random sample drawn from a population an attribute of interest, $A_0/\bar{A}_0$, is observed, together with k 'regressor' attributes $A_\nu/\bar{A}_\nu : \nu = 1,2,\ldots,k$. From the sample evidence we wish to estimate the probability of A_0 occurring given that any particular combination of the k regressor attributes has occurred. The aspect of this problem emphasised in the present article is the obtaining of the maximum likelihood solution when the number, k, of regressor attributes is large and so the 2×2^k table of sample frequencies contains many zeros.

2. The Model

Let i_ν be an index attached to the νth attribute, $A_\nu/\bar{A}_\nu$, taking the value 0 when A_ν occurs and the value 1 when $\bar{A}_\nu$ occurs, for $\nu = 1,2,\ldots,k$. Then the collection of these indices $(i_1,i_2,i_3,\ldots,i_k)$ represents a particular combination of the regressor attributes $A_\nu/\bar{A}_\nu : \nu = 1,2,\ldots,k$. For brevity the typical collection $(i_1,i_2,\ldots,i_k)$ is denoted by the index (i) which enumerates all 2^k possible collections $(i_1, i_2, \ldots, i_k)$ in some order. The conditional population probability of A_0 occurring given that (i) has occurred is denoted by $\pi(0|i)$, and so $1 - \pi(0|i) = \pi(1|i)$ represents the conditional probability of $\bar{A}_0$ occurring given (i). Both $\pi(0|i)$ and $\pi(1|i)$ are assumed positive for each collection (i).

A log-linear model is used following Birch (1963). If $k = 2$ then

$$\pi(0|(0,0)) \propto \exp\{u_0+u_{01}+u_{02}+u_{012}\}$$

$$\pi(0|(1,0)) \propto \exp\{u_0-u_{01}+u_{02}-u_{012}\}$$

$$\pi(0|(0,1)) \propto \exp\{u_0+u_{01}-u_{02}-u_{012}\}$$

$$\pi(0|(1,1)) \propto \exp\{u_0-u_{01}-u_{02}+u_{012}\} .$$

More generally let I denote an index set selected from the integers $(0,1,2,\ldots,k)$ without repetition, in ascending order, and with zero always in first position. Let $s(I;i)$ denote the sum of the indices i_ν belonging to the collection (i) taken over all integers ν belonging to I . {Notice that $\nu = 0$ is excluded since i_o does not belong to the collection $(i) = (i_1, i_2, \ldots, i_k)$}. Then

$$\pi(0|i) \propto \exp\{\sum_I u_1(-1)^{s(I;i)}\}$$

where the summation is over all possible index sets I. If #(I) denotes the number of elements in the index set I, then u_I is a parameter representing the {#(I) - 1}th order interaction between the attributes $A_\nu : \nu \varepsilon I$.

Writing

$$\omega(i) = \tfrac{1}{2}\{\ell n\ \pi(0|i) - \ell n \pi(1|i)\}$$

then

$$\omega(i) = \sum_I u_i(-1)^{s(I;i)}$$

and

$$u_I = 2^{-k} \sum_i \omega(i)(-1)^{s(I;i)}$$

3. Sampling

Let a sample of size N be drawn from the population either completely at random, or with the $A_o/\bar{A}_o$ margin alone fixed (this would imply the fitting of u_o in all subsequent work), or with any collection of margins representing the intersection of margins corresponding to any selection of the attributes $A_\nu/\bar{A}_\nu : \nu = 1,2,\ldots,k$ fixed. The observed frequencies of A_o and $\bar{A}_o$ with (i) are denoted by $n(0,i)$ and $n(1,i)$ respectively. The observed frequency of the collection (i) is $n(i) = n(0,i) + n(1,i)$. The log-likelihood of the observations may be expressed as the sum of three terms. The first term is a function of the observed frequencies alone, the second is a function of the observed frequencies and the marginal probabilities of the collections (i) occurring, and the third is ℓ_1, where

$$\ell_1 = \sum_i \{n(0,i)\ln\pi(0|i) + n(1,i)\ln\pi(1|i)\}$$

in which the summation is over just those collections (i) for which $n(i) > 0$. In the special case when all 2×2^k cell frequencies are positive, we may define $\mu(i)$, $v(i)$ and $h(i)$ by

$$\mu(i) = \tfrac{1}{2}\{\ln n(0,i) - \ln(1,i)\}$$

$$\{h(i)\}^{-1} = v(i) = \tfrac{1}{4}\{\frac{1}{n(0,i)} + \frac{1}{n(1,i)}\}$$

In this case, following Lindley (1964), the expression ℓ_1 differs asymptotically by only a function of the observed frequencies from

$$\ell_a = -\tfrac{1}{2}\Sigma\, h(i)\, \{\omega(i) - \mu(i)\}^2$$

Let J_I denote the set of the collections (i) that have occurred in the sample and for which $s(I;i)$ is even, and let CJ_I denote the set of those (i) that have occurred in the sample and for which $s(I,i)$ is odd. Then

$$\frac{\partial \ell_1}{\partial u_I} = 2\left| \sum_{J_I} \{n(0,i) - n(i)\pi(0|i)\} + \sum_{CJ_I} \{n(1,i) - n(i)\pi(1|i)\} \right| \quad (3.1)$$

$$\frac{\partial^2 \ell_1}{\partial u_{I_1}\partial u_{I_2}} = -4 \sum_i n(i)\pi(0|i)\pi(1|i)(-1)^{s(I_1;i) + s(I_2;i)} ,$$

and, when appropriate,

$$\frac{\partial \ell_a}{\partial u_I} = \Sigma\ h(i)\{\mu(i) - \omega(i)\}\ (-1)^{s(I;i)}$$

$$\frac{\partial^2 \ell_a}{\partial u_{I_1} \partial u_{I_2}} = -\ \Sigma\ h(i)(-1)^{s(I_1;i) + s(I_2;i)} = -\ \tilde{\sigma}(I_1,I_2).$$

4. Estimating Equations

When each of the possible 2×2^k cells has a non-zero observed frequency and every possible interaction is fitted, the maximum likelihood estimate for $\pi(0|i)$ is $n(0,i)/n(i)$. In this situation

$$\hat{u}_I = 2^{-k} \sum_i \mu(i)(-1)^{s(I;i)}$$

together with estimated covariance element

$$\mathrm{cov}(\hat{u}_{I_1} \hat{u}_{I_2}) = 2^{-2k} \sum_i v(i)(-1)^{s(I_1;i) + s(I_2;i)} .$$

For example if $k = 1$

$$\hat{u}_o = \tfrac{1}{4}\left\{\frac{\ell n\ n(0,0)}{n(1,0)} + \frac{\ell n\ n(0,1)}{n(1,1)}\right\} \quad \text{and}$$

$$\hat{u}_{o1} = \tfrac{1}{4}\left\{\frac{\ell n(0,0)}{n(1,0)} - \frac{\ell n\ n(0,1)}{n(1,1)}\right\} .$$

When some interactions are not fitted a minimum logit chi-square solution, Berkson (1955), is given by the solutions of $\partial \ell_a/\partial u_I = 0$ for each u_I , or, equivalently, the solutions of

$$u_{I_1} + \sum_{I_2 \neq I_1} V u_{I_2}\ \tilde{\sigma}(I_1,I_2) = \sum_i Vh(i)\mu(i)(-1)^{s(I_1;i)}$$

for every fitted parameter u_{I_1} , where $V^{-1} = \sum_i h(i)$. If a few observed cell frequencies are zero then these zeroes may be replaced by ½ in order to use these equations. This solution may be used as an initial starting point for a Newton's method interation to the maximum likelihood solution.

When a large number of observed cell frequencies are zero a device suggested by Cox (1966) may be used. Cox suggests the judicious use of an approximation due to Chandler (1962), namely

$$\frac{e^t}{1+e^t} = \begin{cases} 1 & t > 3 \\ \frac{1}{2} + \frac{1}{6}t & -3 \le t \le 3 \\ 0 & t < -3 \end{cases} \tag{4.1}$$

Corresponding to each fitted parameter u_{I_1} from result (3.1) the maximum likelihood equation may be expressed as

$$\sum_{J_{I_1}} n(i)\pi(0|i) + \sum_{CJ_{I_1}} n(i)\pi(1|i) = \sum_{J_{I_1}} n(0,i) + \sum_{CJ_{I_1}} n(1,i) \tag{4.2}$$

Using the central portion of the approximation (4.1) as a first step gives

$$\sum_{I_2} u_{I_2} \left\{ \sum_i n(i)(-1)^{s(I_2;i) + s(I_1;i)} \right\} = \frac{3}{2} \sum_i \{n(0,i) - n(1,i)\}(-1)^{s(I_1;i)}$$

{where the (i) summations are over all observed (i)}, for each fitted parameter u_{I_1}. Using the solutions of these linear equations the approximation (4.1) is reapplied to $\pi(0|i)$ and $\pi(1|i)$ for each observed (i), and the results reinserted in the equations (4.2) which are once more solved and so on.

It is noticed that the maximum likelihood equation (4.2) also implies the corresponding equation

$$\sum_{CJ_{I_1}} n(i)\pi(0|i) + \sum_{J_{I_1}} n(i)\pi(1|i) = \sum_{CJ_{I_1}} n(0,i) + \sum_{J_{I_1}} n(1,i) \tag{4.3}$$

In many situations the parameters are fitted in a hierarchical manner. That is, whenever u_I is fitted then $u_{I'}$ is also fitted for every $I' \subset I$. In this case a further simplification of the maximum likelihood equations occurs. For example if u_0 is fitted the maximum likelihood equation is

$$\Sigma\, n(0,i) = \Sigma n(i)\pi(0|i)$$

and similarly (4.4)

$$\Sigma\, n(1,i) = \Sigma n(i)\pi(1|i)$$

Now also fitting u_{01} leads, after some manipulation, to

$$\sum_{J_{01}} n(0,1) = \sum_{J_{01}} n(i)\pi(0|i)$$

$$\sum_{J_{01}} n(1,i) = \sum_{J_{01}} n(i)\pi(1|i)$$

and

$$\sum_{CJ_{01}} n(0,i) = \sum_{CJ_{01}} n(i)\pi(0|i)$$

$$\sum_{CJ_{01}} n(1,i) = \sum_{CJ_{01}} n(i)\pi(1|i)$$

If we now include u_{02} and u_{012} so that the fitted hierarchical 'pyramid' has 'apex' I = {0,1,2}, then there are four equation pairs of the form

$$\begin{aligned} \Sigma n(0,1) &= \Sigma n(i)\pi(0|i) \\ \Sigma n(1,i) &= \Sigma n(i)\pi(1|i) \end{aligned} \qquad (4.5)$$

where the summations are taken in turn over the four sets of (i) given by

$$J_{01} \cap J_{02},\ CJ_{01} \cap J_{02},\ J_{01} \cap CJ_{02},\ CJ_{01} \cap CJ_{02}\ .$$

In general if a hierarchical pyramid has apex I and $e = \{\#(I)-1\}$ then there are 2^e equation pairs of the form (4.5) in which the summations are taken over the 2^e sets of (i) formed by taking all possible intersections of $J_{0\nu}$ or $CJ_{0\nu}$ where ν is an integer ranging over each non-zero element belonging to I. Let these 2^e sets of (i) be enumerated $\{F_\eta: \eta = 1,2,\ldots,2^e\}$. This hierarchical pyramid of parameters $u_{I'}:I' \subset I$ may be fitted by maximum likelihood if and only if the left hand side of both equations in (4.5) is strictly positive for each of the 2^e sets $\{F_\eta: \eta = 1,2,\ldots,2^e\}$.

5. Iterative Solutions.

The iterative solutions suggested here are based on the techniques described by Mosteller (1968) and Bishop (1969). However, in the present context the Mosteller-Bishop solution requires the storage of 2^{k+1} cells, whilst the solution suggested here requires storage of only twice the number of observed (i) $\{\leq 2N\}$, cells. The proof of the convergence of the suggested iteration to the maximum likelihood solution may be taken from Darroch and Ratcliff (1972).

For each observed (i) the expected frequencies $n(i)\pi(0|i)$ and and $n(i)\pi(1|i)$ are denoted by $m(0,i)$ and $m(1,i)$ respectively. The iteration successively alters the values $m(0,i)$ and $m(1,i)$ so that the equations (4.2) and (4.3) are iteratively satisfied whilst at the same time no extraneous interaction is fitted. Suppose $(\ell - 1)$ parameters of the form u_I are to be fitted, namely $u_{I_1}, u_{I_2}, \ldots, u_{I_{\ell-1}}$. Let the expected frequencies after the rth stage of the $(j + 1)$th cycle of the iteration be denoted by $m^{(\ell j+r)}(0,i)$ and $m^{(\ell j+r)}(1,i)$ respectively. Take $m^{(0)}(0,i) = m^{(0)}(1,i) = 1$ for each observed (i). Then the first stage on each cycle is given by

$$m^{(\ell j+1)}(0,i) = m^{(\ell j)}(0,i) \frac{n(i)}{m^{(\ell j)}(0,i) + m^{(\ell j)}(1,i)}$$

and

$$m^{(\ell j+1)}(1,i) = m^{(\ell j)}(1,i) \frac{n(i)}{m^{(\ell j)}(0,i) + m^{(\ell j)}(1,i)}$$

Let $z(j,r)$ denote

$$\frac{\sum_{J_r} n(0,1) + \sum_{CJ_r} n(1,i)}{\sum_{J_r} m^{(\ell j+r)}(0,i) + \sum_{CJ_r} m^{(\ell j+r)}(1,i)}$$

(where $J_r = J_{I_r}$) for $r = 1,2,\ldots,\ell-1$; and for $j = 0,1,2,\ldots,$ and let $w(j,r)$ denote a similar expression found by interchanging J_r and CJ_r in the summations. Then

$$\left.\begin{aligned} m^{(\ell j+r+1)}(0,i) &= m^{(\ell j+r)}(0,i)\; z(j,r) \\ m^{(\ell j+r+1)}(1,i) &= m^{(\ell j+r)}(1,i)\; w(j,r) \end{aligned}\right\} \text{ for every } i \varepsilon J_r \tag{5.1}$$

and

$$\left.\begin{aligned} m^{(\ell j+r+1)}(0,i) &= m^{(\ell j+r)}(0,i)\; w(j,r) \\ m^{(\ell j+r+1)}(1,i) &= m^{(\ell j+r)}(1,i)\; z(j,r) \end{aligned}\right\} \text{ for every } i \varepsilon CJ_r$$

In order to extend the inference beyond the fitted cells it is necessary to estimate the fitted u_I parameters. This may be done concurrently with the iterative scaling of the expected cell frequencies by using

$$\hat{u}_{I_r} = \tfrac{1}{2} \sum_{j=0} \{\ell n\, z(j,r) - \ell n\, w(j,r)\} \tag{5.2}$$

where the summation is over the cycle index $j = 0,1,2,\ldots$ for each $r = 1,2,\ldots,\ell-1$.

In the description below of the iterative fitting of hierarchical collections of parameters we confine our attention to a single hierarchical pyramid. Each such pyramid would be fitted successively in each cycle. In the concurrent estimation expression given below a certain parameter such as u_o might occur in more than one pyramid in which case its value is successively altered as the iteration moves from pyramid to pyramid inside the same cycle.

Consider the hierarchical pyramid with apex I discussed in the last paragraph of section four. Using the expected frequency notation used above with $m^{(0)}(0,i) = m^{(0)}(1,i) = 1$ for every observed (i), each cycle of the iteration has $d = 2^e + 1$ stages.

The first stage is

$$m^{(dj+1)}(i_o,i) = m^{(dj)}(i_o,i) \frac{n(i)}{m^{(dj)}(0,i) + m^{(dj)}(1,i)}$$

for $i_o = 0,1$ and each observed (i) .

Let $\alpha(j,\eta)$ denote

$$\frac{\Sigma n(0,i)}{\Sigma m^{(dj+\eta)}(0,i)}$$

where the summations are over all observed (i) belonging to F_η, and $\beta(j,\eta)$ denote a similar expression with $n(0,i)$ and $m^{(dj+\eta)}(0,i)$ replaced by $n(1,i)$ and $m^{(dj+\eta)}(1,i)$ respectively, for $\eta = 1,2,\ldots,2^e$ and $j = 0,1,\ldots$. Then the remaining $(d-1)$ stages on the $(j+1)$th cycle are

$$m^{(dj+\eta+1)}(0,i) = m^{(dj+\eta)}(0,i)\ \alpha(j,\eta)$$

and

$$m^{(dj+\eta+1)}(1,i) = m^{(dj+\eta)}(1,i)\ \beta(j,\eta)$$

for $\eta = 1,2,\ldots,d-1$.

Suppose $k \geq 2$ and we fit the hierarchical collection of parameters u_0, u_{01}, u_{02}, and u_{012}. Then we may take $F_1 = J_{01} \cap J_{02}$, $F_2 = CJ_{01} \cap J_{02}$, $F_3 = J_{01} \cap CJ_{02}$ and $F_4 = CJ_{01} \cap CJ_{02}$. It is seen that any of the sets J_0, J_{01}, J_{02}, J_{012} or CJ_0, CJ_{01}, CJ_{02}, CJ_{012} may be expressed as a union of the sets $F_\eta : \eta = 1,2,3,4$. Thus $s(I';i)$, for any $I' \subset I = \{0,1,2\}$, takes the same value for any (i) belonging to a given $F\eta$. This result is true for any general hierarchical situation. Denote this common value by $s(I';\eta)$. Therefore the maximum likelihood parameter values are found concurrently with the cell fitting process by

$$\hat{u}_{I'} = 2^{-(e+1)} \sum_j \left[\sum_{\eta=1}^{d-1} (-1)^{s(I';\eta)} \{\ln \frac{\alpha(j,\eta)}{\beta(j,\eta)}\} \right]$$

for each $I' \subset I$.

6. Related Work.

A log-linear model of this type was fitted using maximum likelihood by Dyke and Patterson (1952). Their method requires each of the 2^{k+1} observed cell values to be non-zero. Bishop (1969) showed that conditioning on the observed $\{n(i)\}$ implies the saturated fitting of interaction parameters between the k regressor variables, and gave an illustration showing that such a procedure may violate Occam's Razor in this sense. However, when k is large this conditioning enables us to keep the problem within solvable bounds. Goodman (1970) gives a variety of solutions for the ill-ustration given by Dyke and Patterson based on fitting differing hierarchical parameter collections. Gokhale (1971) uses the parametrisation suggested by Goodman (1964) and Mantel (1966) to give an iterative scaling procedure for the general unrestricted contingency table, and later, Gokhale (1972) gives another iterative procedure which is more flexible in its model requirements but which does require each observed cell frequency to be non-zero. An algorithm for the fitting of hierarchical log-linear models for both complete and incomplete contingency tables by iterative schaling may be found in Haberman (1972), and a fairly complete bibliography of the iterative scaling technique in Fienberg (1970).

Acknowledgements

The basic equations (4.3) and iterative results (5.1) and (5.2) were developed and programmed during the year 1971 with the financial support of the Canada Council on grant number S69-1437.

References

[1] Berkson, J. (1955), Maximum likelihood and minimum X^2 estimates of the logistic function. J. Amer. Statist. Assoc., 50, 130-162.

[2] Birch, M.W. (1963), Maximum likelihood in three-way contingency tables. J.R. Statist. Soc. B, 25, 220-233.

[3] Bishop, Yvonne M.M. (1969), Full contingency tables, logits and split contingency tables. Biometrics, 25, 383-399.

[4] Chandler, P. (1962), Mixed failure systems. M.Sc. Thesis, University of London.

[5] Cox, D.R. (1966), Some procedures connected with the logistic qualitative response curve. In Festschrift for J. Neyman (ed. F.N. David). New York, Wiley.

[6] Cox, D.R. (1970), Analysis of Binary Data. Methuen: London.

[7] Darroch, J.N. and Ratcliff, D. (1972), Generalised Iterative Scaling for Log-Linear models. Ann. Math. Stat. 43, 1470-1480.

[8] Dyke, G.V. and Patterson, H.D. (1952), Analysis of factorial arrangements when the data are proportions. Biometrics, 8, 1-12.

[9] Fienberg, S.E. (1970), An iterative procedure for estimation in contingency tables. Ann. Math. Statist, 41, 907-917.

[10] Gokhale, D.V. (1971), An iterative procedure for analysing log-linear models. Biometrics, 27, 681-687.

[11] Gokhale, D.V. (1972), Analysis of log-linear models. Jour. Roy. Statist. Soc. B., 34, 371-376.

[12] Goodman, L.A. (1964),Interactions in multidimensional contingency tables, Ann. Math. Statist., 35, 632-646.

[13] Goodman, L.A. (1970). The multivariate analysis of qualitative data; interactions among multiple classifications. Jour. Amer. Statist. Assoc., 65, 226-256.

[14] Haberman, S.J. (1972), Algorithm AS51. Log-linear fit for contingency tables. Appl. Statist, 21, 218-225.

[15] Lindley, D.V. (1964), The Bayesian analysis of contingency tables. Ann. Math. Statist., 35 1622-1643.

[16] Mantel, N. (1966), Models for complex contingency tables and polychotomous dosage response curves. Biometrics, 22, 83-95.

[17] Mosteller, F. (1967), Association and estimation in contingency tables. Jour., Amer. Stat. Assoc., 63, 1-28.

J. B. Garner
Department of Mathematics
Memorial University of Newfoundland
St. John's, Newfoundland
Canada

Applied Statistics, R.P. Gupta, (editor)

BEST LINEAR UNBIASED AND INVARIANT ESTIMATORS FOR REGRESSION PARAMETERS BASED ON ORDERED OBSERVATIONS*

Z. Govindarajulu

ABSTRACT

The best linear unbiased estimators of the parameters in the multiple regression model are obtained when the samples are arbitrarily censored. The homogeneity of variance, polynomial regression and simple linear regression become special cases of the above model. The formulas take simplier forms when the underlying distribution is symmetric and the subsamples are of equal size and are symmetrically censored. Best constant-risk estimators of the parameters α, β and σ in the simple linear regression model are obtained. These results are applied to symmetric uniform and negative exponential cases.

1. INTRODUCTION

Lloyd [4] has given explicit formulas for the ordered least squares estimates of location and scale parameters together with their variances and covariance. Since then numerous contributions have been made to estimation of parameters using ordered observations in random samples (with or without censoring), for instance see Sarhan and Greenberg [8]. When subsamples are censored, the maximum likelihood estimates, in general, are somewhat difficult to compute. In the normal case, Tiku [9] has provided modified likelihood equations which lead to simple estimators; see also references [6, 10, 11]. Using generalized least squares theory, Hamouda and Leone [3] have obtained 0-BLUE estimators of linear regression parameters when the underlying distrubution is symmetric and the subsamples (censored or uncensored) are of the same size. In particular, they tabulate the coefficients for the normal case. In this paper we shall be concerned with 0-BLUE estimation of parameters in multiple linear regression with specialization to (i) homogeneity of variance, (ii) polynomial regression and (iii) simple linear regression. In particular, we further specialize the results to the case when the underlying distribution is symmetric and the subsamples are of equal size and are symmetrically censored. Further, we shall

* Part of this research was supported by the Navy under Office of Naval Research Contract No. N0014-73-A-0385-0001, Task Order NR 042-295. Reproduction in whole or in part is permitted for any purpose of the United States Government.

also consider best constant-risk estimation of parameters in simple linear regression.

2. MULTIPLE REGRESSION MODEL

Let us consider the model

$$Y_{ij} = \sum_{r=1}^{s} \beta_r t_{ir} + \varepsilon_{ij},\ j = 1,\ldots,n_i,\ i = 1,\ldots,k \tag{2.1}$$

where $E\varepsilon_{ij} = 0$ and the ε_{ij} are mutually independent. The ε_{ij}, $j=1,\ldots,n_i$, have the common distribution $F(x/\sigma_i)$.

Let $\underline{Y}_i = (Y_{i,(1)},\ldots,Y_{i,(n_i)})'$ denote the ordered $Y_{i,1},\ldots,Y_{i,n_i}$. Let

$$Z_{i,(j)} = (Y_{i,(j)} - \sum_{r=1}^{s} \beta_r t_{ir})/\sigma_i,\ j=1,\ldots,n_i \ . \tag{2.2}$$

Then $Z_{i,(1)} <\cdots< Z_{i,(n_i)}$ constitute order statistics in a sample of size n_i drawn from $F(x)$. Let

$$\mu_{i,r} = EZ_{i,(r)},\ r=1,\ldots,n_i, \tag{2.3}$$

$$\underline{\mu}_i = (\mu_{i,1},\ldots,\mu_{i,n_i})' \tag{2.4}$$

and let Ω_i = variance-covariance matrix of $Z_{i,(1)}$ $Z_{i,(n_i)}$.

Let $\underline{Y} = (\underline{Y}_1',\ldots,\underline{Y}_k')'$, where Y_i denotes the available portion of the censored sample, like $\underline{Y}_i = (Y_{i,(i_1)},\ldots, Y_{i,(i_{m_i})})'$. Thus we can write the model as

$$E\underline{Y} = Q\underline{\theta} \tag{2.5}$$

where

$$\underline{\theta} = (\beta_1,\ldots,\beta_s,\sigma_1,\ldots,\sigma_k)' \tag{2.6}$$

and

$$Q = \begin{bmatrix} \underline{t}_{11} & \underline{t}_{12} & \underline{t}_{1s} & \underline{\mu}_1 & 0 & \cdots & 0 \\ \underline{t}_{21} & \underline{t}_{22} & \underline{t}_{2s} & 0 & \underline{\mu}_2 & \cdots & 0 \\ \cdot \\ \cdot \\ \cdot \\ \underline{t}_{k1} & \underline{t}_{k2} & \underline{t}_{ks} & 0 & 0 & \cdots & \underline{\mu}_k \end{bmatrix} \tag{2.7}$$

where $\underline{t}_{ij} = t_{ij}\underline{1}_i$, $\underline{1}_i = (1,\ldots,1)'$, having m_i components. Also, the variance-covariance matrix of $\underline{Y}$ is given by

$$\Omega = \begin{pmatrix} \sigma_1^2\Omega_1 & & 0 \\ & \ddots & \\ 0 & & \sigma_k^2\Omega_k \end{pmatrix} . \tag{2.8}$$

Then the ordered best linear unbiased estimators (O-BLUE) of $\underline{\theta}$ is given by

$$\hat{\underline{\theta}} = (Q'\omega Q)^{-1}Q'\omega\underline{Y} \tag{2.9}$$

where

$$\omega = \Omega^{-1} = \begin{bmatrix} \omega_1 & & 0 \\ & \ddots & \\ 0 & & \omega_k \end{bmatrix} = \begin{bmatrix} \sigma_1^{-2}\Omega_1^{-1} & & 0 \\ & \ddots & \\ 0 & & \sigma_k^{-2}\Omega_k^{-1} \end{bmatrix} \tag{2.10}$$

and

$$\underset{(s+k)\times(s+k)}{Q'\omega Q} = \begin{pmatrix} \sum_{i=1}^{k}\underline{t}_{i1}'\omega_i\underline{t}_{i1} & \cdots & \Sigma\underline{t}_{i1}'\omega_i\underline{t}_{is} & \underline{t}_{11}'\omega_1\underline{\mu}_1 & \cdots & \underline{t}_{k1}'\omega_k\underline{\mu}_k \\ \vdots & & & & & \\ \Sigma\underline{t}_{is}\omega_i\underline{t}_{i1} & \cdots & \Sigma\underline{t}_{is}\omega_i\underline{t}_{is} & \underline{t}_{is}'\omega_1\underline{\mu}_1 & \cdots & \underline{t}_{ks}'\omega_k\underline{\mu}_k \\ \underline{\mu}_1'\omega_1\underline{t}_{11} & \cdots & \underline{\mu}_1'\omega_1\underline{t}_{1s} & \underline{\mu}_1'\omega_1\underline{\mu}_1 & & 0 \\ & & & & \ddots & \\ \underline{\mu}_k'\omega_k\underline{t}_{k1} & \cdots & \mu_k'\omega_k\underline{t}_{ks} & 0 & & \mu_k\omega_k\underline{\mu}_k \end{pmatrix} \tag{2.11}$$

and

$$\underset{(s+k)\times 1}{Q'\omega Y} = \left(\sum_{i=1}^{k}\underline{t}_{i1}'\omega_i\underline{Y}_i,\ldots,\sum_{i=1}^{k}\underline{t}'_{is}\omega_i\underline{Y}_i,\underline{\mu}_1'\omega_1\underline{Y}_1,\ldots,\underline{\mu}_k'\omega_k\underline{Y}_k\right) \tag{2.12}$$

Special Case 2.1. The polynomial regression is obtained by setting

$$t_{ik} = t_i^{k-1} = (x_i-\bar{x})^{k-1} , \quad k = 1,\ldots,s \quad .$$

Special Case 2.2. The linear regression is obtained by setting

$$s=2,\ \beta_1=\alpha,\ \beta_2=\beta,\ t_{i1}=1,\ t_{i2}=(x_i-\bar{x}),$$

Special Case 2.3. The homogeneity of variance case is obtained by setting

$$\sigma_1 = \sigma_2 = \ldots = \sigma_k = \sigma \ .$$

Special Case 2.4. If the underlying distribution F is symmetric and the subsample sizes are equal and each subsample is symmetrically censored, that is, $i_{m_i} = n_i - i_1 + 1$ etc., then

$$\underline{1}_i'\omega_i\underline{\mu}_i = 0 \quad \text{for} \quad i=1,\dots,k \quad . \tag{2.13}$$

Hence

$$(Q'\omega Q)^{-1} = \begin{bmatrix} \begin{pmatrix} \Sigma\underline{t}_{i1}'\omega_i\underline{t}_{i1} & \cdots & \Sigma\underline{t}_{i1}'\omega_i\underline{t}_{is} \\ \vdots & & \vdots \\ \Sigma\underline{t}_{is}'\omega_i\underline{t}_{is} & \cdots & \Sigma\underline{t}_{is}'\omega_i\underline{t}_{is} \end{pmatrix} & \bigcirc \\ \bigcirc & \begin{pmatrix} \frac{1}{\underline{\mu}_1'\omega_1\underline{\mu}_1} & & 0 \\ & \ddots & \\ 0 & & \frac{1}{\underline{\mu}_k\omega_k\underline{\mu}_k} \end{pmatrix} \end{bmatrix} \tag{2.14}$$

Then

$$\hat{\theta} = (Q'\omega Q)^{-1}Q'\omega\underline{Y} =$$

$$\begin{bmatrix} \begin{pmatrix} \Sigma\underline{t}_{i1}'\omega_i\underline{t}_{i1} & \cdots & \Sigma\underline{t}_{i1}'\omega_i\underline{t}_{is} \\ \vdots & & \\ \Sigma\underline{t}_{is}'\omega_i\underline{t}_{i1} & \cdots & \Sigma\underline{t}_{is}'\omega_i\underline{t}_{is} \end{pmatrix}^{-1} \begin{pmatrix} \Sigma\underline{t}_{i1}\omega_i\underline{Y}_i \\ \\ \Sigma\underline{t}_{is}\omega_i\underline{Y}_i \end{pmatrix} \\ \begin{pmatrix} \underline{\mu}_1'\omega_1\underline{Y}_1/\underline{\mu}_1'\omega_1\underline{\mu}_1 \\ \vdots \\ \underline{\mu}_k'\omega_k\underline{Y}_k/\underline{\mu}_k\omega_k\underline{\mu}_k \end{pmatrix} \end{bmatrix} \tag{2.15}$$

the summations being on i which ranges from 1 to k.

Special Case 2.5. Let $m_1 = m_2 = \dots = m_k = m$. Then, $\omega_i = \omega_2 = \dots \omega_k = \omega_0$ (say) and consequently

$$\Sigma \underline{t}_{ir}\omega_r\underline{t}_{ir} = (\underline{1}'\omega_0\underline{1}) \sum_{i=1}^{k} t_{ir}^2, \; r=1,\ldots,s$$

$$\Sigma \underline{t}_{ir}'\omega_r\underline{t}_{i\ell} = (\underline{1}'\omega_0\underline{1}) (\sum_{i=1}^{k} t_{ir}t_{i\ell}), \; 1\leq r,\ell\leq s,$$

and

$$\underline{\mu}_r'\omega_r\underline{\mu}_r = \underline{\mu}'\omega_0\underline{\mu}, \; r=1,\ldots,s. \tag{2.16}$$

Consequently,

$$\hat{\underline{\theta}} = \begin{pmatrix} (\underline{1}'\omega_0\underline{1})^{-1} \begin{bmatrix} \Sigma t_{i1}^2 & \cdots & \Sigma t_{i1}t_{is} \\ \vdots & & \\ \Sigma t_{i1}t_{is} & \cdots & \Sigma t_{is}^2 \end{bmatrix}^{-1} \begin{bmatrix} \Sigma t_{i1}\underline{1}'\omega_0\underline{Y}_i \\ \vdots \\ \Sigma t_{is}\underline{1}'\omega_0\underline{Y}_i \end{bmatrix} \\ \underline{\mu}'\omega_0\underline{Y}_1/\underline{\mu}'\omega_0\underline{\mu} \\ \vdots \\ \underline{\mu}'\omega_0\underline{Y}_k/\underline{\mu}'\omega_0\underline{\mu} \end{pmatrix} \tag{2.17}$$

$$\text{Var-Cov } \hat{\underline{\theta}} = \begin{pmatrix} (\underline{1}'\omega_0\underline{1})^{-1} \begin{bmatrix} \Sigma t_{i1}^2 & & \Sigma t_{i1}t_{is} \\ \Sigma t_{i1}t_{is} & \cdots & \Sigma t_{is}^2 \end{bmatrix}^{-1} & \bigcirc \\ \bigcirc & (\underline{\mu}'\omega_0\underline{\mu})^{-1}\underline{1} \end{pmatrix}. \tag{2.18}$$

<u>Special Case 2.6.</u> Let $\sigma_1 = \sigma_2 = \ldots = \sigma_k = \sigma$. Then,

$$\underline{\theta} = (\beta_1,\ldots,\beta_s,\sigma)' \tag{2.19}$$

$$Q = \begin{pmatrix} \underline{t}_{i1} & \cdots & \underline{t}_{1s} & \underline{\mu}_1 \\ \underline{t}_{21} & \cdots & \underline{t}_{2s} & \underline{\mu}_2 \\ \vdots & & & \\ \underline{t}_{k1} & \cdots & \underline{t}_{ks} & \underline{\mu}_k \end{pmatrix} \tag{2.20}$$

$$\Omega = \sigma^2 \begin{bmatrix} \Omega_1 & & 0 \\ & \ddots & \\ 0 & & \Omega_k \end{bmatrix}, \tag{2.21}$$

$$\omega = \sigma^{-2}\begin{pmatrix} \Omega_1 & & 0 \\ & \ddots & \\ 0 & & \Omega_k \end{pmatrix} . \tag{2.22}$$

Hence,

$$Q'\omega Q = \sigma^{-2}\begin{pmatrix} \Sigma \underline{t}'_{i1}\omega_i \underline{t}_{i1} & \cdots & \Sigma \underline{t}'_{i1}\omega_i \underline{t}_{is} & \Sigma \underline{t}'_{i1}\omega_i \underline{\mu}_i \\ \vdots & & & \\ \Sigma \underline{t}'_{is}\omega_i \underline{t}_{i1} & \cdots & \Sigma \underline{t}'_{is}\omega_i \underline{t}_{is} & \Sigma \underline{t}'_{is}\omega_i \underline{\mu}_i \\ \Sigma \underline{\mu}'_i\ \omega_i \underline{t}_{i1} & & \Sigma \underline{\mu}'_i\ \omega_i \underline{t}_{is} & \Sigma \underline{\mu}'_i\ \omega_i \underline{\mu}_i \end{pmatrix} , \tag{2.23}$$

$$(Q'\omega Q)^{-1} = \sigma^2 \left(\qquad \right)^{-1} . \tag{2.24}$$

Also,

$$Q'\omega Y = \sigma^{-2}\begin{pmatrix} \Sigma \underline{t}'_{i1}\omega_i \underline{Y}_i \\ \vdots \\ \Sigma \underline{t}'_{is}\omega_i \underline{Y}_i \\ \Sigma \underline{\mu}'_i\ \omega_i \underline{Y}_i \end{pmatrix} . \tag{2.25}$$

So

$$\underset{(s+1)\times 1}{\hat{\theta}} = (Q'\omega Q)^{-1}Q'\omega Y$$
$$= \left(\qquad \right)^{-1} \begin{pmatrix} \Sigma \underline{t}'_{i1}\omega_i \underline{Y}_i \\ \vdots \\ \Sigma \underline{t}'_{is}\omega_i \underline{Y}_i \\ \Sigma \underline{\mu}'_i\ \omega_i \underline{Y}_i \end{pmatrix} . \tag{2.26}$$

Special Case 2.6.1. If F is symmetric and we have a symmetrically censored sample with $m_1 = m_2 = \ldots = m_k = m$ (say) and $\omega_1 = \ldots = \omega_k = \omega_0$, then

$$(Q'\omega Q)^{-1} = \sigma^2 \begin{pmatrix} (\underline{1}'\omega_0\underline{1})^{-1}\begin{pmatrix} \Sigma t_{i1}^2 & \cdots & \Sigma t_{i1}t_{is} \\ \Sigma t_{i1}t_{is} & \cdots & \Sigma t_{is}^2 \end{pmatrix}^{-1} & \bigcirc \\ \bigcirc & 1/k\mu'\omega_0\underline{\mu} \end{pmatrix} . \tag{2.27}$$

Then

$$\underline{\hat{\theta}} = \left[\begin{array}{c} (\underline{1}'\omega_0\underline{1})^{-1} \begin{pmatrix} \Sigma t_{i1}^2 & \cdots & \Sigma t_{i1}t_{is} \\ \vdots & & \vdots \\ \Sigma t_{i1}t_{is} & \cdots & \Sigma t_{is}^2 \end{pmatrix}^{-1} \begin{pmatrix} \Sigma t_{i1}\underline{1}'\omega_0\underline{Y}_i \\ \\ \Sigma t_{is}\underline{1}'\omega_0\underline{Y}_i \end{pmatrix} \\ \Sigma\underline{\mu}'\omega_0\underline{Y}_i/k\underline{\mu}'\omega_0\underline{\mu} \end{array}\right]. \tag{2.28}$$

In a typical case we need to invert an $s\times s$ matrix. In certain cases, because of its special structure, we can explicitly invert such a matrix. Towards this, we have the following lemma.

Lemma 2.1. Consider a matrix A of the form

$$A = \left(\begin{array}{c|c} A_{11} & C' \\ \hline C & D \end{array}\right) . \tag{2.29}$$

Then

$$A^{-1} = \left[\begin{array}{c|c} I_2 & -(A_{11}-C'D^{-1}C)^{-1}C' \\ \hline 0 & I + D^{-1}C(A_{11}-C'D^{-1}C)^{-1}C' \end{array}\right] L^{-1} \tag{2.30}$$

where

$$L = \left(\begin{array}{c|c} A_{11} & 0 \\ \hline C & D \end{array}\right) \tag{2.31}$$

and

$$L^{-1} = \left(\begin{array}{c|c} A_{11}^{-1} & 0' \\ \hline -D^{-1}CA_{11}^{-1} & D^{-1} \end{array}\right) . \tag{2.32}$$

So

$$A^{-1} = \begin{bmatrix} A_{11}^{-1} + (A_{11}-C'D^{-1}C)^{-1}C'D^{-1}CA_{11}^{-1} & -(A_{11}-C'D^{-1}C)^{-1}C'D^{-1} \\ -D^{-1}CA_{11}^{-1} + D^{-1}C(A_{11}-C'D^{-1}C)^{-1}C'D^{-1}CA_{11}^{-1}, & D^{-1}\{I+D^{-1}C(A_{11}-C'D^{-1}C)^{-1}C'\} \end{bmatrix} . \tag{2.33}$$

Proof: See Fairweather and Govindarajulu [1].

3. LINEAR REGRESSION

In the linear regression case

$$\underline{\hat{\theta}} = \begin{pmatrix} \hat{\alpha} \\ \hat{\beta} \\ \hat{\sigma} \end{pmatrix} = (Q'\omega Q)^{-1}Q'\omega Y \tag{3.1}$$

where

$$(Q'\omega Q)^{-1} = \begin{bmatrix} \Sigma \underline{1}_i'\omega_i\underline{1}_i & \Sigma(\underline{1}_i'\omega_i\underline{1}_i)(x_i-\bar{x}) & \Sigma\underline{1}_i'\omega_i\underline{\mu}_i \\ \Sigma(\underline{1}'\omega_i\underline{1}_i)(x_i-\bar{x}) & \Sigma(\underline{1}_i'\omega_i\underline{1}_i)(x_i-\bar{x})^2 & \Sigma(\underline{1}_i'\omega_i\underline{\mu}_i)(x_i-\bar{x}) \\ \Sigma\underline{1}_i'\omega_i\underline{\mu}_i & \Sigma(\underline{1}_i'\omega_i\underline{\mu}_i)(x_i-\bar{x}) & \Sigma\mu_i'\omega_i\underline{\mu}_i \end{bmatrix} \tag{3.2}$$

$$Q'\omega\underline{Y} = (\Sigma\underline{1}_i'\omega_i\underline{Y}_i,\ \Sigma(\underline{1}_i'\omega_i\underline{Y}_i)(x_i-\bar{x}),\ \Sigma\underline{\mu}_i'\omega\underline{Y}_i)' . \tag{3.3}$$

Special Case 3.1. Let $m_1 = m_2 = \ldots = m_k = m$. Then $\omega_1 = \ldots = \omega_k = \omega_0$ and consequently

$$Q'\omega Q = \begin{bmatrix} k\underline{1}'\omega_0\underline{1} & 0 & k\underline{1}'\omega_0\underline{\mu} \\ 0 & (\underline{1}'\omega_0\underline{1})\Sigma(x_i-\bar{x})^2 & 0 \\ k\underline{1}'\omega_0\underline{\mu} & 0 & k\underline{\mu}'\omega_0\underline{\mu} \end{bmatrix} . \tag{3.4}$$

Also

$$Q'\omega\underline{Y} = (\underline{1}'\omega_0(\Sigma\underline{Y}_i)\quad \underline{1}'\omega_0\{\Sigma\underline{Y}_i(x_i-\bar{x})\}\quad \underline{\mu}'\omega_0(\Sigma\underline{Y}_i))'. \tag{3.5}$$

Applying the lemma for inverting $Q'\omega Q$, straightforward computations yield

$$(Q'\omega Q)^{-1} = k^{-1}\begin{bmatrix} \underline{\mu}'\omega_0\underline{\mu}/\Delta & & -\underline{1}'\omega_0\underline{\mu}/\Delta \\ 0 & 1/s^2(\underline{1}'\omega_0\underline{1}) & 0 \\ -\underline{1}'\omega_0\underline{\mu}/\Delta & 0 & \underline{1}'\omega_0\underline{1}/\Delta \end{bmatrix} , \tag{3.6}$$

where

$$\Delta = (\underline{1}'\omega_0\underline{1})(\underline{\mu}'\omega_0\underline{\mu}) - (\underline{1}'\omega_0\underline{\mu})^2 , \tag{3.7}$$

and

$$s^2 = (1/k)\sum_1^k (x_i-\bar{x})^2 . \tag{3.8}$$

Hence

$$\hat{\alpha} = \frac{1}{k\Delta}[(\underline{\mu}'\omega_0\underline{\mu})\{\underline{1}'\omega_0(\Sigma\underline{Y}_i)\}-(\underline{1}'\omega_0\underline{\mu})\{\mu'\omega_0(\Sigma\underline{Y}_i)\}], \tag{3.9}$$

$$\hat{\beta} = \underline{1}'\omega_0\{(x_1-\bar{x})\underline{Y}_1+\ldots+(x_k-\bar{x})\underline{Y}_k\}/ks^2(\underline{1}'\omega_0\underline{1}) , \tag{3.10}$$

and

$$\hat{\sigma} = \frac{1}{k\Delta}[(\underline{1}'\omega_0\underline{1}')\{\underline{\mu}'\omega_0(\Sigma\underline{Y}_i)\}-(\underline{1}'\omega_0\underline{\mu})\{\underline{1}'\omega_0(\Sigma\underline{Y}_i)\}] . \tag{3.11}$$

In particular, if $F(x)$ is symmetric about zero then $\underline{1}'\omega_0\underline{\mu} = 0$ and consequently

$$\hat{\alpha} = \underline{1}'\omega_0(\Sigma\underline{Y}_i)/k(\underline{1}'\omega_0\underline{1}) , \tag{3.12}$$

$$\hat{\beta} = \text{same as in (3.10)}, \tag{3.13}$$

and

$$\hat{\sigma} = \underline{\mu}'\omega_0(\Sigma\underline{Y}_i)/k(\underline{\mu}'\omega_0\underline{\mu}), \tag{3.14}$$

$$\text{var } \hat{\alpha} = \sigma^2/k(\underline{1}'\omega_0\underline{1}), \text{ var } \hat{\beta} = \sigma^2/ks^2(\underline{1}'\omega_0\underline{1}') \tag{3.15}$$

$$\text{var } \hat{\sigma} = \sigma^2/k(\underline{\mu}'\omega_0\underline{\mu}) \tag{3.16}$$

and all the covariances are zero. Expressions (3.12)-(3.14) have been obtained by Hamouda and Leone [3].

4. BEST LINEAR CONSTANT RISK ESTIMATORS

Constant risk estimators have been considered by Mann [5] and Vännman [12]. In the following we shall obtain the best regression model with straightforward generalizations to the multiple regression model. Let

$$R(\underline{\theta},\hat{\underline{\theta}}) = \theta_{s+1}^{-2}E[(\hat{\underline{\theta}}-\underline{\theta})(\hat{\underline{\theta}}-\underline{\theta})'] \tag{4.1}$$

where $\theta_{s+1} = \sigma$, and $\hat{\underline{\theta}}$ is any estimator of $\underline{\theta}$. An estimator of $\underline{\theta}$ is a constant risk estimator if every entry of its risk function matrix is free of $\underline{\theta}$. The best constant risk estimator of $\underline{\theta}$ is defined as the unique constant risk estimator of $\underline{\theta}$ that minimizes each diagonal entry of the risk function matrix.

Let D denote the class of estimators with expectations equal to linear functions of the unknown parameters and K be a linear space of estimators. We say that an estimator $\hat{\underline{\theta}} = (\hat{\theta}_1,\ldots,\hat{\theta}_s,\hat{\sigma})'$ belongs to $K\cap D$ if each $\hat{\theta}_i$, $i=1,\ldots,s$ and $\hat{\sigma}$ belongs to $K\cap D$. Then we have the following result pertaining to the best constant risk estimators.

RESULT 4.1. The best constant risk estimator $\tilde{\underline{\theta}}$ of $\underline{\theta}$ is given by

$$\tilde{\underline{\theta}} = \hat{\underline{\theta}} - \hat{\sigma}\tau \tag{4.2}$$

with risk function

$$R(\underline{\theta},\tilde{\theta}) = V^* - (1+v_{s+1,s+1})\underline{\tau}\,\underline{\tau}' \tag{4.3}$$

where $V^* = ((v_{ij}))$ = variance-covariance matrix of $\hat{\underline{\theta}}$, the best linear unbiased estimator of $\underline{\theta}$, and

$$\underline{\tau} = (1+v_{s+1,s+1})^{-1}(v_{1,s+1},\ldots,v_{s+1,s+1})'\ . \tag{4.4}$$

Proof. It readily follows from Corollary 1 of Vännman [12] (see her Remark on page 10).

REMARK 4.1. If σ is known, then

$$\underset{\sim}{\tilde{\theta}} = \underset{\sim}{\hat{\theta}}. \tag{4.5}$$

When the subsamples are of equal size and are symmetrically censored, using Result 4.1 and the expressions in (3.9) - (3.16) we obtain the constant-risk estimators for α, β and σ:

$$\tilde{\alpha} = \hat{\alpha} + \hat{\sigma}(\underset{\sim}{1}'\omega_0\underset{\sim}{\mu})/\{k\Delta+\underset{\sim}{1}'\omega_0\underset{\sim}{1}\}, \tag{4.6}$$
$$\tilde{\beta} = \hat{\beta}, \tag{4.7}$$
$$\tilde{\sigma} = \hat{\sigma}.\ k\Delta/\{k\Delta+\underset{\sim}{1}'\omega_0\underset{\sim}{1}\} \tag{4.8}$$

and

$$R(\underset{\sim}{\theta},\underset{\sim}{\tilde{\theta}}) = \frac{1}{k\Delta}\begin{bmatrix} \underset{\sim}{\mu}'\omega_0\underset{\sim}{\mu} - \dfrac{(\underset{\sim}{1}'\omega_0\underset{\sim}{\mu})^2}{k\Delta+\underset{\sim}{1}'\omega_0\underset{\sim}{1}} & 0 & -1'\omega_0\underset{\sim}{\mu} + \dfrac{(\underset{\sim}{1}'\omega_0\underset{\sim}{1})(\underset{\sim}{1}'\omega_0\underset{\sim}{\mu})}{k\Delta+\underset{\sim}{1}'\omega_0\underset{\sim}{1}} \\ 0 & \dfrac{\Delta}{s^2(\underset{\sim}{1}'\omega_0\underset{\sim}{1})} & 0 \\ -\underset{\sim}{1}'\omega_0\underset{\sim}{\mu} + \dfrac{(\underset{\sim}{1}\omega_0\underset{\sim}{\mu})(1'\omega_0\underset{\sim}{1})}{k\Delta+\underset{\sim}{1}'\omega_0\underset{\sim}{1}} & 0 & \dfrac{(\underset{\sim}{1}'\omega_0\underset{\sim}{1})k\Delta}{k\Delta+\underset{\sim}{1}'\omega_0\underset{\sim}{1}} \end{bmatrix} \tag{4.9}$$

Further, if F is symmetric about zero, then

$$\tilde{\alpha} = \hat{\alpha},\ \tilde{\beta} = \hat{\beta},\ \tilde{\sigma} = \hat{\sigma}\ .k/k+(\underset{\sim}{\mu}'\omega_0\underset{\sim}{\mu})^{-1}\} \tag{4.10}$$

and

$$R(\underset{\sim}{\theta},\underset{\sim}{\tilde{\theta}}) = \text{diag}\left(\frac{1}{k\underset{\sim}{1}'\omega_0\underset{\sim}{1}}\quad \frac{1}{ks^2(\underset{\sim}{1}'\omega_0\underset{\sim}{1})}\quad \frac{1}{1+k\underset{\sim}{\mu}'\omega_0\underset{\sim}{\mu}}\right)\ . \tag{4.11}$$

5. GENERAL PROPERTIES OF ESTIMATORS

Govindarajulu [2] has studied certain general properties of the ordered best linear estimators (O-BLUE) of the location and scale parameters. In the following we shall give certain general properties of the O-BLUE estimators of the parameters in the simple linear regression model. For the sake of simplicity, we shall further assume that F is symmetric about zero and the subsamples are of equal size and are symmetrically censored. Then we have the following result.

RESULT 5.1. Let

$$\sum_1^k \underset{\sim}{Y}_i/k = \underset{\sim}{\bar{Y}} = (\bar{Y}_1,\ldots,\bar{Y}_m),\ \bar{Y}_\ell = k^{-1}\sum_{i=1}^k Y_{i,(\ell)},\ \ell=1,\ldots,m\ ,$$

and

$$\underset{\sim}{g}' = (g_{i1},\ldots,g_{im})$$

denote the best coefficients for estimating α . That is, $\hat{\alpha} = \underline{g}'\underline{\bar{Y}}$. Then

$$\underline{g}'\underline{1} = 1 \quad \text{and} \quad \underline{g}'\underline{\mu} = 0 \ . \tag{5.1}$$

If $\underline{g}'$ denotes the vector of best coefficients for estimating σ then

$$\underline{g}'\underline{\mu} = 1 \quad \text{and} \quad \underline{g}'\underline{1} = 0 \ . \tag{5.2}$$

If $\underline{g}'$ denotes the coefficient of

$$\frac{\sum_1^k (x_i - \bar{x})\underline{Y}_i}{ks^2}$$

then

$$\underline{g}'\underline{1} = 1 \quad \text{and} \quad \underline{g}'\underline{\mu} = 0 \ . \tag{5.3}$$

Proof. (5.1)-(5.3) readily follows from (3.12)-(3.14). An analogous result can be gotten for the constant-risk estimators of α, β and σ after using (4.6)-(4.8).

6. EXAMPLES

In this section we shall apply the preceding results to specific distributions. Again let us confine ourselves to a simple linear regression model and assume that the subsample sizes are equal to n and they are uncensored.

Example 6.1. Let $F'(x) = f(x) = \begin{cases} 1/2 & -1 \le y \le 1 \\ 0 & \text{elsewhere.} \end{cases}$

Then

$$E(Z_{i,(r)}) = \mu_{i,r} = (2r-n-1)/(n+1), \quad i \le r \le n, \tag{6.1}$$

$$\mathrm{Cov}(Z_{i(\ell)}, Z_{i,(r)}) = 4\ell(n-r+1)/(n+1)^2(n+2), \quad 1 \le \ell \le r \le n. \tag{6.2}$$

Lloyd [4] has obtained an explicit expression for ω_0 given by

$$\omega_0 = \frac{(n+1)(n+2)\sigma^2}{4} \begin{pmatrix} 2 & -1 & & & & \\ -1 & 2 & -1 & & & \\ & -1 & 2 & -1 & & \\ & & -1 & 2 & -1 & \\ & & & -1 & 2 & -1 \\ & & & & -1 & 2 \end{pmatrix} . \tag{6.3}$$

Then we obtain

$$\underline{1}'\omega_0 = \frac{(n+1)(n+2)}{4}(1,0,\ldots,0,1),\quad \underline{\mu}'\omega_0 = \frac{(n+1)(n+2)}{4}(-1,0,\ldots,0,1)$$

$$\underline{1}'\omega_0\underline{1} = (n+1)(n+2)/2, \quad \text{and} \quad \underline{\mu}'\omega_0\underline{\mu} = (n-1)(n+2)/2. \tag{6.4}$$

The 0-BLUE estimators are given by

$$\hat{\alpha} = (\bar{Y}_n - \bar{Y}_1)/2,\quad \hat{\beta} = \{\sum_{i=1}^{k}(x_i-\bar{x})(Y_{i,(n)}-Y_{i,(1)})\}/2\sum_1^k(x_i-\bar{x})^2,$$

$$\hat{\sigma} = (n+1)(\bar{Y}_n-\bar{Y}_1)/2(n-1), \tag{6.5}$$

$$\operatorname{var}\hat{\alpha} = 2\sigma^2/k(n+1)(n+2),\quad \operatorname{var}\hat{\beta} = 2\sigma^2/\{(n+1)(n+2)\sum_1^k(x_i-\bar{x})^2\},$$

and

$$\operatorname{var}\hat{\sigma} = 2\sigma^2/k(n-1)(n+2), \tag{6.6}$$

where

$$\bar{Y}_\ell = \sum_{i=1}^{k} Y_{i,(\ell)}/k,\quad \ell=1,\ldots,n.$$

The constant-risk estimators and their variances are given by

$$\tilde{\alpha} = \hat{\alpha},\quad \tilde{\beta} = \hat{\beta}$$

$$\tilde{\sigma} = k\hat{\sigma}/\{k+2(n-1)^{-1}(n+2)^{-1}\} \tag{6.7}$$

$$\sigma^{-2}E(\tilde{\sigma}-\sigma)^2 = 2/\{2+k(n-1)(n+2)\}. \tag{6.8}$$

Example 6.2. Let $F'(x) = f(x) = \begin{cases} e^{-x}, & x \geq 0 \\ 0, & x < 0 \end{cases}$.

Then it is well known that

$$E(Z_{i,(r)}) = \mu_{ir} = \sum_{\ell=n-r+1}^{n} \ell^{-1}$$

$$\operatorname{Cov}(Z_{i,(\ell)}, Z_{i,(r)}) = \sum_{m=n-\ell+1}^{n} m^{-2} \quad 1 \leq \ell \leq r \leq n. \tag{6.9}$$

Sarhan [7] has obtained ω_0 and is given by

$$\omega_0 \begin{pmatrix} n^2+(n-1)^2 & -(n-1)^2 & 0 & \ldots & 0 \\ -(n-1)^2 & (n-1)^2+(n+2)^2 & -(n-2)^2 & \ldots & 0 \\ 0 & & & & -1 \\ \vdots & & & & \\ 0 & 0 & \ldots & -1 & 1 \end{pmatrix} \sigma^2 \tag{6.10}$$

Routine computations yield

$$\underline{1}'\omega_0 = (n^2,0,\ldots,0)\sigma^{-2},\quad \underline{1}'\omega_0\underline{1} = n^2\sigma^{-2}$$

$$\underline{\mu}'\omega_0 = (1,\ldots,1)\sigma^{-2},\quad \underline{\mu}'\omega_0\underline{\mu} = n\,\sigma^{-2},\quad \underline{1}'\omega_0\underline{\mu} = n\,\sigma^{-2}, \tag{6.11}$$

and

$$\Delta = n^2(n-1). \tag{6.12}$$

Substituting these in Equations (3.9)-(3.11) we obtain

$$\hat{\alpha} = \{n^2\bar{Y}_1 - (\bar{Y}_1 + \ldots + \bar{Y}_n)\}/n(n-1), \tag{6.13}$$

$$\hat{\beta} = \sum_{i=1}^{k} (x_i - \bar{x}) Y_{i,(1)} / \sum_{1}^{k} (x_i - \bar{x})^2 , \tag{6.14}$$

$$\hat{\sigma} = (\bar{Y}_1 + \ldots + \bar{Y}_n - n\bar{Y}_1)/(n-1) \tag{6.15}$$

and the variance-covariance matrix of $\hat{\alpha}$, $\hat{\beta}$ and $\hat{\sigma}$ is given by

$$\frac{\sigma^2}{k(n-1)} \begin{pmatrix} n^{-1} & 0 & -n^{-1} \\ 0 & \frac{n-1}{n^2 s^2} & 0 \\ -n^{-1} & 0 & 1 \end{pmatrix} .$$

REFERENCES

[1] Fairweather, G. and Govindarajulu, Z. (1974). Inversion of a certain patterned matrix (In preparation).

[2] Govindarajulu, Z. (1968). Certain general properties of unbiased estimates of location and scale parameters based on ordered observations. SIAM J. Appl. Math., 16, 533-551.

[3] Hamouda, E.M. and Leone, F.C. (1974). The 0-BLUE estimators for complete and censored samples in linear regression. Technometrics (To appear).

[4] Lloyd, E.H. (1952). Least squares estimation of location and scale parameters using order statistics. Biometrika 39, 89-95.

[5] Mann, R.N. (1970). Estimation of Location and Scale Parameters Using Various Models of Censoring and Truncation. Aerospace Research Laboratories Report ARL70-0026. Wright Patterson Air Force Base, Ohio, p. 65.

[6] Moeschberger, M.L. and David, H.A. (1971). Life tests under competing causes of failure and the theory of competing risks. Biometrics 27, 909-33.

[7] Sarhan, A.E. (1955). Estimation of the mean and standard deviation by order statistics, Part III. Ann. Math. Statist. 26, 576-592.

[8] Sarhan, A.E. and Greenberg, B. G. (1962). Contributions to Order Statistics. John Wiley, New York.

[9] Tiku, M.L. (1973). Testing group effects from type II censored normal samples in experimental design. Biometrics 29, 25-33.

[10] Tiku, M.L. (1968). Estimating the parameters of log-normal distribution from censored samples. J. Amer. Stat. Assn. 63, 134-40.

[11] Tiku, M.L. (1968). Estimating the parameters of normal and logistic distributions from censored samples. Aust. J. Stat. 10, 64-74.

[12] Vannman, K. (1973). Constant Risk Estimators with Applications to Linear Functions of Order Statistics. Technical Report No. 6, University of Umeå, Sweden.

Z. Govindarajulu
Department of Statistics
University of Kentucky
Lexington, Kentucky 40506
U.S.A.

Applied Statistics, R.P. Gupta, (editor)

ON A CERTAIN MANOVA MODEL

R. P. Gupta

SUMMARY

Leefland and Praag [3] consider a certain univariate analysis of variance model which is useful for predicting football league results and elsewhere. Here the model is generalized to the multivariate analysis of variance (MANOVA) case.

1. THE MODEL

Consider the MANOVA model

$$z_{ij} = \rho_i - \pi_j + \alpha + \varepsilon_{ij} \ , \ i,j = 1,\ldots,N, i \neq j \ , \tag{1}$$

where each ε_{ij} is a p component random column vector normally distributed with mean vector zero and covariance matrix Σ . ρ_i is the p component column vector of the i-th block effects, i.e., on each plot of the i-th block we have one vector observation, and π_j is the p component column vector of j-th column effect, α is the general mean vector, and z_{ij} is the observed yield, z_{ii} is assumed to be a null vector. Thus this is a two way classification, without interaction, normal MANOVA model with one observation per cell. This is an orthogonal model except that z_{ii} is equal to zero. If

$$Z = (z_{12}, z_{13}, \ldots, z_{1N}, z_{21}, \ldots, z_{2N}, \ldots, z_{N-1N}) \ , \tag{2}$$

then the total model (1) may be written as

$$Z = [\rho_1, \rho_2, \ldots, \rho_N, -\pi_1, -\pi_2, \ldots, -\pi_N, \alpha] X + E \ , \tag{3}$$

E is $p \times N(N-1)$, and $(2N+1) \times (N-1)N$ X is

$$X' = \begin{bmatrix} J & 0 & 0 & \cdots & 0 & M_1 & J \\ 0 & J & 0 & \cdots & 0 & M_2 & J \\ \cdot & \cdot & \cdot & \cdots & \cdot & \cdot & \cdot \\ 0 & 0 & 0 & \cdots & J & M_N & J \end{bmatrix} \tag{4}$$

where J denotes an $(N-1)$ component column vector of unities everywhere and M_i is the $(N-1) \times N$ matrix obtained from $I, N \times N$, by deleting its i-th row. We have $(2N+1)$ parameter vectors

$[\rho,-\pi,\alpha]$, $\uparrow$ $p\times N$, π $p\times N$, α $p\times 1$, and we have exactly $(2N-1)$ independent normal equations. Since block totals and column totals add to the grand total the matrix XX' is such that its column total and row totals add to the same quantity. It follows that XX', and hence X is a matrix of rank $(2N-1)$. However, for the unique estimation of parameters we must have $(2N+1)$ independent normal equations, and hence we must impose two linear restrictions on the parameters. Following the general practice in MANOVA we take these restrictions to be

$$e'\rho' = 0 \tag{5}$$

and

$$e'\pi' = 0 \ , \tag{6}$$

where e is a column vector of N components of unities everywhere. The error sum of products matrix (SPM) is

$$\sum_{i\neq j=1}^{N}\sum (z_{ij}-\rho_i+\pi_j-\alpha)\,(z_{ij}-\uparrow_i+\pi_j-\alpha)'$$

$$= (Z-BX)(Z-BX)' \tag{7}$$

where $B = (\rho,-\pi,\alpha)$. The conditions (5) and (6) are written as

$$\begin{pmatrix} e' & 0 & 0 \\ 0 & e' & 0 \end{pmatrix} B' = FB' = 0 \ . \tag{8}$$

Thus we have come to linear hypothesis MANOVA model where the parameters are linearly restricted by design. To differentiate (7), or a suitable scalar function of (7), subject to the conditions (8), we may use the following results collected by Kabe [2].

Let y be an N component column vector, D a given $q\times N$ matrix of rank $q(<N)$, H a given $N\times N$ positive definite symmetric matrix, μ a column vector of N components. Then the minimum of $(y-\mu)'H(y-\mu)$, subject to $Dy = v$ is obtained as

$$u = (y-\mu)'H(y-\mu) - 2\lambda'(Dy-v) \ , \tag{9}$$

where λ is a q component column vector of Lagrange's multipliers, we find that $\partial u/\partial y = 0$, $\partial u/\partial\lambda = 0$ readily yield

$$H(y-\mu) - D\lambda = 0 \ , \tag{10}$$

$$Dy-v = 0 \tag{11}$$

$$\lambda = (DH^{-1}D')^{-1}(v-D\mu) \ , \tag{12}$$

$$y-\mu = H^{-1}D'(DH^{-1}D')^{-1}(v-D\mu) \tag{13}$$

$$u = (v-D\mu)'(DH^{-1}D')^{-1}(v-D\mu) \quad . \tag{14}$$

When Y is $p\times N$, Ω is $p\times N$, $N \geq q+p$, then $\partial U/\partial Y$, $\partial U/\partial\Lambda$ equated to zero for

$$U = tr[(Y-\Omega)H(Y-\Omega)' - 2\Lambda(DY'-V')] , \tag{15}$$

readily yield that

$$\Lambda' = (DH^{-1}D')^{-1}(V-\Omega D')' , \ \Lambda \ p\times q , \tag{16}$$

$$Y-\Omega = (V-\Omega D')(DH^{-1}D')^{-1}DH^{-1} , \tag{17}$$

and the minimum of $(Y-\Omega)H(Y-\Omega)$, subject to $DY' = V'$ is

$$(V-\Omega D')(DH^{-1}D')^{-1}(V-\Omega D')' \tag{18}$$

When H is of rank t, $q < t < N$, then the above results hold good provided H^{-1} is replaced by its Moore-Penrose inverse defined by $H^{-} = R'(RHR')^{-1}R$, where R is the $t\times N$ matrix of the first t latent vectors of H corresponding to the first t nonzero roots of H.

From MANOVA theory it is known that

$$(Y-BX)(Y-BX)' = (Y-\hat{B}X)(Y-\hat{B}X)' + (\hat{B}-B)XX'(\hat{B}-B)' , \tag{19}$$

where $\hat{B} = YX'(XX')^{-}$. In case B is subjected to g linear restrictions $FB' = 0$, F is $g\times q$ and of rank $g(<q)$, then we have to find

$$\text{Min(over B)}(B-\hat{B})XX'(B-\hat{B})' \quad \text{subject to} \quad FB' = 0 \quad . \tag{20}$$

A solution $\hat{\hat{B}}$ of (20) is obtained by using (17), and we find that

$$\hat{\hat{B}}-\hat{B} = - \hat{B}F'(F(XX')^{-}F')^{-1}F(XX')^{-} \quad . \tag{21}$$

If F is chosen such that the right hand side of (21) vanishes then $\hat{\hat{B}} = \hat{B}$. In MANOVA models F is so chosen by design that the particular estimates $\hat{B}$ coincide with the unique estimates $\hat{\hat{B}}$. In case X is of full rank then $\hat{B}F'$ does not vanish. When X is not of full rank the vanishing of the right hand side of (21) is tested by noting that the Lagrange multiplier matrix vanishes.

This is done as follows;

On differentiating

$$(Z-BX)(Z-BX)' - 2\Lambda FB' , \tag{22}$$

with respect to B and Λ we find that

$$ZX' - \hat{B}XX' - \hat{\Lambda}F = 0 \, , \tag{23}$$

and

$$F\hat{B}' = \begin{pmatrix} e' & 0 & 0 \\ 0 & e' & 0 \end{pmatrix} \hat{B}' = 0 \, . \tag{24}$$

Thus from (23) and (24) we find that

$$\begin{aligned} \hat{\Lambda} &= ZX'(XX')^{-}F'(F(XX')^{-}F')^{-1} \\ &= \hat{B}F'(F(XX')^{-}F')^{-1} = 0 \, , \text{ as } \hat{B}F' = 0 \, . \end{aligned} \tag{25}$$

Thus $\hat{\Lambda}$ vanishes, i.e., $\hat{B}F' = 0$ and hence from (21) we conclude that for our model particular estimates and unique estimates coincide. The particular estimates are easy to obtain. We know that α is estimated by the grand mean vector $\bar{z} = (N(N-1))^{-1}\Sigma\Sigma z_{ij}$ and hence

$$\sum_{i \neq j = 1}^{N} \sum (z_{ij}-\rho_i+\pi_j-\bar{z})(z_{ij}-\rho_i+\pi_j-\bar{z})' \, , \tag{26}$$

with respect to ρ_i and equating the derivative to zero we find that

$$\sum_j (z_{ij}-\bar{z}-\rho_i+\pi_j) = 0 = \sum_j z_{ij}-(N-1)\bar{z}-(N-1)\uparrow_i \, , \tag{27}$$

i.e. ρ_i is given by

$$\rho_i = \frac{1}{(N-1)} \sum_j z_{ij} - \bar{z} = \frac{1}{(N-1)} z_{i\cdot} - \bar{z} \, , \tag{28}$$

and similarly we find that

$$-\hat{\pi}_j = \frac{1}{(N-1)} z_{\cdot j} - \bar{z} \, . \tag{29}$$

We arrange the data as

blocks \ columns	1	2		j		N	Totals
1	0	z_{12}		z_{1j}		z_{1N}	$z_{1\cdot}$
2	z_{21}	0		z_{2j}		z_{2N}	$z_{2\cdot}$
i				z_{ij}			
N						0	$z_{N\cdot}$
Totals	$z_{\cdot 1}$			z_{ij}		$z_{\cdot N}$	$\Sigma\Sigma z_{ij} = G$

(30)

and write the MANOVA table as

Source	d.f.	SPM	Test Statistic
Blocks	$N-1$	$\sum_i z_{i\cdot}z'_{i\cdot} - \frac{GG'}{N} = (N-1)W$	$\frac{\lvert (N^2-3N+1)\Sigma \rvert^n}{\lvert (N^2-3N+1)\Sigma + N\rvert^n}$
Columns	$N-1$	$\sum_j z_{\cdot j}z'_{\cdot j} - \frac{GG'}{N} = (N-1)T$	$\frac{\lvert T\rvert}{\lvert \hat{\Sigma}+T\rvert}$ (31)
Error	N^2-3N+1	$\hat{\Sigma} = \frac{\text{by subtraction}}{N^2-3N+1}$	
Total	$N(N-1)-1$	$\sum\sum z_{ij}z'_{ij} - \frac{GG'}{N(N-1)}$	

The test statistics are the usual Wilks' likelihood ratio statistics which reduce to F criteria in the univariate case considered by Leefland and Praag. In case a new set of values of X is available say X_0, then $\hat{Z}_0 = \hat{B}X_0$ predicts future Z.

For the generalized least squares model in the univariate case Leefland and Praag ([3], p. 73) remark that the exact proof that the Lagrange multiplier λ vanishes is tiresome. We give a simple proof in the next section.

2. UNIVARIATE GENERALIZED LEAST SQUARES MODEL

In this section we use the notation given by Leefland and Praag ([3], p. 72, Section 4), whose problem is to minimize

$$\text{Min(over } \gamma)(\bar{z}-Q\gamma)'\hat{\Omega}^{-1}(\bar{z}-Q\gamma), \text{ subject to } i'\gamma = 0 . \tag{32}$$

Now write

$$(\bar{z}-Q\gamma)'\hat{\Omega}^{-1}(\bar{z}-Q\gamma) = (\bar{z}-Q\hat{\gamma})\hat{\Omega}^{-1}(\bar{z}-Q\hat{\gamma})' + (\hat{\gamma}-\gamma)Q'\hat{\Omega}^{-1}Q(\hat{\gamma}-\gamma), \tag{33}$$

where $\hat{\gamma} = (Q'\hat{\Omega}^{-1}Q)^{-1}Q'\hat{\Omega}^{-1}\bar{z}$. Thus it remains to minimize

$$(\hat{\gamma}-\gamma)'Q'\hat{\Omega}^{-1}Q(\hat{\gamma}-\gamma) , \text{ subject to } i'\gamma = 0 . \tag{34}$$

The problem (34) is the same as (9) and hence from (12) we find that

$$\hat{\lambda} = i'\hat{\gamma}/(i'(Q'\hat{\Omega}^{-1}Q)^{-1}i) = 0 , \text{ as } i'\hat{\gamma} = 0 , \tag{35}$$

and hence from (21) we observe that particular estimates and unique estimates coincide. The particular estimates may be computed from

$$\sum_{i \neq j=1}^{n} \sum (\bar{z}_{ij} - \tau_i + \pi_j)^2 / \hat{\sigma}_i^2 \theta_j^2 \hat{k}^2 , \qquad (36)$$

where $\theta_j = j\delta$ in the notation of [3]. From (36) we note that we have $2n$ equations in the $2n$ unknowns $\tau_i, \ldots, \tau_n, \pi_1, \ldots, \pi_n$. The analysis of variance testing theory for generalized least squares model is difficult, as testing problems are generalized Fisher-Behrens problems.

In (36) if the conditions on parameters are taken as $\Sigma\theta_j^{-2}\pi_j = 0$ and $\Sigma\sigma_i^{-2}\tau_i = 0$, then the parameters are easily estimated.

With respect to model (1) we may remark that when the yield is zero, in MANOVA we assume the yield to be δ , a small quantity. Then analyse the design in the usual way and then we take the limits as $\delta \to 0$.

REFERENCES

[1] Anderson, T. W., An Introduction to Multivariate Statistical Analysis, John Wiley, New York, 1958.

[2] Kabe, D. G., Minima of vector quadratic forms with applications to statistics. Metrika 12 (1968), 155-160.

[3] Leefland, P.S.H. and Praag, B.M.S. van. A procedure to estimate realtive powers in binary contacts and an application to Dutch football league results. Statistica Neerlandica 25 (1971), 63-84.

R. P. Gupta
Department of Mathematics
Dalhousie University
Halifax, Nova Scotia
Canada

Applied Statistics, R.P. Gupta, (editor)

ON SUBSET SELECTION PROCEDURES FOR POISSON POPULATIONS AND SOME APPLICATIONS TO THE MULTINOMIAL SELECTION PROBLEMS

by Shanti S. Gupta & Deng-Yuan Huang

1. Introduction

We discuss the problem of selecting a subset of k Poisson populations including the best which is associated with the largest value of the parameters. Goel [2] has shown that the usual type of selection rules do not exist for some values of the probability P^* of correct selection. We propose some subset selection procedures which exist for all P^*. These rules are based on a result of Chapman [1] who showed that there is no unbiased estimator of the ratio $\frac{\theta_1}{\theta_2}$ with finite variance, where θ_1, θ_2 are expected values of X_1, X_2 with Poisson distributions, and X_1, X_2 are independent, but that the estimator $\frac{X_1}{X_2+1}$ is "almost unbiased."

Let $\pi_1,\pi_2,\ldots,\pi_k$ be k independent Poisson populations, i.e. π_i has a Poisson distribution with parameter λ_i, $i = 1,2,\ldots,k$. Suppose that we have equal sample size from each population. Without loss of generality, one can assume the sample size to be one. Let $\lambda_{[1]} \leq \lambda_{[2]} \leq \cdots \leq \lambda_{[k]}$ and let us assume λ's are unknown.

Given any P^*, $\frac{1}{k} < P^* < 1$, we want to select a subset of these k populations such that the subset contains the population corresponding to the parameter $\lambda_{[k]}$ with probability at least P^*, no matter what the configuration of $\lambda_1,\lambda_2,\ldots,\lambda_k$ is. We denote this notation by CS. Therefore we are interested in some selection rules R such that

$$\inf_{\Omega} P(CS|R) \geq P^* \tag{1.1}$$

where, Ω is the set of all k-tuples $(\lambda_1,\lambda_2,\ldots,\lambda_k)$, $\lambda_i > 0$, $i = 1,2,\ldots,k$.

Let $X_1,X_2,\ldots,X_k$ denote the sample from $\pi_1,\pi_2,\ldots,\pi_k$, respectively. Let $X_{(i)}$ be associated with $\lambda_{[i]}$.

* This research was supported by the Office of Naval Research under Contract N00014-67-A-00014 at Purdue University. Reproduction in whole or part is permitted for any purpose of the United States Government.

In Section 2, we discuss a subset selection rule to satisfy the probability requirement (1.1), and to find an upper bound for the expected subset size. In Section 3, we are interested in deriving a conditional procedure based on the total sum of observations. In Section 4, we discuss methods for finding values of constants associated with the unconditional and conditional procedures, respectively. In Section 5, we show some applications of the conditional procedure of Section 3 to the multinomial distribution for the selection problem of the largest cell probability.

2. The Unconditional Selection Procedure R_1

2.1. The Rule R_1 and Probability of Correct Selection

R_1: Select the population π_i in the subset if and only if $X_i + 1 \geq c_1 \max_{1\leq j\leq k} X_j$, where $0 < c_1 \leq 1$ is to be chosen to satisfy the basic probability requirement.

We have

$$P(CS|R_1) = P(X_{(k)} + 1 \geq c_1 \max_{1\leq j\leq k-1} X_{(j)})$$

$$= \sum_{x=0}^{\infty} e^{-\lambda_{[k]}} \frac{\lambda_{[k]}^x}{x!} \{ \prod_{j=1}^{k-1} \sum_{i=0}^{[(x+1)/c_1]} e^{-\lambda_{[j]}} \frac{\lambda_{[j]}^i}{i!} \}$$

$$= \sum_{x=0}^{\infty} e^{-\lambda_{[k]}} \frac{\lambda_{[k]}^x}{x!} \{ \prod_{j=1}^{k-1} \int_{\lambda_{[j]}}^{\infty} \frac{1}{[\frac{x+1}{c_1}]} y^{[\frac{x+1}{c_1}]} e^{-y} dy\} .$$

Hence we have the following lemma.

Lemma 2.1.1.

$$\inf_{\Omega} P(CS|R_1) = \inf_{\Omega_0} P(CS|R_1) ,$$

where $\Omega_0 = \{\underline{\lambda} : \lambda_1 = \lambda_2 = \ldots = \lambda_k = \lambda\}$.

Under the parameter space Ω, we know that the joint distribution of $X_1,\ldots,X_k$, given $\sum_1^k X_i = t$, is a multinomial distribution with parameters $t ; \theta_1,\ldots,\theta_k; \theta_j = \frac{\lambda_j}{\sum_1^k \lambda_i}$, $j = 1,\ldots,k$, i.e.

$$P(X_1 = x_1, \ldots, X_k = x_k \mid \sum_1^k X_i = t)$$

$$= \frac{t!}{x_1! \ldots x_k!} \theta_1^{x_1} \ldots \theta_k^{x_k} .$$

Lemma 2.1.2. For any t, $t \geq 0$, $0 < c_1(t) \leq 1$,

$$P_{\Omega_0}(X_k + 1 \geq c_1(t) \max_{1 \leq j \leq k-1} X_j \mid \sum_1^k X_i = t)$$

$$= \sum \frac{t!}{x_1! x_2! \ldots x_k!} (\tfrac{1}{k})^{x_1} \ldots (\tfrac{1}{k})^{x_k} , \text{ where}$$

the summation is over all x_i's such that $x_k + 1 \geq c_1(t) \max_{1 \leq j \leq k-1} x_j$, $x_j \geq 0$, for $j = 1,2,\ldots,k$, $\sum_1^k X_i = t$.

Theorem 2.1.1. For given P^*, let $P_1^* = (P^*)^{\frac{1}{k-1}}$, any r, $r \geq 0$, let $c_1(r)$ be the largest value such that

$$P_{\Omega_0}(X_1 \leq \frac{r+1}{1+c_1(r)} \mid X_1 + X_2 = r) \geq P_1^* .$$

If, $c_1 = \inf\{c_1(r): r \geq 0\}$, then

$$\inf_{\Omega} P(CS \mid R_1) \geq P^* .$$

Proof.

$$P_{\Omega_0}(CS \mid R_1) = P_{\Omega_0}(X_{(k)} + 1 \geq c_1 \max_{1 \leq j \leq k-1} X_{(j)}) = \sum_{x=0}^{n} \frac{\lambda^t e^{-\lambda}}{x!} \left\{ \sum_{i=0}^{[\frac{x+1}{c}]} \frac{e^{-\lambda} \lambda^2}{i!} \right\}^{k-1}$$

$$\geq \left\{ \sum_{x=0}^{n} \frac{\lambda^x e^{-\lambda}}{x!} \left(\sum_{i=0}^{[\frac{x+1}{c}]} \frac{e^{-\lambda} \lambda^2}{i!} \right) \right\}^{k-1}$$

$$= \left\{ P_{\Omega_0}(X_2 + 1 \geq c_1 X_1) \right\}^{k-1}$$

$$= \left\{ \sum_{r=0}^{\infty} P_{\Omega_0}(X_2 + 1 \geq c_1 X_1 \mid X_1 + X_2 = r) P_{\Omega_0}(X_1 + X_2 = r) \right\}^{k-1}$$

$$\geq \left\{ \sum_{r=0}^{\infty} P_{\Omega_0}(X_2 + 1 \geq c_1(r) X_1 \mid X_1 + X_2 = r) P_{\Omega_0}(X_1 + X_2 = r) \right\}^{k-1}$$

$$= \left\{ \sum_{r=0}^{\infty} P_{\Omega_0}(X_1 \leq \frac{r+1}{1+c_1(r)} \mid X_1 + X_2 = r) P_{\Omega_0}(X_1 + X_2 = r) \right\}^{k-1}$$

$\geq (P_1^*)^{k-1} = P^*$. And by Lemma 2.1.1, the proof is complete.

2.2. An Upper Bound on the Expected Subset Size Associated with R_1

For the special case, $\lambda_{[1]} = \ldots = \lambda_{[k-1]} = \lambda$, $\lambda_{[k]} = \delta\lambda$, $\delta > 1$, $\lambda > \lambda_0 > 0$. Let us denote the space of all slippage configurations of the type discussed above by Ω_1 . We discuss the expected subset size as follows.

Theorem 2.2.1

$$E_{\Omega_1}(S|R_1) \leq \sup_{r\geq 1} g(r) + (2 - \sup_{r\geq 1} g(r))e^{-(1+\delta)\lambda_0}$$

$$+ (k-2)\{\sup_{r\geq 1} h(r) + (1 - \sup_{r\geq 1} h(r))e^{-2\lambda_0}\} ,$$

where

$$g(r) = \{\sum_{s=0}^{[\frac{r+1}{1+c_1}]} \binom{r}{s} (\frac{\delta}{1+\delta})^s (\frac{\delta}{1+\delta})^{r-s}$$

$$+ \sum_{r=r-[\frac{r+1}{1+c_1}]}^{r} \binom{r}{s} (\frac{1}{1+\delta})^s (\frac{\delta}{1+\delta})^{r-s}\} , \text{ and}$$

$$h(r) = \sum_{s=0}^{[\frac{r+1}{1+c_1}]} \binom{r}{s} \frac{1}{2^r} .$$

Proof. $E_{\Omega_1}(S|R_1) = \sum_{i=1}^{k} P(X_{(i)} + 1 \geq c_1 \max_{\substack{1\leq j\leq k\\ j\neq i}} X_{(j)})$

$$\leq \frac{1}{k-1} \sum_{i=1}^{k} \sum_{j\neq i} P(X_{(i)} + 1 \geq c_1 X_{(j)})$$

$$= \frac{1}{k-1} \sum_{i=1}^{k} \sum_{j\neq i} \sum_{r=0}^{\infty} P(X_{(i)}+1 \geq c_1 X_{(j)} | X_{(i)}=r)\ P(X_{(i)}+X_{(j)}=r)$$

$$= \frac{1}{k-1} \sum_{i=1}^{k} \sum_{j\neq i} \sum_{r=0}^{\infty} P(X_{(j)} \leq \frac{r+1}{1+c_1} \quad X_{(i)}+X_{(j)}=r)\ P(X_{(i)}+X_{(j)}=r)$$

$$= \frac{1}{k-1} \sum_{i=1}^{k} \sum_{j\neq i} \sum_{r=0}^{\infty} \sum_{s=0}^{[\frac{r+1}{1+c_1}]} \binom{r}{s} (\frac{\lambda_{[j]}}{\lambda_{[i]}+\lambda_{[j]}})^s (\frac{\lambda_{[i]}}{\lambda_{[i]}+\lambda_{[j]}})^{r-s} P(X_{(i)}+X_{(j)}=r)$$

$$= \frac{1}{k-1} \{ \sum_{j=1}^{k-1} \sum_{r=0}^{\infty} \sum_{s=0}^{[\frac{r+1}{1+c_1}]} \binom{r}{s} (\frac{1}{1+\delta})^s (\frac{\delta}{1+\delta})^{r-s} P(X_{(k)}+X_{(j)}=r)$$

$$+ \sum_{i=1}^{k-1} \sum_{j\neq i,k} \sum_{r=0}^{\infty} \sum_{s=0}^{[\frac{r+1}{1+c_1}]} \binom{r}{s} \frac{1}{2^r} P(X_{(i)}+X_{(j)}=r)$$

$$+ \sum_{i=1}^{k-1} \sum_{r=0}^{\infty} \sum_{s=0}^{[\frac{r+1}{1+c_1}]} \binom{r}{s} (\frac{\delta}{1+\delta})^s (\frac{\delta}{1+\delta})^{r-s} P(X_{(i)}+X_{(k)}=r) \}$$

$$= \frac{1}{k-1} [\sum_{j=1}^{k-1} \sum_{r=0}^{\infty} \{ \sum_{s=0}^{[\frac{r+1}{1+c_1}]} \binom{r}{s} (\frac{1}{1+\delta})^s (\frac{\delta}{1+\delta})^{r-s}$$

$$+ \sum_{s=r-[\frac{r+1}{1+c_1}]}^{r} \binom{r}{s} (\frac{1}{1+\delta})^s (\frac{\delta}{1+\delta})^{r-s} \} \times P(X_{(j)}+X_{(k)}=r)$$

$$+ \sum_{i=1}^{k-1} \sum_{j\neq i,k} \sum_{r=0}^{\infty} \sum_{s=0}^{[\frac{r+1}{1+c_1}]} \binom{r}{s} \frac{1}{2^r} P(X_{(i)}+X_{(j)}=r)]$$

$$= \sum_{r=0}^{\infty} [\sum_{s=0}^{[\frac{r+1}{1+c_1}]} \binom{r}{s} (\frac{1}{1+\delta})^s (\frac{\delta}{1+\delta})^{r-s} + \sum_{s=r-[\frac{r+1}{1+c_1}]}^{r} \binom{r}{s} (\frac{1}{1+\delta})^s (\frac{\delta}{1+\delta})^{r-s}]$$

$$\times \frac{[(1+\delta)\lambda]^r e^{-(1+\delta)\lambda}}{r!}$$

$$+ (k-2) \{ \sum_{r=0}^{\infty} [\sum_{s=0}^{[\frac{r+1}{1+c_1}]} \binom{r}{s} \frac{1}{2^r}] \frac{(2\lambda)^r e^{-2\lambda}}{r!} \} .$$

The proof is completed by taking supremum of each square bracket for all $r(\geq 1)$ terms.

3. The Conditional Procedure R_2

R_2 : Select the population π_i in the subset if and only if

$$X_i + \geq c_2(t) \max_{1\leq j\leq k} X_j \text{ , given } T = \sum_1^k X_i = t \text{ ,}$$

where $0 \leq t$ and $0 < c_2(t) \leq 1$ is chosen to satisfy the probability requirement. For this rule R_2, we obtain an exact result for $k = 2$ in Theorem 3.1.1. For $k \geq 3$, we have a lower bound for the probability of a correct selection in Theorem 3.1.2.

3.1. The Probability of a Correct Selection for R_2

Theorem 3.1.1. For a given P^*, $(\frac{1}{k} < P^* < 1)$, $k = 2$, and any $r \geq 0$, let $c_2(r)$ be the largest value such that

$$P_{\Omega_0}(X_1 \leq \frac{r+1}{1+c_2(r)} \mid X_1 + X_2 = r) \geq P^* .$$

Then, $\inf_{\Omega} P(CS|R_2) = \inf_{\Omega_0} P(CS|R_2) \geq P^*$.

Proof.

$$P(CS|R_2) = P(X_{(2)}+1 \geq c_2(r)X_{(1)} | X_{(1)}+X_{(2)} = r)$$

$$= P(X_{(1)} \leq \frac{r+1}{1+c_2(r)} \mid X_{(1)}+X_{(2)} = r)$$

$$(3.1.1) \quad = \sum_{x=0}^{[\frac{r+1}{1+c_2(r)}]} \binom{r}{s} \left(\frac{\lambda_{[1]}}{\lambda_{[1]}+\lambda_{[2]}}\right)^x \left(1 - \frac{\lambda_{[1]}}{\lambda_{[1]}+\lambda_{[2]}}\right)^{r-x} .$$

For fixed $\lambda_{[2]}$, $\dfrac{1}{1 + \frac{\lambda_{[2]}}{\lambda_{[1]}}}$ increases with $\lambda_{[1]}$ to $\frac{1}{2}$, and the expression on the right hand side of (3.1.1) decreases as $p = \dfrac{\lambda_{[1]}}{\lambda_{[1]}+\lambda_{[2]}}$ increases to the value $\frac{1}{2}$.

Hence, $\inf_{\Omega} P(CS|R_2) = \inf_{\Omega_0} P(CS|R_2)$.

This completes the proof.

Theorem 3.1.2. For $k \geq 3$, and any P^*, let $P_2^* = 1 - \frac{1-P^*}{k-1}$, $0 \leq r \leq t$; also let $c_2(r)$ be the largest value such that

$$P_{\Omega_0}(X_2+1 \geq X_1c_2(r) | X_1+X_2 = r) \geq P_2^* .$$

If, $c_2(t) = \min\{c_2(r);\ 0 \le r \le t\}$, then

$$\inf_{\Omega} P(CS|R_2) \ge P^* .$$

Proof.

$$P(CS|R_2) = P\{X_{(k)}+1 \ge c_2(t) \max_{1\le j\le k-1} X_{(j)} \mid \sum_1^k X_i = t\}$$

(3.1.2) $$\ge 2-k + \sum_{j=1}^{k-1} P(X_{(k)}+1 \ge c_2(t)X_{(j)} \mid \sum_1^k X_i = t) .$$

For any j , $1 \le j \le k-1$,

$$P(X_{(k)}+1 \ge c_2(t)X_{(j)} , \sum_1^k X_{(i)} = t)$$

$$= \sum_{r=0}^{t} P(X_{(k)}+1 \ge c_2(t)X_{(j)}, X_{(k)}+X_{(j)}=r, \sum_{i\ne j,k} X_{(i)} = t)$$

$$= \sum_{r=0}^{t} P(X_{(k)}+1 \ge c_2(t)X_{(j)}|X_{(k)}+X_{(j)}=r)P(X_{(k)}+X_{(j)}=r, \sum_1^k X_i=t)$$

$$\ge \sum_{r=0}^{t} P(X_{(k)}+1 \ge c_2(r)X_{(j)}|X_{(k)}+X_{(j)}=r)P(X_{(k)}+X_{(j)}=r, \sum_1^k X_i=t)$$

$$\ge \sum_{r=0}^{t} P(X_2+1 \ge c_2(r)X_1|X_1+X_2 = r) P(X_{(k)}+X_{(j)} = r , \sum_1^k X_i=t)$$

(3.1.3) $$\ge P_2^* \cdot P(\sum_1^k X_i = t) .$$

From (3.1.2), (3.1.3) and

$$P(X_{(k)}+1 \ge c_2(t)X_{(j)} \mid \sum_1^k X_i = t)$$

$$= P(X_{(k)}+1 \ge c_2(t)X_{(j)} , \sum_1^k X_i = t)/P(\sum_1^k X_i = t) .$$

Thus, we have proved the result.

3.2. An Upper Bound on the Expected Subset Size for R_2

For the special case, $\lambda_{[1]} = \dots = \lambda_{[k-1]} = \lambda$, $\lambda_{[k]} = \delta\lambda$, $\delta > 1$, $\lambda > 0$. Let us denote the space of all slippage configurations of the type discussed as above by Ω_2. We discuss the expected subset size as follows.

Theorem 3.2.1.

$$E_{\Omega_2}(S|R_2) \leq \sum_{r=0}^{t} \left\{ \sum_{s=0}^{[\frac{r+1}{1+c_2(t)}]} \binom{r}{s}\delta^{r-s} + \sum_{r-[\frac{r+1}{1+c_2(t)}]}^{r} \binom{r}{s}\delta^{r-s} \right\}$$

$$\times \frac{t!}{r!(t-r)!} \frac{(k-2)^{t-r}}{(k-1+\delta)^{t}} + (k-2) \sum_{r=0}^{t} \left(\sum_{s=0}^{[\frac{r+1}{1+c_2(t)}]} \binom{r}{s} \frac{(k-2)^{t-r}}{k^{t}} \right)$$

Proof.

$$E_{\Omega_2}(X|R_2) = \sum_{i=1}^{k} P(X_{(i)}+1 \geq c_2(t) \max_{\substack{1\leq j\leq k\\ j\neq i}} X_{(j)} \mid \sum_1^k X_i = t)$$

$$\leq \frac{1}{k-1} \sum_{i=1}^{k} \sum_{j\neq i} P(X_{(i)}+1 \geq c_2(t)X_{(j)} \mid \sum_1^k X_i = t) .$$

$$= \frac{1}{(k-1)P(T=t)} \sum_{i=1}^{k} \sum_{j\neq i} \sum_{r=0}^{t} P(X_{(i)}+1 \geq c_2(t)X_{(j)} | X_{(i)}+X_{(j)} = r)$$

$$P(X_{(i)}+X_{(j)} = r , \sum_1^k X_i = t)$$

$$= \frac{1}{(k-1)P(T=t)} \sum_{i=1}^{k} \sum_{j\neq i} \sum_{r=0}^{t} \sum_{s=0}^{[\frac{r+1}{1+c_2(t)}]} \binom{r}{s} \left(\frac{\lambda_{[j]}}{\lambda_{[i]}+\lambda_{[j]}}\right)^{s} \left(\frac{\lambda_{[i]}}{\lambda_{[i]}+\lambda_{[j]}}\right)^{r-s}$$

$$P(X_{(i)}+X_{(j)} = r , \sum_1^k X_i = t) ,$$

the same reason as in (2.2.1).

$$= \frac{1}{(k-1)P(T=t)} \sum_{j=1}^{k-1} \sum_{r=0}^{t} \sum_{s=0}^{[\frac{r+1}{1+c_2(t)}]} \binom{r}{s} \left(\frac{1}{1+\delta}\right)^{s} \left(\frac{\delta}{1+\delta}\right)^{r-s} P(X_{(k)}+X_{(j)}=r,\sum_1^k X_i=t)$$

$$+ \sum_{i=1}^{k-1} \sum_{j\neq i,k} \sum_{r=0}^{t} \sum_{s=0}^{[\frac{r+1}{1+c_2(t)}]} \binom{r}{s} \frac{1}{2^r} P(X_{(i)}+X_{(j)} = r , \sum_1^k X_i = t)$$

$$+ \sum_{i=1}^{k-1} \sum_{r=0}^{t} \sum_{s=0}^{[\frac{r+1}{1+c_2(t)}]} \binom{r}{s} \left(\frac{\delta}{1+\delta}\right)^{s} \left(\frac{1}{1+\delta}\right)^{r-s} P(X_{(i)}+X_{(k)} = r , \sum_1^k X_i = t)\}$$

$$= \frac{1}{(k-1)P(T=t)} \sum_{j=1}^{k-1} \sum_{r=0}^{t} \left[\sum_{s=0}^{[\frac{r+1}{1+c_2(t)}]} \binom{r}{s} \left(\frac{1}{1+\delta}\right)^s \left(\frac{\delta}{1+\delta}\right)^{r-s} \right.$$

$$\left. + \sum_{s=r-[\frac{r+1}{1+c_2(t)}]}^{r} \binom{r}{s} \left(\frac{1}{1+\delta}\right)^s \left(\frac{\delta}{1+\delta}\right)^{r-s} \right] P(X_{(j)}+X_{(k)} = r\ ,\ T = t)$$

$$+ \sum_{i=1}^{k-1} \sum_{j \neq i,k} \sum_{r=0}^{t} \sum_{s=0}^{[\frac{r+1}{1+c_2(t)}]} \binom{r}{s} \frac{1}{2^r} P(X_{(i)}+X_{(j)} = r\ ,\ T = t)\}$$

$$+ \sum_{r=0}^{t} \{ \sum_{s=0}^{[\frac{r+1}{1+c_2(t)}]} \binom{r}{s} \left(\frac{1}{1+\delta}\right)^s \left(\frac{\delta}{1+\delta}\right)^{r-s} + \sum_{s=r-[\frac{r+1}{1+c_2(t)}]}^{r} \binom{r}{s} \left(\frac{1}{1+\delta}\right)^s \left(\frac{\delta}{1+\delta}\right)^{r-s} \}$$

$$\times \frac{t!}{r!(t-r)!} \cdot \frac{(1+\delta)^r (k-2)^{t-r}}{(k-1+\delta)^t}$$

$$+ (k-2) \sum_{r=0}^{t} \sum_{s=0}^{[\frac{r+1}{1+c_2(t)}]} \binom{r}{s} \frac{1}{2^r} \cdot \frac{2^r \cdot (k-2)^{t-r}}{k^t} .$$

By simplifying, we have the result.

4. A Method to Find the c-Values for the Procedures R_1 and R_2

The conditional distribution of X_1, given $X_1+X_2 = r$ is a binomial random variable Y with $b(r, \frac{1}{2})$. We know that

$$(4.1) \qquad P(X_1 \leq \frac{r+1}{1+c_i(r)} \mid X_1+X_2=r) = \sum_{x=0}^{[\frac{r+1}{1+c(r)}]} \binom{r}{x} \frac{1}{2^r} \geq P_i^* \ ,\ i = 1,2 .$$

(1) For every r, $(1 \leq r \leq 150)$, we can find the largest c-values satisfying (4.1) from the tables of the cumulative binomial populations [4] for $p = 0.5$ and $r = 1(1)150$.

(2) For $r \geq 20$, by using normal approximation with continuity correction, we have

$$P(X_1 \leq \frac{r+1}{1+c_i(r)} \mid X_1+X_2 = r)$$

$$= P(Y \leq \frac{r+1}{1+c_i(r)}) \approxeq P(Z \leq \frac{r+1}{1+c_i(r)} + \frac{1}{2} - \frac{r}{2})/\frac{1}{2}\sqrt{r})$$

$$= P(Z \leq \frac{(1-c_i(r))r+3+c_i(r)}{\sqrt{r}(1+c_i(r))})$$

$$= \Phi(\frac{(1-c_i(r))r+3+c_i(r)}{\sqrt{r}(1+c_i(r))}) \, , \, i = 1,2 \, ,$$

where Φ is the cdf of the standard normal distribution. Since for $r \geq 20$, $0 < c \leq 0.99$,

$\frac{(1-c)r+3+c}{\sqrt{r}(1+c)}$ is increasing in r, hence,

$$\Phi(\frac{(1-c)r+3+c}{\sqrt{r}(1+c)}) \geq \Phi(\frac{20(1-c)+3+c}{\sqrt{20}(1+c)})$$

$$= \Phi(\frac{23-19c}{\sqrt{20}(1+c)}) \, .$$

Let c_i^*, $0 < c_i^* \leq 0.99$, be the largest value such that

$$\Phi(\frac{23-19c_i^*}{\sqrt{20}(1+c_i^*)}) \geq P_i^* \, , \, i = 1,2.$$

For every r, $r \geq 20$, we choose $c_i(r) = c_i^*$, $i = 1,2.$ Then the c-values can be obtained as follows:

$$c_i = \min\{ \min_{1 \leq r \leq 19} c_i(r) \, , \, c_i^*\} \, , \, i = 1,2.$$

5. Subset Selection Procedures for Multinomial Distributions

We wish to consider selection problems for the multinomial distribution. In some situations the experimenter may be interested in selecting a subset containing the cell with the highest or the lowest probability based on a set of observations. The main problem is to define a selection rule which selects a small, non-empty subset such that the probability of including the cell with the largest or the smallest cell-probability is at least equal to a preassigned number P^*.

Let $p_1, p_2, \ldots, p_k$ be the unknown cell-probabilities in the multinomial distribution with $\sum_1^n p_i = 1$. Let $X_1, \ldots, X_k$ be the respective observations in the k cells of the distribution with $\sum_1^k X_i = N$. Let the ordered cell-probabilities be given by

$$0 < p_{[1]} \le p_{[2]} \le \cdots \le p_{[k]}\ , \text{ and } \Omega' = \{(p_1,\ldots,p_k) : \sum_{i=1}^k p_i = 1\}\ .$$

Let $X_{(i)}$ be associated with $p_{[i]}$. The goal of the experimenter is to select a subset containing the cell corresponding to $p_{[k]}$. A correct selection {CS} is defined as the selection of any subset of the k cells which contains the cell with the largest probability. Gupta and Nagel [3] have proposed procedures for the problem. We are interested in deriving a different procedure.

The rule R_3 for selecting the subset to contain the cell with the largest probability is as follows:

R_3: Select the cell with observed X_i if and only if $X_i + 1 \ge c_3 \max_{1\le j\le k} X_j$, where $0 < c_3 \le 1$.

From Lemma 2.1.2, we know that

$$P(CS|R_3) = P(X_{(k)} + 1 \ge c_3 \max_{1\le j\le k} X_j)$$

$$= P(Y_{(k)} + 1 \ge c_3 \max_{1\le j\le k} Y_j \mid \sum_1^k Y_i = N)\ ,$$

where $Y_1,\ldots,Y_k$ are independent observations have Poisson distributions with parameter $p_1,p_2,\ldots,p_k$ and $\sum_1^k Y_i = N$. Let $Y_{(i)}$ be associated with $p_{[i]}$, $(i = 1,\ldots,k)$. Hence, the rule R_3 is equivalent to the conditional procedure for Poisson distributions as follows:

R_3': Select the Poisson population π_i with parameter p_i if and only if

$$Y_i + 1 \ge c_3 \max Y_j\ , \text{ given } \sum_1^k Y_i = N\ .$$

We have discussed how to compute the fule R_3' in Section 3. Similarly, Gupta and Nagel [3] procedure

R_4: Select the cell with observed X_i iff $X_i \ge X_{max} - D_N$, where $X_{max} = \max(X_1,\ldots,X_k)$, $\sum_1^k X_i = N$ and D_N is a given non-negative integer, is equivalent to the rule

R_4': Select the Poisson population π_i with parameter p_i if and only if

$$Y_i \geq Y_{max} - D_N \ , \text{ given } \sum_1^k Y_i = N \ .$$

For rule R_4' , the discussion is similar as in Section 3. Thus we have the result for the rule R_4 as follows:

Theorem 5.1. For given P^* , let $P_2^* = 1 - \frac{1-P^*}{k-1}$, $0 \leq r \leq N$; also let $D(r)$ be the smallest value such that

$$P(Y_2 \geq Y_1 - D(r) \mid Y_1 + Y_2 = r) \geq P_2^* \ .$$

If, $D_N = \max\{D(r): \ 0 \leq r \leq N\}$, then

$$\inf_{\Omega'} P(CS|R_4) \geq P^* \ .$$

Note that it is an exact result, when $k = 2$.

Theorem 5.2. Let $\Omega_1' = \{p_{[1]} = \cdots = p_{[k-1]} = p \ , \ p_{[k]} = \delta p \ , \ p > 0\}$, $\delta > 1$.

$$E_{\Omega_1'}(S|R_4) \leq \sum_{r=0}^{N} \left\{ \sum_{s=0}^{[\frac{r+D(r)}{2}]} \binom{r}{s} \delta^{r-s} + \sum_{r-[\frac{r+D(r)}{2}]}^{r} \binom{r}{s} \delta^{r-s} \right\}$$

$$\times \frac{N!}{r!(N-r)!} \frac{(k-2)^{N-r}}{(k-1+\delta)^N} + (k-2) \sum_{r=0}^{N} \left(\sum_{s=0}^{[\frac{r+D(r)}{2}]} \binom{r}{s} \frac{(k-2)^{N-r}}{k^N} \right) .$$

Remark: (i) We know that if we consider the sample size of the multinomial distribution as random variables with Poisson distribution associated with the same cell probabilities, then the multinomial distribution is changed into the joint distribution of independent Poisson associated with the parameters corresponding to the same cell probabilities. For this case, the procedure R_4 is reduced to the usual type procedure as discussed by Goel [2]. From this point of view, we prefer R_3 to R_4 for the multinomial distribution. (ii) Gupta and Nagel [3] discussed the infimum of the probability of a correct selection associated with R_4 . By using a similar argument, we obtain a similar result for R_3 and connect it with R_2 .

6. Acknowledgement

The authors wish to thank Mr. Ray Carroll for his critical reading and helpful comments. Thanks are also due to Professor Milton Sobel of the University of Minnesota for making some helpful comments.

References

[1] Chapman, D. G. (1952). On tests and estimates for the ratio of Poisson means. Ann. Inst. Statist. Math., 4, 45-49.

[2] Goel, P. K. (1972). A note on the non-existence of subset selection procedure for Poisson populations. Mimeo Series No. 303, Dept. of Statistics, Purdue University, W. Lafayette, Indiana.

[3] Gupta, S. S. and Nagel, K. (1967). On selection and ranking procedures and order statistics from the multinomial distribution. Sankhyā, Ser. B, 29, 1-34.

[4] Tables of the cumulative binomial probabilities (1952). U. S. Office of the Chief of Ordance.

Shanti S. Gupta
Department of Statistics
Purdue University
Lafayette, Indiana 47907
U.S.A.

Dueng-Yuan Huang
Department of Statistics
Purdue University
Lafayette, Indiana 47907
U.S.A.

Applied Statistics, R.P. Gupta, (editor)

A BAYESIAN APPROACH TO SOME PROBLEMS INVOLVING THE DETECTION OF SPURIOSITY

by Irwin Guttman* and C.G. Khatri**

1. INTRODUCTION

One of us (Guttman (1973)) has investigated the problem of detection of spuriosity in the following setting. A random sample of n independent observations $\underset{\sim}{Y}_1,\ldots,\underset{\sim}{Y}_n$ is thought to be generated from a m-variate $N(\underset{\sim}{\mu},\Sigma)$ distribution. It is feared however, that one of the $\underset{\sim}{Y}_j$ is generated from the spurious source $N(\underset{\sim}{\mu}+\underset{\sim}{a},\Sigma)$. The posterior of the mean shift $\underset{\sim}{a}$ is then shown to be of the form

$$(1.1)\qquad p(\underset{\sim}{a}|\underset{\sim}{y},\ldots,\underset{\sim}{y}_n) = \sum_{j=1}^{n} c_j h_m(\underset{\sim}{a}|\underset{\sim}{\eta}_j;B_m^{(j)};n-m-1)$$

that is, a weighted combination of generalized m-variate t-densities defined by

$$(1.1a)\qquad h_m(\underset{\sim}{\omega}|\underset{\sim}{\eta};B;v) = \frac{\Gamma(\frac{m+v}{2})\,|B|^{\frac{1}{2}}}{\Gamma(\frac{v}{2})\,(\pi v)^{m/2}}\left[1+\frac{(\underset{\sim}{\omega}-\underset{\sim}{\eta})'B(\underset{\sim}{\omega}-\underset{\sim}{\eta})}{v}\right]^{-(m+v)/2}$$

with the weights c_j of (1.1) given by

$$(1.1b)\qquad c_j = \left|A_m^{(j)}\right|^{-(n-2)/2} \Big/ \sum_{\ell=1}^{n}\left|A_m^{(\ell)}\right|^{-(n-2)/2} \quad ;\ j=1,\ldots,n$$

with the $(m\times m)$ matrices $A_m^{(j)}$ defined by

$$(1.1c)\qquad A_m^{(j)} = \sum_{\ell\neq j}(\underset{\sim}{y}_\ell-\bar{\underset{\sim}{y}}^{(j)})(\underset{\sim}{y}_\ell-\bar{\underset{\sim}{y}}^{(j)})'$$

and $\bar{y}^{(j)} = (n-1)^{-1}\sum_{\ell\neq j}\underset{\sim}{y}_\ell$. The quantities appearing in the density h_m in (1.1) are given by

$$(1.1d)\qquad \underset{\sim}{\eta}_j = \frac{n}{n-1}(\underset{\sim}{y}_j-\bar{\underset{\sim}{y}})\ ,\quad B_m^{(j)} = \left[n\,A_m^{(j)}\Big/(n-1)(n-m-1)\right]^{-1} .$$

---

* Work supported by a grant from the National Research Council of Canada, Grant No. A-8743.

** On leave of absence from Gujarat University; work supported by a grant from the National Research Council of Canada, Grant No. A-4060.

The use of the above result is discussed, interpreted and illustrated with numerical examples in Guttman (1973).

In this paper, we extend the above to the discussion of a model which includes the case of spuriosity of the scale-change variety which may be combined with spuriousness of the mean-shift variety, and to the situation where one or two observations may be spurious. We will confine our discussion to the univariate $(m=1)$ situation.

2. THE MIXED-SHIFT MODEL: ONE SPURIOUS OBSERVATION

In this section, we will be concerned with the analysis of the following situation: it is hoped that $Y_1,\ldots,Y_n$ are n independent observations from $N(\mu,\sigma^2)$, but it is feared, however, that one of the Y_j has been generated from the spurious source $N\left(\mu+a\frac{\sigma}{\delta};\frac{\sigma^2}{\delta^2}\right)$, $\delta>0$, that is, the spurious source has a distribution whose variance is different from the intended $N(\mu,\sigma^2)$ source by the "inflation" factor δ^2, and has mean shifted by "a" standard deviations (of the spurious source) away from μ, the unknown mean of the intended source. Such occurrences are common in the physical sciences -- for example, in the determination of the amount of methanol in a certain solution, it could happen that every once in a while the apparatus slips and that the subsequent observation on the amount of methanol is generated from a distribution whose mean and variance are different than the mean and variance of observations generated when the apparatus is correctly adjusted.

We wish then, to find the posterior of "a" for fixed δ^2, and to see how we may use this in the detection of spuriousness of the kind just indicated. We will then allow δ^2 to vary to see what effect, if any, the assumption of known δ^2 has on our method, i.e., we will perform a sensitivity analysis.

Now assuming that one observation is from a spurious source and that it is equally likely that any one observation be generated from the spurious source as any other, then the likelihood of (μ,a,σ^2) for fixed δ^2 is of the form

$$(2.1)\quad \ell(\mu,a,\sigma^2|y_1,\ldots,y_n;\delta^2) = \sum_{j=1}^{n}\frac{1}{n}\left\{\prod_{i\neq j}(2\pi\sigma^2)^{-\frac{1}{2}}\exp -\frac{1}{2\sigma^2}(y_i-\mu)^2\right\} \times\left[\frac{2\pi\sigma^2}{\delta^2}\right]^{-\frac{1}{2}}\exp\left\{-\frac{\delta^2}{2\sigma^2}(y_j-\mu-a\frac{\sigma}{\delta})^2\right\}.$$

The factor "1/n" is present in the above likelihood because the probability that y_j comes from $N\left[\mu+a\frac{\sigma}{\delta};\frac{\sigma^2}{\delta^2}\right]$ is identically $1/n$, for all j. If we further assume that before obtaining the data, only vague information is available on μ and σ^2, and indeed about "likely values of a", then we are led to the use of the non-informative prior for (μ,σ^2,a) given by

$$p(\mu,\sigma^2,a) \propto (\sigma^2)^{-1} \quad . \tag{2.2}$$

It is shown in Appendix I of Guttman and Khatri (1974) that if (2.2) is combined with (2.1) by Bayes' Theorem, that the posterior of "a" takes the form

$$p(a|\delta^2;y_1,\ldots,y_n) = \sum_{j=1}^{n} d_j\, g_j(a_1) \quad , \quad a_1 = a\sqrt{\frac{n-1}{n-1+\delta^2}} \quad , \tag{2.3}$$

where the weights d_j are given by

$$d_j = (A^{(j)})^{-(n-1)/2} \Big/ \sum_{\ell=1}^{n} (A^{(\ell)})^{-(n-1)/2} \quad , \quad \sum_{j=1}^{n} d_j = 1 \tag{2.3a}$$

with

$$A^{(j)} = \sum_{\ell\neq j} (y_\ell-\bar{y}^{(j)})^2 \quad , \quad \bar{y}^{(j)} = (n-1)^{-1} \sum_{\ell\neq j} y_\ell \tag{2.3b}$$

and where the function $g_j(a_1)$ is defined by

$$g_j(a_1) = \sum_{t=0}^{\infty} (1-\rho_j^2)^{(n-1)/2} \rho_j^t \frac{\Gamma(\frac{n+t-1}{2})}{\Gamma(\frac{t}{2}+1)\Gamma(\frac{n-1}{2})} h_t(a_1) \quad . \tag{2.3c}$$

The quantities ρ_j are given by

$$\rho_j = \delta\sqrt{n-1}\,(y_j-\bar{y}^{(j)}) \Big/ \sqrt{(n-1+\delta^2)[A^{(j)}+k(y_j-\bar{y}^{(j)})^2]} \quad , \tag{2.3d}$$

where $k = (n-1)\delta^2/(n-1+\delta^2)$, and the function h_t is of the form

$$h_t(a_1) = \frac{\Gamma(\frac{t}{2}+1)2^{t/2}}{\sqrt{2\pi}\,\Gamma(t+1)} \sqrt{\frac{n-1}{n-1+\delta^2}}\, a_1^t \exp(-a_1^2/2) \quad . \tag{2.3e}$$

For convenience, we will denote $(y_1,\ldots,y_n)$ by $\underset{\sim}{y}$. Using the above, it is shown in Appendix I of Guttman and Khatri (1974) that the posterior expectation and variance of a are given by

$$(2.4)\quad E(a|\delta^2;\underset{\sim}{y}) = \frac{\sqrt{2}\,\Gamma(\frac{n}{2})}{\Gamma(\frac{n-1}{2})}\,\delta \sum_{j=1}^{n} d_j\,(y_j-\bar{y}^{(j)})\,/\,(A^{(j)})^{\frac{1}{2}} \quad ,$$

and

$$(2.5)\quad \mathrm{Var}(a|\delta^2;\underset{\sim}{y}) = \frac{n-1+\delta^2}{n-1} + (n-1)\delta^2 \sum_{j=1}^{n} \frac{d_j(y_j-\bar{y}^{(j)})^2}{(A^{(j)})} - \{E(a|\delta^2;\underset{\sim}{y})\}^2 .$$

The results (2.3), (2.4) and (2.5) are very interesting for a variety of reasons. Firstly, we note that the posterior of "a" is a weighted combination of functions g_j, and, regarding g_j as a function of a, it is shown in Appendix I of Guttman and Khatri (1974) that the mean value for the density g_j is

$$(2.6)\quad \frac{\sqrt{2}\,\Gamma(\frac{n}{2})}{\Gamma(\frac{n-1}{2})}\,\delta\,\frac{(y_j-\bar{y}^{(j)})}{(A^{(j)})^{\frac{1}{2}}} = \frac{\sqrt{2}\,\Gamma(\frac{n}{2})}{\Gamma(\frac{n-1}{2})}\,\frac{n}{n-1}\,\frac{\delta(y_j-\bar{y})}{(A^{(j)})^{\frac{1}{2}}} \quad .$$

Further, inspection of the weights d_j, as given by (2.3a), shows that the j-th weight d_j is inversely proportional to the power $(n-1)/2$ of the internal sum of squares of all the observations but y_j.

With the above in mind, suppose we consider the following extreme case; that is, suppose one observation has indeed been generated from a spurious source with "extremely large mean shift parameter a", and for the sake of conversation, suppose it is y_j. Then with high probability, y_j will be an *outlier*, that is, it will lie "far away" from the remaining $(n-1)$ observations. Hence, $A^{(j)}$ will tend to be smallest among the n values of $A^{(\ell)}$, $\ell = 1,\ldots,n$, so that the d_j will be largest among the n weights, and for our extreme case, have value very near one. That is, the posterior of "a" will tend to be dominated by g_j, which itself is "located at" the quantity (2.6). Now for this case, $y_j - \bar{y}$ is a source of information for $[(\mu+a\sigma/\delta) - (\mu+a\sigma/n\delta)]$ or $(n-1)a\sigma/n\delta$. Further, $(A^{(j)})^{\frac{1}{2}}$ is a source of information about $[\sqrt{2}\,\Gamma(\frac{n-1}{2})/\Gamma(\frac{n-2}{2})]\sigma$, so that, roughly speaking, (2.6) is a source of information about

$$(2.7)\quad \begin{aligned} &\frac{\sqrt{2}\,\Gamma(\frac{n}{2})}{\Gamma(\frac{n-1}{2})}\,\frac{n}{n-1}\,\delta\,\frac{n-1}{n}\,\frac{a\sigma}{\delta} \Bigg/ \frac{\sqrt{2}\,\Gamma(\frac{n-1}{2})}{\Gamma(\frac{n-2}{2})}\,\sigma \\ &= e_n a \quad , \text{ where } \quad e_n = 1 + o\left(\frac{1}{\sqrt{n}}\right) \quad , \end{aligned}$$

that is, (2.6) is a source of information essentially for a , and for this case, then, so is (2.4), so that the posterior of a would be located at a quantity which gives us some information about a . We note that δ^2 does not come into play and so we would expect that if we vary δ^2 (i.e., do a so-called sensitivity analysis) that the role of δ^2 will not be influential (see below).

To further help us in the examination of the statement "no mean shift has occurred", examination of the weights d_j may be undertaken. If indeed $a = 0$, then the weights should all be approximately "the same". This may be coupled with examination of the posterior odds that "$a > 0$" . Guttman and Khatri (1974) show (in their Appendix I) that

$$(2.8) \quad P(a > 0|\delta^2;\underset{\sim}{y}) = \frac{1}{2} \sum_{j=1}^{n} \sum_{t=0}^{\infty} d_j \, (1-\rho_j^2)^{(n-1)/2} \rho_j^t \frac{\Gamma(\frac{n+t-1}{2})}{\Gamma(\frac{t}{2}+1)\,\Gamma(\frac{n-1}{2})} ,$$

so that the posterior odds in favour of a exceeding zero is

$$(2.9) \qquad \gamma = P(a > 0|\delta^2;\underset{\sim}{y})/P(a < 0|\delta^2;\underset{\sim}{y})$$

$$= P(a > 0|\delta^2;y)/[1-P(a > 0|\delta^2;\underset{\sim}{y}] \ .$$

We now illustrate the above with some numerical examples. We would first like to point out that since we are dealing with the possibility of one spurious observation occurring in n , that for reasons discussed and outlined in Guttman (1973), we would use the method described and implied by the above results for samples of size $n \leq 10$.

To begin with, we generated a set of data (authors' data set 2.1A) comprising six observations from the $N(0,1)$ distribution. The observations (when ordered) turned out to be

(2.10) -.395 , -.312 , .111 , .332 , .566 , .789 .

The corresponding weights d_j (see (2.3a)) turn out to be

(2.10a) .245 , .174 , .083 , .087 , .125 , .286 .

The posterior expectation of a (from (2.4)) and variance (see (2.5)) turn out to be

$$(2.10b) \qquad E(a|\delta^2;\underset{\sim}{y}) = E(a|1;\underset{\sim}{y}) = .017$$

(2.10c) $$\mathrm{Var}(a|\delta^2;\underset{\sim}{y}) = \mathrm{Var}(a|1;\underset{\sim}{y}) = 3.663$$

and the posterior of a (see (2.3)) is graphed in Figure 2.1. Examination of (2.3) and the weights (2.10a) shows that $p(a|1;\underset{\sim}{y})$ is mainly an amalgam of g_1 and g_6, and to a lesser extent, g_2 and g_5, while in comparison wiht these, g_3 and g_4 play a small role. The composite is as graphed in Figure 2.1. As is easily seen from the graph, there is a "lot of probability" in the vicinity of zero, and indeed it turns out that γ, computed from (2.9) is

(2.10d) $$\gamma = 1.002$$

A closer inspection of the weights shows that all are really well behaved -- using the reasoning in Guttman (1973), and in particular, the Table 3.1 of that paper, we would expect d_j's to have value $1/n = 1/6$ and fluctuate in the interval $(0, .449)$ when all observations are from the same source. Also, we note again that the posterior expectation has value $.017$, and the standard deviation is $\sqrt{3.663} = 1.914$. All these results would indicate that there is no spuriousness, and as we know that indeed $a = 0$, the method is working well. The results for $n = 10$ (authors' data set 2.11A) yields, as expected, similar results. (We wish to point out that the program for $n \leq 10$ and the results obtained by the authors for their data sets 2.1A - 2.20A with $n = 6$ and 10 are available on request. Generally, when $n = 10$, the results are "even nicer than for $n = 6$" -- we cannot reproduce all the results here, so are confining ourselves for the present illustration to the $n = 6$ case).

We turn now to a non-null case and a sensitivity analysis. To begin with, we mention how the sample of size n is generated for a non-null case, and, simply, we begin by generating n independent $N(0,1)$'s, say $N_t(0,1)$, $t = 1,\dots,n$. Arbitrarily, we then choose the first observation, $N_1(0,1)$, and generate a spurious observation, say N_{sp}, by letting

(2.11) $$N_{sp} = [N_1(0,1) + a]/\delta \sim N(a/\delta\ ,\ 1/\delta^2)\ .$$

The sample is then taken to be N_{sp}, $N_2(0,1),\dots,N_n(0,1)$. For illustration, we let $n = 6$, $a = 5$, $\delta^2 = 7$, yielding the sample (authors' data set 2.8A), which when ordered has observations

(2.12) $$-1.771,\ -1.131,\ -.3120,\ 1.350,\ 1.624,\ 6.895\ .$$

As $a = 5$, it is no surprise that the outlier 6.895 is that

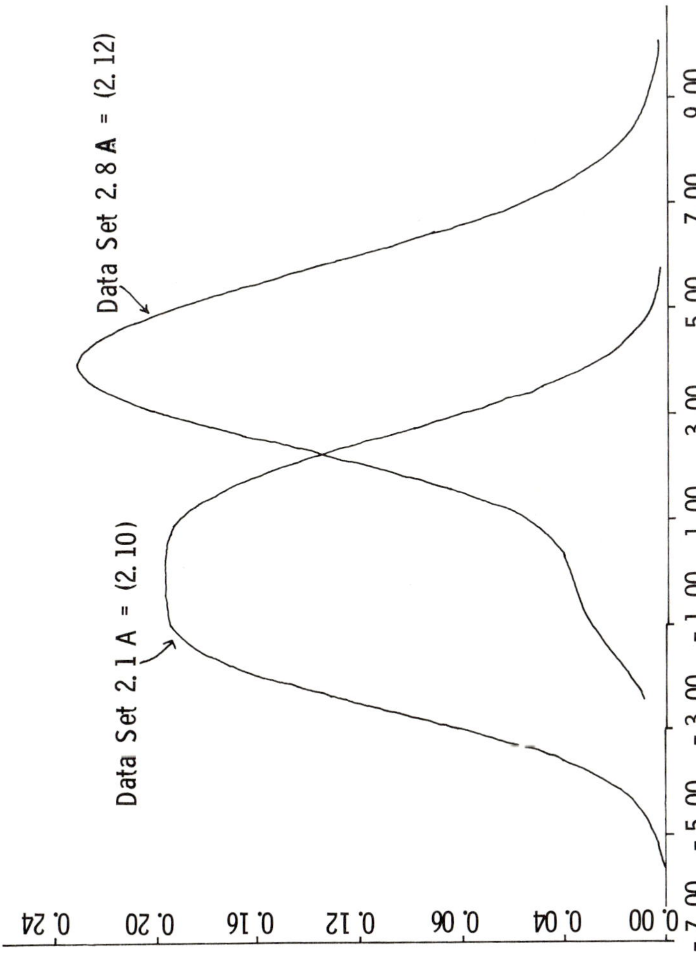

Figure 2.1 - The posteriors of a based on the data sets 2.1 A = (2.10), a null case, and 2.8 A = (2.12), a non-null case.

observation generated from the spurious source. Doing the various computations yields

(2.12a) $E(a|.7\ ;\ \underset{\sim}{y}) = 3.752 \qquad Var(a|.7\ ;\ \underset{\sim}{y}) = 4.395$

and from the graph of $p(a|.7\ ;\ \underset{\sim}{y})$ which we also give in Figure 2.1, we graphically determine

(2.12b) $$\text{Mode of } p(a|.7\ ;\ \underset{\sim}{y}) = 4.07\ .$$

The weights d_j corresponding to the observations (2.12) turned out to be

(2.12c) .023 , .018 , .015 , .013 , .013 , .918 .

That is, the weight d_6 is very much greater than the rest, and this in turn means that the posterior of a is virtually dominated by g_6. A small chunk of probability lies in the neighbourhood of zero, and in fact

(2.12d) $$\gamma = 16.147\ .$$

Examination of (2.12a, b and c), together with (2.12d) would lead to no other conclusion that 6.895 has been generated from a spurious source. As we know this to be the case, $(a = 5\ ,\ \delta^2 = .7)$, this is as should be. As mentioned earlier, we have repeated the above with $n = 10$, and even "nicer" results than for $n = 6$ have been obtained for the cases $a = 1$, 3 and 5 .

Now in the above, we have assumed that δ^2 , the "damping" or "inflating" factor of the variance of the spurious source is known. This could well arise in practice -- from bitter experience, an experimenter may keep track of detected spurious observations, and after a long period of time gain information which says that with high accuracy, that the variance of the spurious source is $1/\delta^2$ times the variance of the intended source. The question as to how sensitive our results would be to δ^2 can now be looked at. To examine this question, we have decided to see what happens when a given set of $n = 10$ observations is analyzed as above if δ^2 is allowed to vary over the values 7/10 , 1 , 10/7 that is, a deflation factor running from 70% to 143% , and to do this for the value of $a = 5$. To this end, we first generated a set of 10 independent $N(0,1)$'s , say $N_1(0,1),\ldots,N_{10}(0,1)$, which in order of appearance were

$$1.1840 \ , \quad 1.5317 \ , \quad -.3996 \ , \quad -2.0174 \ , \quad -1.9089$$

(2.13)

$$.9718 \ , \quad -2.2187 \ , \quad 1.1882 \ , \quad .5390 \ , \quad .8738$$

Again, arbitrarily choosing $N_1(0,1) = 1.1840$, we find N_{sp} using (2.11) for a fixed δ^2, consider the sample as $N_{sp}, N_2(0,1), \ldots, N_{10}(0,1)$, which is then ordered, and produce the "same computations" as in the previous two examples above. When we do this for $\delta^2 = 7/10$, 1, 10/7, we obtain the data sets 2.18B, 2.19B and 2.20B respectively, and summarize the computations for these samples in the following table.

Table 2.1

δ^2	$E(a\|\delta^2;\underset{\sim}{y})$	$Var(a\|\delta^2;\underset{\sim}{y})$	max d_j	next largest d_j	γ	Modes of $p(a\|\delta^2;\underset{\sim}{y})$
.7	4.567	2.6904	.9840	.0025	33.13	4.51
1	4.403	3.5952	.9498	.0086	28.83	4.56
10/7	3.957	5.6945	.8622	.0246	9.96	4.74

The modes are graphically determined from Figure 2.2

The graphs of the resulting posteriors $p(a|\delta^2 ; y)$ are given in Figure 2.2. Note how they resemble each other. Their modes are very nearly the same, as is the case with their expectations. Note the behaviour of the d_j - one is very large compared to all the others. Note the high odds γ in favour of " $a > 0$ ", ranging from 10 to 33. In each case, the decision called for is that spuriousness is present, and though the damping factor δ^2 varies from one of 70% to 143%, this is as should be, for indeed we know that $a = 5$. That is, the method is quite insensitive to changes in δ^2. Put another way, whether N_{sp} is generated with $\delta^2 = 7/10$, 1, or 10/7, its presence with $N_2, \ldots, N_{10}$ leads to its detection.

3. A DIFFERENT MIXED SHIFT MODEL: TWO SPURIOUS OBSERVATIONS

In this section, our model for spuriousness will be different to that of the previous section, namely, if spuriousness occurs, we will assume that the spurious observation is from a source whose mean is shifted by " a " absolute units, and whose variance is

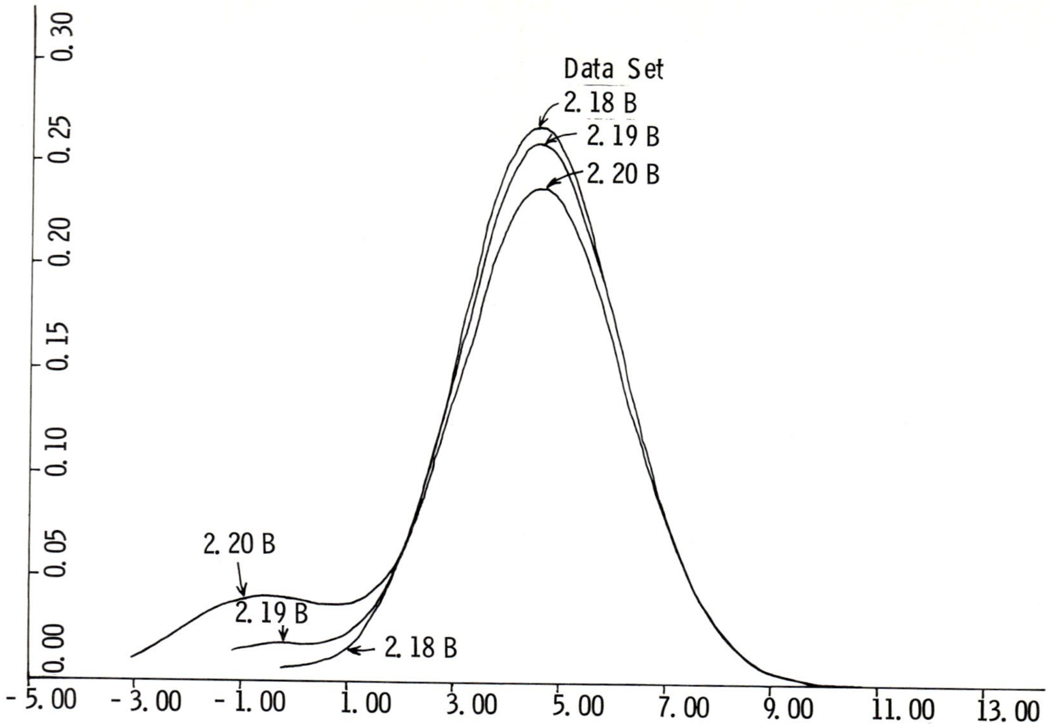

Figure 2.2 - The posteriors $p(a|\delta^2; y)$, where y comprises the data sets 2.18 B - 2.20 B, with respective values of $\delta^2 = .7, 1, 10/7$.

"damped" by δ^2 , i.e., $N(\mu + a , \sigma^2/\delta^2)$, δ^2 known. If $\delta^2 = 1$, the case of 1 spurious observation amongst the n observations has been dealt with in Guttman (1973), and so we pass directly to the case where it is hoped that $Y_1,\ldots,Y_n$ are generated from $N(\mu , \sigma^2)$, but it is feared that two of the observations may be spurious of the type mentioned above. We first point out however, that since our model of this section calls for the possibility of two spurious observations, we assume that the sample size n is such that this is a realistic assumption. An extension of the discussion of this point in the introduction of Guttman (1973) leads us here to assume that $10 \leq n \leq 20$.

3a. Mean Shifts Unrelated

In this sub-section, we assume that $Y_1,\ldots,Y_n$ are n independent observations, hopefully from $N(\mu , \sigma^2)$, but that it may be the case that two of the observations are generated from a spurious source. Specifically, if Y_i and Y_j are spurious, then $Y_i \sim N(\mu + a_1 , \sigma^2/\delta_1^2)$, $Y_j \sim N(\mu + a_2 , \sigma^2/\delta_2^2)$. We note again that the mean shifts are in absolute terms and not scaled (as in section 2) in terms of the standard deviation of the spurious source. We assume that if spuriousness occurs, then any pair of observations is equally likely to be spurious as any other pair, which will explain the factor $1/n(n-1)$ in the likelihood, which, if spuriousness does indeed occur is of the form (for given δ_1^2 and δ_2^2)

(3.1)

$$\ell(\mu,\sigma^2,a_1,a_2|\delta_1^2,\delta_2^2;\underset{\sim}{y}) \propto \sum_{i\neq j}\sum \frac{1}{n(n-1)} \left\{ \prod_{\ell\neq i,j} (\sigma^2)^{-\frac{1}{2}} \exp - \frac{1}{2\sigma^2} (y_\ell-\mu)^2 \right\}$$

$$\times \left[(\sigma^2)^{-\frac{1}{2}} \delta_1 \exp - \frac{\delta_1^2}{2\sigma^2} (y_i-\mu-a_1)^2\right]\left[(\sigma^2)^{-\frac{1}{2}} \delta_2 \exp - \frac{\delta_2^2}{2\sigma^2} (y_j-\mu-a_2)^2\right] .$$

If we further assume that only vague information is available for μ and σ^2 before obtaining the data, and also for "likely values of a_1 and a_2 " , then we will use the prior

$$p(\mu,\sigma^2,a_1,a_2) \propto (\sigma^2)^{-1} . \tag{3.1a}$$

It is proved in Appendix II of Guttman and Khatri (1974) that combining (3.1) with (3.1a) leads to the posterior for (a_1,a_2) , given $(\delta_1^2 , \delta_2^2)$, as

(3.2) $$p(a_1,a_2|\delta_1^2,\delta_2^2;\underset{\sim}{y}) = \sum_{i\neq j}\sum d_{ij}^{(1)} \; h_2(a_1,a_2|\underset{\sim}{\eta}_{ij};\; S_{ij};\; n-3)$$

where the function h_2 is given in (1.1a) on putting $m = 2$. Further, the vector $\underset{\sim}{\eta}_{ij}$ is given by

(3.2a) $$\underset{\sim}{\eta}_{ij}' = (y_i-\bar{y}^{(ij)},\; y_j-\bar{y}^{(ij)}),\quad \bar{y}^{(ij)} = (n-2)^{-1} \sum_{\ell\neq i,j} y_\ell$$

and the matrix S_{ij} is given by

(3.2b) $$S_{ij} = \frac{n-3}{A^{(ij)}} M_1,\quad M_1 = \frac{1}{n-2+\delta_1^2+\delta_2^2}\begin{bmatrix} \delta_1^2(n-2+\delta_2^2) & -\delta_1^2\delta_2^2 \\ -\delta_1^2\delta_2^2 & \delta_2^2(n-2+\delta_1^2) \end{bmatrix}$$

with

(3.2c) $$A^{(ij)} = \sum_{\ell\neq i,j} (y_\ell - \bar{y}^{(ij)})^2 ,$$

and the weights $d_{ij}^{(1)}$ are given by

(3.2d) $$d_{ij}^{(1)} = \left[A^{(ij)}\right]^{-(n-3)/2} \Big/ \sum_{i\neq j}\sum \left[A^{(ij)}\right]^{-(n-3)/2},\quad \sum_{i\neq j}\sum d_{ij}^{(1)} = 1,$$
$$d_{ij}^{(1)} = d_{ji}^{(1)} .$$

Using properties of the bivariate generalized Student - t distribution h_2 that appears in (3.2), it is also shown in Appendix II of Guttman and Khatri (1974), that

(3.3)
$$E(a_1|\delta_1^2,\delta_2^2;\underset{\sim}{y}) = \sum_{i\neq j}\sum d_{ij}^{(1)} \; (y_i - \bar{y}^{(ij)})$$

$$\mathrm{Var}(a_1|\delta_1^2,\delta_2^2;\underset{\sim}{y}) = \sum_{i\neq j}\sum d_{ij}^{(1)}\left[(y_i-\bar{y}^{(ij)})^2 + \frac{A^{(ij)}}{n-5}\,\frac{n-2+\delta_1^2}{(n-2)\delta_1^2}\right] - \left[\sum_{i\neq j}\sum d_{ij}^{(1)} (y_i-\bar{y}^{(i)})\right.$$

with the obvious similar expressions for the posterior expectation and variance of a_2. We note that the $E(a_s|\delta_1^2,\delta_2^2;\underset{\sim}{y})$'s are equal and that when $\delta_1^2 = \delta_2^2$, that the $\mathrm{Var}(a_s|\delta^2,\delta^2;\; \underset{\sim}{y})$'s are equal. Further, it is also shown that

$$P(a_1 > 0 | \delta_1^2 , \delta_2^2 ; \underset{\sim}{y}) = \sum_{i \neq j}\sum d_{ij}^{(1)} G_{n-3} \left| \sqrt{D_{11}} (y_i - \bar{y}^{(ij)}) \right|$$

(3.4)

$$D_{11} = (n-3)(n-2)\delta_1^2 \Big/ \left[A^{(ij)} (n-2 + \delta_1^2) \right]$$

where $G_{n-3}(x)$ is the cumulative distribution function of the Student - t distribution wtih (n-3) degrees of freedom. A similar obvious expression holds for $P(a_2 > 0 | \delta_1^2 , \delta_2^2 ; \underset{\sim}{y})$. We note here too that when $\delta_1^2 = \delta_2^2 = \delta^2$, that the $P(a_s > 0 | \delta^2 , \delta^2 , \underset{\sim}{y})$'s are equal. From (3.4), we may, for example, look at the posterior odds of " $a_s > 0$ " by computing

$$\text{(3.4a)} \qquad \gamma_s = P(a_s > 0 | \delta_1^2 , \delta_2^2 ; \underset{\sim}{y}) \Big/ |1 - P(a_s > 0 [\delta_1^2 , \delta_2^2 ; \underset{\sim}{y})] , \; s=1 \text{ or } 2.$$

We turn now to an illustrative example. As previously mentioned, we assume that the sample size n here is such that $10 \leq n \leq 20$, say, so that we are indeed in situations where it is realistic to assume that at most two spurious observations will occur in the sample. The model of this sub-section takes this into account. We have generated many examples for $n = 10$, 16 and 20 , $(\delta_1^2 , \delta_2^2)$ one of the nine pairs (.7, .7) , (.7, 1) , (.7, 10/7),..., (10/7, 10/7) , with (a_1 , a_2) one of nine pairs (1,1) , (1,3) , (1,5),...,(5,5) for non-null cases, and of course $(a_1 , a_2) = (0,0)$, $(\delta_1^2 , \delta_2^2) = (1,1)$ for the null case. We do not reproduce all of these here, but their typical behaviour is seen in the examples we now quote. (The complete output for the above cases is available on request). We begin with a null case; specifically, we generated $n = 10$ independent observations (data set 3a.1A), each $N(0,1)$, and these were, when ordered

(3.5)

$$\begin{array}{ccccc} -.877 , & -.568 , & -.409 , & -.298 , & -.297 \\ -.178 , & .440 , & .569 , & .755 , & 1.081 . \end{array}$$

We recall that $d_{ij}^{(1)} = d_{ji}^{(1)}$, and we thus tabulate these weights as follows:

Table 3.1

$d_{ij}^{(1)}$ for $i < j$

(indices corresponding to the ordered observations (3.5))

Row i \ j=	i+1	i+2	i+3	i+4	i+5	i+6	i+7	i+8	i+9
1	.0185	.0134	.0114	.0114	.0100	.0102	.0118	.0163	.0437
2	.0067	.0059	.0059	.0054	.0057	.0066	.0087	.0204	
3	.0048	.0048	.0044	.0048	.0055	.0073	.0165		
4	.0043	.0040	.0044	.0051	.0067	.0150			
5	.0040	.0044	.0050	.0067	.0150				
6	.0042	.0048	.0063	.0142					
7	.0062	.0086	.0229						
8	.0106	.0308							
9	.0568								

For example, $d_{27}^{(1)}$ is found in the second row (i=2), fifth column (7 = 2 + 5 = i + 5) of the above table, and has value $d_{27}^{(1)} = .0057$. We note that all entries are quite small, that for any row i, the last entry in the row is largest, but still small, and that the value of $d_{9,10}^{(1)}$ is not very much greater than the entry for $d_{8,10}^{(1)}$. (Compare this behaviour with Table 3.2, which tabulates $d_{ij}^{(1)}$'s for a non-null case). This behaviour described here for Table 3.1 is typical of what we have found for the $d_{ij}^{(1)}$'s for null cases.

Further, for the sample (3.5) = data set 3a.1A, we may calculate (3.3) which yields

$$E(a_s|\delta_1^2 , \delta_2^2 ; \underset{\sim}{y}) = E(a_s|1 , 1 ; \underset{\sim}{y}) = .1984 , \qquad s = 1,2$$

(3.6)

$$\text{Var}(a_s|\delta_1^2 , \delta_2^2 ; \underset{\sim}{y}) = \text{Var}(a_s|1 , 1 ; \underset{\sim}{y}) = 1.219 \qquad s = 1,2$$

and also that (see 3.4b)

$$\gamma_s = 1.336 , \qquad s = 1,2 \tag{3.6a}$$

which are certainly not divergent from the odds of 1 : 1 , which we expect when the true values of a_1 and a_2 are zero. Inspection of the above calculations would lead an experimenter to conclude that there is no spuriousness present in the sample.

We turn now to a non-null case, and for illustrative purposes here we choose the following case (data set 3a.245A). For this example, we first generated $n = 20$ $N(0,1)$'s , say $N_1, N_2, \ldots, N_{20}$, and, arbitrarily, chose the first two, say N_1 , N_2 and introduced spuriousness by replacing these with

$$N_{sp,t} = N_t/\delta_t + a_t \sim N(a_t \; , \; 1/\delta_t^2) \quad . \tag{3.7}$$

By following the above procedure, we arrived at the following sample, when $(\delta_1^2 \; , \; \delta_2^2) = (10/7, \; 1)$ and $(a_1 \; , \; a_2) = (3,4)$ -- we have ordered the observations for convenience:

$$\begin{array}{l} -2.005 \; , \; -1.507 \; , \; -1.346 \; , \; -1.272 \; , \; -1.148 \; , \; -.8392 \; , \\ -.7682, \; -.7599, \; -.7010, \; -.4659, \; -.4292, \; -.1391 \; , \\ -.0594, \; .1785, \; .4956, \; 1.216, \; 1.579, \; 1.667 \; , \\ 2.900 \; , \; 3.972 \; . \end{array} \tag{3.8}$$

(The above sample is data set 3a.245A).

We are involved with 190 distinct $d_{ij}^{(1)}$'s , and we do not list them all here (although our output does have these weights tabulated). But, we comment that most are pretty small and the last three rows of a table of $d_{ij}^{(1)}$'s looks like the following:

Table 3.2

Some $d_{ij}^{(1)}$'s for $i < j$

corresponding to the ordered observations (3.8)

i \ j		i+1	i+2	i+3
Row i	17	.00046	.00186	.01857
	18	.00202	.02077	
	19	.25630		

Notice the last, i.e. the (19,20) entry. It is very much bigger than the rest (cf. with Table 3.1). Further, the entries in the last place of each row are almost ten times that of the preceding ones in that row -- this does not happen in Table 3.1. Our experience shows that this behaviour is typical. Returning to the sample 3a.245A = (3.8), we calculate (3.3) for this sample which yields

$$E(a_s|10/7\ ,\ 1\ ;\ \underset{\sim}{y}) = 2.721 \qquad s = 1,2$$

(3.8a)

$$\mathrm{Var}(a_1|10/7\ ,\ 1\ ;\ \underset{\sim}{y}) = 5.158\ ,\ \mathrm{Var}(a_2|10/7\ ,\ \underset{\sim}{y}) = 5.642\ .$$

Note that a two-sided $100(1-\alpha)\%$ confidence interval for a_2 would be longer in length than the one for a_1. Finally, it turns out that the odds for $a_s > 0$ are

(3.8b) $$\gamma_1 = 5.39\ ,\ \gamma_2 = 5.43\ .$$

Conclusion of spuriousness is warranted -- the odds in favour of $a_1 > 0$, for example, are very high (>5), the $d^{(1)}_{19,20}$ entry is some 125 times the $d^{(1)}_{18,19}$ entry etc. Since we know that $(a_1\ ,\ a_2) = (3,4)$, this is as should be and the method is doing the job of detecting spuriousness well.

We wish at this point to emphasize that an experimenter would put this procedure into practice as follows. If he fears spuriousness, and if as mentioned earlier "he knows" that spurious observations have variance $1/\delta^2_t$ times the variance of the intended source, he sets the δ^2_t to the required value in the various formulae and computes the weights $d^{(1)}_{ij}$, the odds γ_s, the expectations and variances. If, for example, $a_1 = a_2 = 0$ truly, and he has set the δ^2_t equal to one, he will obtain results as in the first example of this subsection. If he has set δ^2_t to some value, and if a_1 and a_2 do not equal zero, he will obtain results as in the seocnd example of this section. The question that now is raised, is how sensitive is this method to the values of δ^2_t that the experimenter has chosen? (Parathentially, we remark that the implied philosophy and procedure expressed in this paragraph holds for all section of this paper).

To see this, we have chosen the following case: $n = 16$, $(a_1\ ,\ a_2) = (5,5)$, and nine data sets (data sets 3a.120B - 3a.128B) were generated as follows. First, $n = 16$ independent $N(0,1)$

observations were generated, say N_t , $t = 1,\ldots,16$. These in order of appearance were

$$\begin{array}{llllll} -.6660 , & .4990 , & -.7258 , & .2464 , & 1.236 , & -1.735 , \\ 1.213 , & -.0565 , & 1.109 , & 1.280 , & 1.050 , & .3791 , \\ 1.374 , & -.5007 , & -.7515 , & -.1585 . & & \end{array} \tag{3.9}$$

The nine samples were then generated as follows. For a given pair $(\delta_1^2 , \delta_2^2)$, two spurious observations $N_{sp,t}$ were generated by arbitrarily selecting N_1 and N_2 (i.e., -.6660 and .4990) , and letting

$$N_{sp,t} = \frac{1}{\delta_t} N_t + a_t \sim N_t(a_t , 1/\delta_t^2) . \tag{3.9a}$$

The sample is then $N_{sp,1}$, $N_{sp,2}$, $N_3,\ldots,N_{16}$. This sample is then ordered, the corresponding $d_{ij}^{(1)}$ calculated, as well as the expectation, variance and γ_s , and the results tabulated in the appropriate line of Table 3.3. As mentioned before, nine pairs of $(\delta_1^2 , \delta_2^2)$ were chosen, with values for the δ_t^2 ranging over .7 , 1 , 10/7 , that is, the damping factors vary from 70% to 143%.

Inspection of Table 3.3 shows that the results are affected very little by changes in the δ_t^2 . The same behaviour as in the non-null example previously discussed in this section occurs throughout. The odds γ_t , which are very high, the behaviour of the weights (e.g. $\max d_{ij}^{(1)}/d_{14,15}^{(1)}$ is very large), the expectations having value away from zero, would point, for any one case, to the fact that spuriousness has occurred. That is, our method seems to be robust to variation in the assumed known δ_t^2 . Put another way, the presence of $N_{sp,1}$ and $N_{sp,2}$ are detected in the samples $N_{sp,1}$, $N_{sp,2}$, N_3 ,..., N_{16} no matter what values of $(\delta_1^2 , \delta_2^2)$ are used.

3b. Mean Shifts Related: $a_1 = -a_2 = a$

In this sub-section, we assume that if spuriousness occurs, then the two observations are spurious of the type described in Section 3a, with the added assumption, however, that one of the spurious observations has mean shifted by a units above μ , while the other has mean shifted by a units below μ . Thus, the likelihood is as in (3.1), with $a_1 = a$, $a_2 = -a$. If we make use of the prior $p(\mu , \sigma^2 , a) \propto (\sigma^2)^{-1}$, then by following the

Table 3.3

Calculations based on the data sets 3a.120B - 128B
A Sensitivity Study $(a_1 = 5, a_2 = 5)(n = 16)$

(δ_1^2, δ_2^2)	$E(a_1 \mid \delta_1^2, \delta_2^2; \underset{\sim}{y})$ $E(a_2 \mid \delta_1^2, \delta_2^2; \underset{\sim}{y})$	$Var(a_1 \mid \delta_1^2, \delta_2^2; \underset{\sim}{y})$ $Var(a_2 \mid \delta_1^2, \delta_2^2; \underset{\sim}{y})$	γ_1 γ_2	$\max d_{ij}^{(1)}$ $= d_{15, 16}^{(1)}$	$d_{14, 15}^{(1)}$
(.7, .7)	4.236 4.236	4.499 4.499	19.567 19.567	.4319	$.342 \times 10^{-2}$
(.7, 1)	4.175 4.175	4.471 3.907	18.960 19.092	.4304	$.341 \times 10^{-2}$
(.7, 10/7)	4.122 4.122	4.456 3.492	18.405 18.471	.4291	$.340 \times 10^{-2}$
(1, .7)	4.355 4.355	3.557 4.110	23.823 23.800	.4436	$.276 \times 10^{-2}$
(1, 1)	4.294 4.294	3.536 3.536	22.967 22.967	.4422	$.275 \times 10^{-2}$
(1, 10/7)	4.234 4.234	3.528 3.137	22.218 22.176	.4408	$.274 \times 10^{-2}$
(10/7, .7)	4.449 4.449	2.891 3.820	27.839 27.762	.4518	$.230 \times 10^{-2}$
(10/7, 1)	4.389 4.389	2.876 3.261	26.740 26.892	.4504	$.229 \times 10^{-2}$
(10/7, 10/7)	4.337 4.337	2.872 2.872	25.771 25.771	.4491	$.229 \times 10^{-2}$

arguments of Appendix II with $a_1 = -a_2 = a$ (see Appendix IIb) we find that the posterior of a is given by

$$p(a|\delta_1^2, \delta_2^2; \underset{\sim}{y}) = \sum_{i \neq j}\sum t_{ij} h_1(a|\eta(ij); f_{ij}; n-2) \tag{3.10}$$

where the function h_1 is defined in (1.1a) on putting $m = 1$, and where

(3.10a)

$$\eta(ij) = \frac{[\delta_1^2(n-2) + 2\delta_1^2\delta_2^2](y_i - \bar{y}^{(ij)}) - [\delta_2^2(n-2) + 2\delta_1^2\delta_2^2](y_j - \bar{y}^{(ij)})}{(n-2)(\delta_1^2 + \delta_2^2) + 4\delta_1^2\delta_2^2}$$

with

$$f_{ij} = \left\{ \frac{u_{ij}(n - 2 + \delta_1^2 + \delta_2^2}{(n-2)[(n-2)(\delta_1^2 + \delta_2^2) + 4\delta_1^2\delta_2^2} \right\}^{-1} \tag{3.10b}$$

and where

(3.10c)

$$u_{ij} = A^{(ij)} + \frac{\delta_1^2\delta_2^2(n-2)}{(n-2)(\delta_1^2 + \delta_2^2) + 4\delta_1^2\delta_2^2} \left[(y_i - \bar{y}^{(ij)}) + (y_j - \bar{y}^{(ij)})\right]^2.$$

The weights t_{ij} are given by

(3.10d)

$$t_{ij} = (u_{ij})^{-(n-2)/2} \Big/ \sum_{i \neq j}\sum (u_{ij})^{-(n-2)/2}, \quad \sum_{i \neq j}\sum t_{ij} = 1, \quad t_{ij} = t_{ji}.$$

Using properties of h_1, it is easy to see that

$$E(a|\delta_1^2, \delta_2^2; \underset{\sim}{y}) = \sum_{i \neq j}\sum t_{ij}\, \eta(ij). \tag{3.11}$$

(3.11a)

$$\operatorname{Var}(a|\delta_1^2, \delta_2^2; \underset{\sim}{y}) = \sum_{i \neq j}\sum t_{ij} \left[\eta^2(ij) + \frac{n-2}{n-4} f_{ij}^{-1}\right] - \left[\sum_{i \neq j}\sum t_{ij}\, \eta(ij)\right]^2,$$

and that

$$P(a > 0|\delta_1^2, \delta_2^2; \underset{\sim}{y}) = \sum_{i \neq j}\sum t_{ij}\, G_{n-2}(\sqrt{f_{ij}}\, \eta(ij)). \tag{3.11b}$$

We wish now to make a few comments on the above results. If $\delta_1^2 = \delta_2^2 = \delta^2$, then it is easy to see that

$$\eta(ij) = -\eta(ji) = (y_i - y_j)/2 \tag{3.12}$$

$$u_{ij} = u_{ji} = A^{(ij)} + \frac{(n-2)\delta^2}{2[n-2+\delta^2]}\left[(y_i - \bar{y}^{(ij)}) + (y_j - \bar{y}^{(ij)})\right] \tag{3.12a}$$

$$t_{ij} = t_{ji} \quad \text{and} \quad f_{ij} = f_{ji} . \tag{3.12b}$$

This automatically means that

$$E(a|\delta^2 , \delta^2 ; \underset{\sim}{y}) = 0 \tag{3.13}$$

and that, as is easily seen,

$$P(a > 0|\delta^2, \delta^2, \underset{\sim}{y}) = \left(\sum_{i<j}\sum + \sum_{i>j}\sum\right)\left[t_{ij}\, G_{n-2}(\sqrt{f_{ij}}\, \eta(ij))\right] \tag{3.13a}$$

$$= \sum_{i>j}\sum t_{ij}\left[1 - G_{n-2}(\sqrt{f_{ij}}\, \eta(ij)\right] + \sum_{i>j}\sum t_{ij}\, G_{n-2}(\sqrt{f_{ij}}\, \eta(ij))$$

$$= \sum_{i>j}\sum t_{ij} = \frac{1}{2}$$

so that, if $\delta_1^2 = \delta_2^2 = \delta^2$, then

$$\gamma = P(a > 0|\delta^2 , \delta^2 ; \underset{\sim}{y})/P(a < 0|\delta^2 , \delta^2 ; \underset{\sim}{y}) = 1 , \tag{3.13b}$$

independent of $\underset{\sim}{y}$. Further, as shown in the examples below, as $(\delta_1^2 , \delta_2^2)$ vary away from equality, the expectation of a remains close to zero and γ remains close to 1. That is to say, for the model of this sub-section, we must (apart from examining the behaviour of the weights t_{ij} - see below) look elsewhere to detect spuriousness of the type described here.

This comes rather easily, when one looks at $p(a|\delta_1^2 , \delta_2^2 ; \underset{\sim}{y})$ and more importantly, its graph. When indeed two observations are spurious, it turns out that, typically, the graph of $p(a , \delta^2 , \delta^2 ; \underset{\sim}{y})$ is bi-modal, symmetric about zero or nearly so, and has the appearance given in Figure 3.1, that is, it is U-shaped

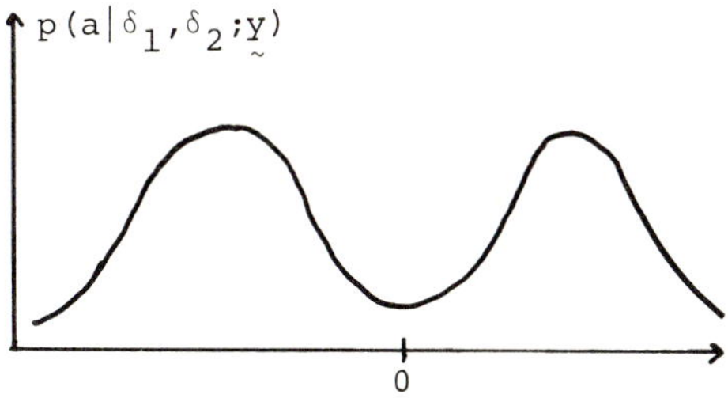

Figure 3.1. Typical shape of the posterior of a when two observations are spurious.

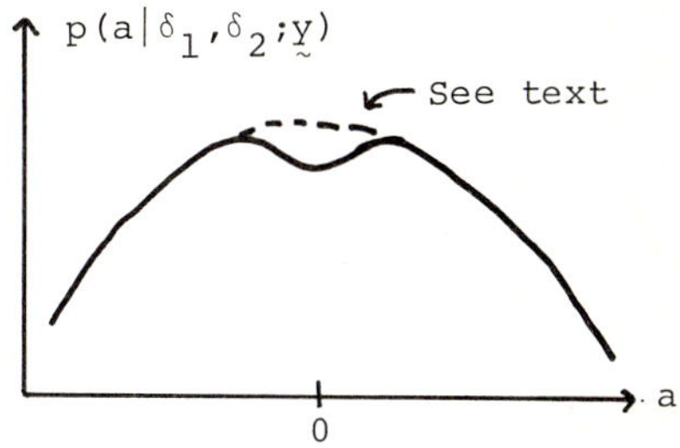

Figure 3.2. Typical shape of the posterior of a when no observations are spurious.

at or near zero. Indeed the mean is zero, or very nearly so, and little probability is concentrated near zero. On the other hand, if there is no spuriousness, then the posterior of a takes on a typical shape portrayed in Figure 3.2 -- indeed, sometimes the slight cusp at or near zero is sometimes absent so that the posterior of a here is unimodal. (This is the reason for the dotted portion of the curve in Figure 3.2). The mean is zero, or approximately so, and there is a sizeable chunk of probability near zero. Hence, it is appropriate to look at the graph of the posterior, and perhaps, compute

$$P(|a| < k\sqrt{\mathrm{Var}(a)} \mid \delta_1^2, \delta_2^2 ; \underset{\sim}{y}) \tag{3.13c}$$

for let us say, $k = \frac{2}{3}$. (This choice is made because in normal theory, the "probable error" is $\frac{2}{3}$ times the standard deviation). If there is no spuriousness, $(a = 0)$ we would expect (3.13c) with $k = \frac{2}{3}$, to be near $.5$, and if spuriousness were present, we would expect (3.13c) to be a long way from $.5$. But, we repeat, the shape of the graph and the behaviour of the t_{ij} will be very informative. We now would like to illustrate how this works in practice. We have generated many examples and the full output is available on request.

We begin with a null case. Sixteen independent $N(0,1)$ observations were generated (i.e., $a = 0$, $\delta_1^2 = \delta_2^2 = 1$) and found to be, when ordered (data set 3b.29A)

(3.14)
$$\begin{array}{lllll} -1.978, & -1.426, & -1.300, & -.7025, & -.6068, \\ -.3423, & -.2739, & -.2353, & -.0311, & .1653, \\ .4754, & .6327, & .7063, & .9453, & 1.425, \\ 1.619. \end{array}$$

On the basis of this sample, we may compute

(3.14a)
$$E(a|1\ ,\ 1\ ;\ \underset{\sim}{y}) = .135 \times 10^{-14} \simeq 0$$
$$\mathrm{Var}(a|1\ ,\ 1\ ;\ \underset{\sim}{y}) = 1.972\ .$$

Further, the t_{ij} (tabulated in Table 3.4) are all small, with no sudden large inflation from, for example, $t_{1,15}$ to $t_{1,16}$ ($t_{1,15} = .0447$, $t_{1,16} = .0747$) contrary to the behaviour in non-null cases (see the example below). The posterior $p(a|1\ ,\ 1\ ;\ \underset{\sim}{y})$ of this example is sketched in Figure 3.3. Note the symmetry about zero. Indeed it turns out that for this example

(3.14b)
$$P(|a| < \tfrac{2}{3}\sqrt{\mathrm{Var}(a)}) = .4506\ .$$

Put another way, the posterior odds are 55 to 45 or 11 to 9 that a lies outside the invetval $(-\frac{2}{3}\sqrt{\mathrm{Var}(a)}\ ,\ \frac{2}{3}\sqrt{\mathrm{Var}(a)})$ etc., a not very remarkable ratio. All these facts would easily lead to the decision that there is no spuriousness.

In contrast, let us consider the following non-null case, based on the authors' data set 3b.48A. This sample was generated as in Section 3a, with the help of (3.9a), with $a_1 = 5$, $a_2 = -5$ and $(\delta_1^2\ ,\ \delta_2^2) = (.7\ ,\ .7)$. The sample so generated, when ordered, turned out to be

(3.15)
$$\begin{array}{lllll} -3.358, & -2.792, & -1.227, & -1.193, & -.5209, \\ -.2395, & -.1019, & .1510, & .4333, & .4669, \\ .7963, & .8584, & 1.160, & 1.415, & 1.447, \\ 4.731. \end{array}$$

On the basis of this sample, we find

(3.15a)
$$E(a|.7\ ,\ .7\ ;\ \underset{\sim}{y}) = .210 \times 10^{-13} \simeq 0$$

Table 3.4

The weights t_{ij} for the data of the null case example of Section 3b (Data Set 3b.29A)

i \ j=	i+1	i+2	i+3	i+4	i+5	i+6	i+7	i+8	i+9	i+10	i+11	i+12	i+13	i+14	i+15
1	.1761	.1825	.2402	.2554	.3105	.3288	.3399	.4122	.5093	.7535	.9473	1.062	1.600	4.465	.7475
2	.1649	.1853	.1917	.2157	.2237	.2286	.2595	.2997	.3940	.4634	.5030	.6744	1.404	2.021	
3	.1783	.1834	.2031	.2097	.2137	.2394	.2726	.3501	.4065	.4384	.5749	1.134	1.590		
4	.1646	.1692	.1713	.1727	.1822	.1954	.2269	.2495	.2620	.3136	.5000	.6324			
5	.1669	.1685	.1696	.1772	.1884	.2156	.2351	.2460	.2907	.4500	.5614				
6	.1645	.1647	.1680	.1743	.1916	.2046	.2119	.2421	.3473	.4184					
7	.1644	.1665	.1717	.1869	.1986	.2051	.2323	.3272	.3908						
8	.1659	.1704	.1845	.1954	.2016	.2273	.3167	.3766							
9	.1658	.1742	.1818	.1861	.2048	.2707	.3143								
10	.1679	.1727	.1757	.1889	.2379	.2704									
11	.1652	.1663	.1728	.2023	.2227										
12	.1645	.1680	.1898	.2057											
13	.1664	.1849	.1991												
14	.1731	.1823													
15	.1657														
Multiplier	10^{-2}	10^{-2}	10^{-2}	10^{-2}	10^{-2}	10^{-2}	10^{-2}	10^{-2}	10^{-2}	10^{-2}	10^{-2}	10^{-2}	10^{-2}	10^{-2}	10^{-1}

e.g. $t_{59} = .1772 \times 10^{-2}$

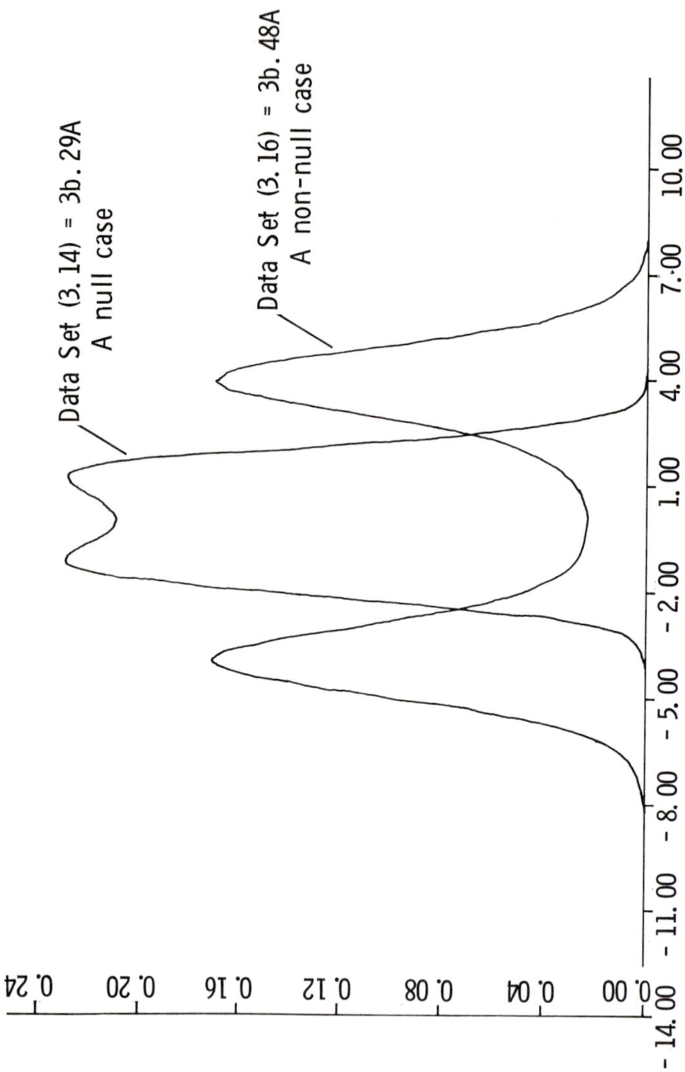

Figure 3.3 - The posteriors of a based on a null case data set 3b.29A = (3.14) and a non-null case data set 3b.48A = (3.16)

(3.15b) $$\mathrm{Var}(a \mid .7\ ,\ .7\ ;\ \underset{\sim}{y}) = 15.201$$

and a posterior whose graph, given in Figure 3.3, is decidely U-shaped in the vicinity of the origin, and with very little probability near zero -- for example,

(3.15c) $$P(|a| < \tfrac{2}{3}\sqrt{\mathrm{Var}(a)}) = .3218\ ,$$

a long way from the $.5$ we would expect if the true value of $a = 0$ (c.f. with the null case example, for which this probability turns out to be $.4506$). Further, the weights t_{ij} behave rather differently here than for the null case example, as given in Table 3.4. We do not give the full table of the t_{ij} here, but mention, in extract, that

(3.15d) $$t_{1,15} = .0018\ ,\quad t_{1.16} = .3302$$

that is, $t_{1,16}$ is over 180 times $t_{1,15}$. This behaviour of the weights along with the graph of $p(a \mid .7\ ,\ .7\ ;\ \underset{\sim}{y})$ leads to the decision that there is spuriousness of the outliers -3.358 and 4.731, which we know to be the case.

Now underlying the results and examples of this section is the same philosophy discussed in Section 3a. This means that we investigate how sensitive the above method is to the assumption that the δ_t^2 are known. We again do this by letting the "damping" factors δ_t^2 vary over the values $.7$ to 1 to $10/7$ and make up nine pairs $(\delta_1^2\ ,\ \delta_2^2)$ in the obvious way, as we did in Section 3a. In the examples below we let $a_1 = -a_2 = a = 5$. (Other examples were generated with $a = 1$ and 3, and similar results to those below were achieved). We generated the samples as follows. We first generated $n = 20$ independent $N(0,1)$'s, say N_t, $t=1,\ldots,20$. Arbitrarily, we chose the first two, and generated 2 spurious observations $N_{sp,t}$ by using (3.9a) with $a_1 = -a_2 = a = 5$, and letting $(\delta_1^2\ ,\ \delta_2^2)$ be one of the nine pairs referred to above. For each pair, the sample is $N_{sp,1}, N_{sp,2}, N_3, \ldots, N_{20}$, which is then ordered and calculations performed as in the example above. (The nine samples comprise authors' data sets 3b.76B - 3b.84B. Other cases with different n and a were also performed, and the output is available on request). The results of these calculations are given in Table 3.5, and graphs of the resulting posteriors are found in Figure 3.4. A glance at this figure and Table 3.5 shows

Table 3.5

Calculations based on the data sets 3b.76B - 3b.84B, $n = 20$, $a_1 = -a_2 = 5$

(δ_1^2, δ_1^2)	$E(a\|\delta_1^2,\delta_2^2;\underset{\sim}{y})$	$Var(a\|\delta_1^2,\delta_2^2;\underset{\sim}{y})$	$P(\|a\| < \frac{2}{3}\sqrt{Var(a)})$	$\max t_{ij} = t_{1,20}$	$t_{ij} = t_{1,19}$
(.7,.7)	$.3675 \times 10^{-13}$	29.476	.2756	.4990	$.5515 \times 10^{-5}$
(.7,1)	-.1426	29.959	.2723	.4994	$.4709 \times 10^{-5}$
(.7,10/7)	-.2590	30.353	.2685	.4996	$.4014 \times 10^{-5}$
(1,.7)	.1300	28.131	.2736	.4991	$.7105 \times 10^{-5}$
(1,1)	$.3578 \times 10^{-13}$	28.634	.2711	.4995	$.5715 \times 10^{-5}$
(1,10/7)	-.1124	29.046	.2686	.4996	$.4566 \times 10^{-5}$
(10/7,.7)	.2251	27.015	.2697	.4992	$.8568 \times 10^{-5}$
(10/7,1)	.1046	27.534	.2690	.4995	$.6440 \times 10^{-5}$
(10/7,10/7)	$-.3543 \times 10^{-13}$	27.961	.2675	.4997	$.4764 \times 10^{-5}$

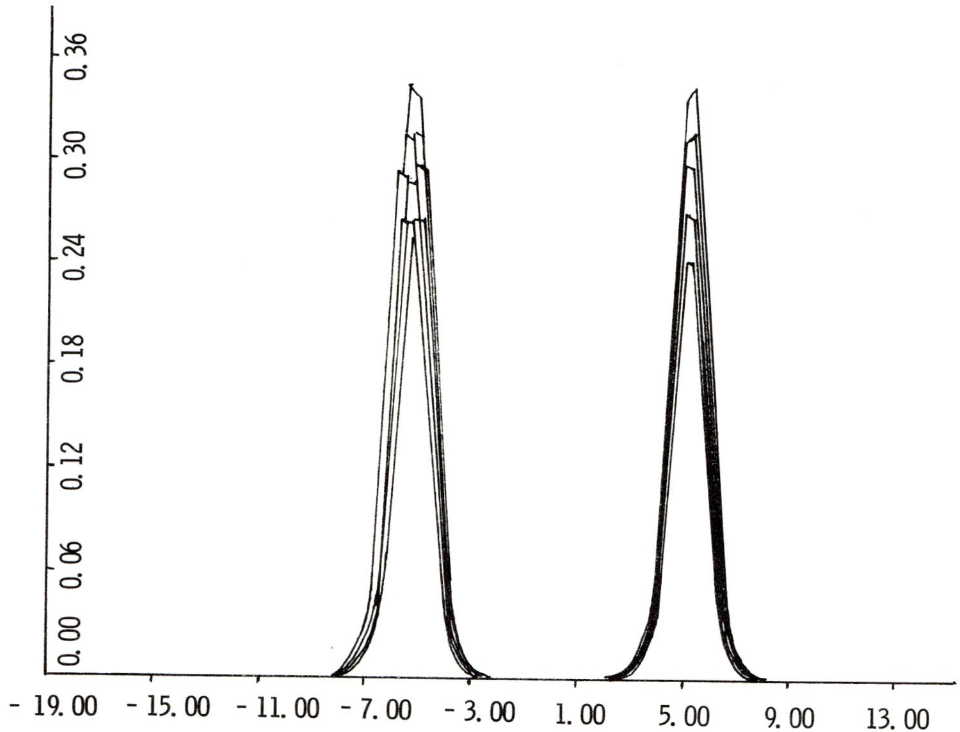

Figure 3.4 - The graphs of the posteriors of a based on the nine samples 3b.76B - 3b.84B $\left(n = 20,\ a = 5 \text{ and } [\delta_1^2, \delta_2^2] \text{ vary over the nine pairs } \{(.7, 1, 10/7) \times (.7, 1, 10/7)\}\right)$

that for all intents and purposes, the posteriors are the same. They are bunched so closely together that we have not labelled which posterior corresponds to which data set. Each is bimodal (near the values ± 5) and U-shaped near the origin, with very little probability near the origin (see the fourth column of Table 3.5, for example.) There is no doubt that the method is insensitive to changes in (δ_1^2, δ_2^2), at least over the range that the damping factors vary over here, that is, from 70 to 143 percent.

Indicentally, the 20 independent $N(0,1)$'s used to construct the nine samples were (in order of appearance)

$$\begin{array}{lllll} 1.178, & .5650, & -1.544, & .6997, & .7276, \\ .0567, & .2913, & 1.088, & -.0128, & -1.213, \\ -.7173, & -.4593, & .8487, & -.0950, & -1.949, \\ .2795, & .8221, & 1.214, & .8112, & .8640. \end{array} \tag{3.16}$$

4. THE MIXED-SHIFT MODEL: TWO SPURIOUS OBSERVATIONS

4a. Mean Shifts Unrelated

In this sub-section, we allow for the same type of spuriosity as in Section 1, but now deal with the case that we may fear that two of the n independent observations are spurious, that is, specifically, generated from $N(\mu + a_s\sigma/\delta_s^2)$, $s = 1,2$, with the δ_s known, while the other $(n-2)$ observations are generated from $N(\mu,\sigma^2)$. The likelihood would thus be of the form

(4.1)

$$\ell(\mu,\sigma^2,a_1,a_2|\delta_1^2,\delta_2^2;\underset{\sim}{y}) = \sum_{i\neq j}^{n}\sum \frac{1}{n(n-1)}\left\{\prod_{\ell\neq i,j}(2\pi\sigma^2)^{-\frac{1}{2}}\exp - \frac{1}{2\sigma^2}(y_\ell-\mu)^2\right\} \times$$

$$\left[(2\pi\sigma^2)^{-\frac{1}{2}}\delta_1\exp - \frac{\delta_1^2}{2\sigma^2}(y_i-\mu-a_1\sigma/\delta_1)^2\right]\left[(2\pi\sigma^2)^{-\frac{1}{2}}\delta_2\exp \frac{\delta_2^2}{2\sigma^2}(y_j-\mu-a_2\sigma/\delta_2)^2\right].$$

The factor " $1/n(n-1)$ " appears in the above likelihood because we are assuming that the probability of $y_i \sim N(\mu + a_1\sigma/\delta_1, \sigma^2/\delta_1^2)$, $y_j \sim N(\mu + a_2\sigma/\delta_2, \sigma^2/\delta_2^2)$, or $y_i \sim N(\mu + a_2\sigma/\delta_2, \sigma^2/\delta_2^2)$, $y_j \sim N(\mu + a_1\sigma/\delta_1, \sigma^2/\delta_1)$ is identically $1/n(n-1)$ for all i and j. If we further assume that only vague information is available for μ and σ^2, before obtaining the data, and also for "likely values of a_1 and a_2 ", then we will use the prior

$$p(\mu , \sigma^2 , a_1 , a_2) \propto (\sigma^2)^{-1} . \tag{4.2}$$

Guttman and Khatri (1974) show (see Appendix III of that paper) that combining (4.2) with (4.1) by Bayes' Theorem leads to the posterior for (a_1 , a_2) , given δ_1 , δ_2 , as

$$p(a_1 , a_2 | \delta_1^2 , \delta_2^2 ; \underset{\sim}{y}) = \sum_{\substack{i \neq j \\ 1}}^{n} \sum d_{ij}^{(2)} h_{ij} (a_1 , a_2) \tag{4.3}$$

where the weights $d_{ij}^{(2)}$ are given by

(4.3a)

$$d_{ij}^{(2)} = \left[A^{(ij)}\right]^{-(n-1)/2} \bigg/ \sum_{i \neq j} \sum \left[A^{(ij)}\right]^{-(n-1)/2} ,$$

$$\sum_{i \neq j} \sum d_{ij}^{(2)} = 1 , \quad d_{ij}^{(2)} = d_{ji}^{(2)} ,$$

with

$$A^{(ij)} = \sum_{\ell \neq i,j} (y_\ell - \bar{y}^{(ij)})^2 , \quad \bar{y}^{(ij)} = (n-2)^{-1} \sum_{\ell \neq i,j} y_\ell . \tag{4.3b}$$

The function $h_{ij} (a_1 , a_2)$ is defined by

(4.3c)

$$h_{ij}(a_1 , a_2) = \sum_{t=0}^{\infty} \frac{2^{t/2} \Gamma(\frac{n-1+t}{2}) |M|^{\frac{1}{2}}}{t! \Gamma(\frac{n-1}{2}) 2\pi} \left[\rho_{1(ij)} a_1 + \rho_{2(ij)} a_2\right]^t \times$$

$$(1 - \varepsilon_{ij})^{(n-1)/2} \exp - Q/2$$

where

$$Q = (a_1,a_2)M(a_1,a_2)' , \quad M = I_2 - \frac{1}{n-2+\delta_1^2+\delta_2^2} \begin{pmatrix} \delta_1 \\ \delta_2 \end{pmatrix} (\delta_1,\delta_2) ,$$

(4.3d)

$$M = 1 - (\delta_1^2+\delta_2^2)/(n-2+\delta_1^2+\delta_2^2) = (n-2)/(n-2+\delta_1^2+\delta_2^2) ,$$

and

(4.3e) $$\varepsilon_{ij} = B_{ij} / (A^{(ij)} + B_{ij}) \quad , \quad 0 < \varepsilon_{ij} < 1$$

(4.3f) $$\begin{aligned} B_{ij} &= (c_{1(ij)} \, , \, c_{2(ij)}) M^{-1} (c_{1(ij)} \, , \, c_{2(ij)})' \\ &= (\bar{y}^{(ij)} \, , \, y_i \, , \, y_j) \; T \; (\bar{y}^{(ij)} \, , \, y_i \, , \, y_j)' \geq 0 \, , \end{aligned}$$

with

$$M^{-1} = I_2 + \frac{1}{n-2} \begin{pmatrix} \delta_1 \\ \delta_2 \end{pmatrix} (\delta_1 , \delta_2) \quad , \text{ and}$$

(4.3g) $$T = \left(\begin{bmatrix} n-2 & 0 & 0 \\ 0 & \delta_1^2 & 0 \\ 0 & 0 & \delta_2^2 \end{bmatrix} - \frac{1}{n-2+\delta_1^2+\delta_2^2} \begin{pmatrix} n-2 \\ \delta_1^2 \\ \delta_2^2 \end{pmatrix} (n-2 , \, \delta_1^2 , \, \delta_2^2) \right) ,$$

and where

(4.3h) $$\rho_{s(ij)} = c_{s(ij)} (1 - \varepsilon_{ij})^{\frac{1}{2}} \, / \, (A^{(ij)})^{\frac{1}{2}} \quad , \quad s = 1,2,$$

with

(4.3i) $$\begin{cases} c_{1(ij)} = \dfrac{\delta_1}{n-2+\delta_1^2+\delta_2^2} \left[-(n-2)\bar{y}^{(ij)} + (n-2+\delta_2^2) y_i - \delta_2^2 \, y_j \right] \\[2ex] c_{2(ij)} = \dfrac{\delta_2}{n-2+\delta_1^2+\delta_2^2} \left[-(n-2)\bar{y}^{(ij)} - \delta_1^2 \, y_i + (n-2+\delta_1^2) y_j \right] . \end{cases}$$

Guttman and Khatri (1974) also prove (Appendix III) that

(4.4) $$\begin{aligned} E(a_s | \delta_1^2, \delta_2^2, \underset{\sim}{y}) &= \frac{\sqrt{2} \; \Gamma(\frac{n}{2})}{\Gamma(\frac{n-1}{2})} \sum_{i \neq j} \sum \frac{d_{ij}^{(2)}}{(1-\varepsilon_{ij})^{\frac{1}{2}}} z_{s(ij)} \\ \mathrm{Var}(a_s | \delta_1^2, \delta_2^2, \underset{\sim}{y}) &= \left(1 + \frac{\delta_s^2}{n-2}\right) + (n-1) \sum_{i \neq j} \sum d_{is}^{(2)} \; z^2_{s(ij)}/)1-\varepsilon_{ij}) \\ &\quad - \left[E(a_s | \delta_1^2 \, , \, \delta_2^2 \; ; \; \underset{\sim}{y}) \right]^2 \end{aligned}$$

where

(4.4a) $$z_{s(ij)} = \left[\rho_{s(ij)} + \frac{\delta_s}{n-2}\left(\delta_1 \rho_{1(ij)} + \sigma_2\rho_{2(ij)}\right)\right] .$$

They also show that

(4.5)

$$P(a_s > 0 \mid \delta_1^2 , \delta_2^2 ; \underset{\sim}{y}) = \frac{1}{2} + \sum_{i \neq j}\sum d_{ij}^{(2)} \sum_{m=0}^{\infty} \frac{2^{m+1/2}\,\Gamma(\frac{n}{2} + m)}{(2m+1)!\,\Gamma(\frac{n-1}{2})} \cdot$$

$$\cdot\; C_{2m+1}^{(s)} (i,j) \left[1-\varepsilon_{ij}\right]^{(n-1)/2}$$

where

(4.5a) $$C_{2m+1}^{(s)} (i,j) = \sum_{\beta=0}^{m} \frac{(-1)^{\beta}\, v_s^{2\beta+1}\, \varepsilon_{ij}^{m-}\, (2m+1)!}{\sqrt{2\pi}\, 2^m\, (m-\beta)!\, \beta!\, (2\beta+1)}$$

$$= \frac{2^{m+1/2}}{\pi}\, \Gamma(m + \tfrac{3}{2}) \int_0^{v_s} (\varepsilon_{ij} - x^2)^m\, dx ,$$

and where $\quad v_{s(ij)} = v_s = z_{s(ij)} \Big/ \sqrt{1 + \delta_s^2/(n-2)}$.

Guttman and Khatri (1974) also show (Appendix III) that the probability of the joint event $(a_1 > 0 , a_2 > 0)$ is given by

$$\frac{1}{4} + \frac{1}{2\pi} \sum_{i \neq j}\sum d_{ij}^{(2)} \cdot \int_0^{\rho} \left[1 + \frac{\tau_1^2 - 2z\tau_1\tau_2 + \tau_2^2}{1 - z^2}\right]^{\frac{n-1}{2}} \frac{dz}{\sqrt{1 - z^2}}$$

$$+ \frac{n-1}{2\pi} \sum_{i \neq j}\sum d_{ij}^{(2)} \int_0^{\tau_1}\int_0^{\tau_2} (1 + z_1^2 + z_2^2)^{-(n+1)/2}\, dz_2 dz_1$$

$$+ \frac{\Gamma(\frac{n}{2})}{2\Gamma(\frac{1}{2})\,\Gamma(\frac{n-1}{2})} \sum_{i \neq j}\sum d_{ij}^{(2)} \left[\int_0^{\tau_1} (1 + z^2)^{-n/2}\, dz\right.$$

$$\left. + \int_0^{\tau_2} (1 + z^2)^{-n/2}\, dz \right] ,$$

where $\tau_s = \tau_{s(ij)} = v_{s(ij)}/\sqrt{1-\varepsilon_{ij}}$, $\rho = \delta_1\delta_2/\sqrt{(n-2+\delta_1^2)(n-2+\delta_2^2)}$, and where $v_s = v_{s(ij)}$ is defined in (4.5a), with $d_{ij}^{(2)}$ and ε_{ij} defined in (4.3a) and (4.3e) respectively.

From (4.5) for example, we may examine the statement $a_s = 0$ (s=0 or 1), by computing $\gamma_s = P(a_s > 0|\delta_1^2, \delta_2^2; \underset{\sim}{y})/P(a_s < 0|\delta_1^2, \delta_2^2; \underset{\sim}{y})$, that is, the posterior odds in favour of " $a_s > 0$ " . Examination of the weights $d_{ij}^{(2)}$ are also useful. The result (4.4) is also interesting in this respect. If a_1 and a_2 are "very large", and y_i and y_j are generated from the corresponding spurious sources $N(\mu + a_1\sigma/\delta_1, \sigma^2/\delta_1^2)$ and $N(\mu + a_2\sigma/\delta_2, \sigma^2/\delta_2^2)$, respectively, then with high probability y_i and y_j will be outliers and $A^{(ij)}$ will be least among the $n(n-1)/2$ distinct $A^{(ij)}$'s . Hence $d_{ij}^{(2)}$ will be very close to 1 . To concentrate on a_1 for a moment, we have from (4.4), that for our extreme case

$$E(a_1|\delta_1, \delta_2; \underset{\sim}{y}) \cong \frac{\sqrt{2}\,\Gamma(\frac{n}{2})}{\Gamma(\frac{n-1}{2})} \frac{z_{s(ij)}}{(1-\varepsilon_{ij})^{\frac{1}{2}}} . \tag{4.6}$$

Now using definition (4.4a) along with (4.3h) and (4.3i), it is straightforward (but tedious) to see that we may write (4.6) as

$$E(a_1|\delta_1^2, \delta_2^2; \underset{\sim}{y}) \simeq \frac{\sqrt{2}\,\Gamma(\frac{n}{2})}{\Gamma(\frac{n-1}{2})} \frac{\delta_1}{\sqrt{A^{(ij)}}} (y_i - \bar{y}^{(ij)}) . \tag{4.7}$$

Now for our extreme case, y_i gives information on $\mu + a_1\delta_1/\sigma$, $\bar{y}^{(ij)}$ on μ , and $\sqrt{A^{(ij)}}$ on $\sqrt{2}\,\Gamma(\frac{n-2}{2})\sigma/\Gamma(\frac{n-3}{2})$, so that the quantity on the right-hand side of (4.7) gives information on

$$\left\{\left[\frac{\sqrt{2}\,\Gamma(\frac{n}{2})}{\Gamma(\frac{n-1}{2})}\right]\Bigg/\left[\frac{\sqrt{2}\,\Gamma(\frac{n-2}{2})}{\Gamma(\frac{n-3}{2})}\right]\right\} \frac{\delta_1(\mu + a_1\sigma/\delta_1 - \mu)}{\sigma}$$

$$= e_n' a_1 \tag{4.7a}$$

where $e_n' = 1 + o\left(\frac{1}{\sqrt{n}}\right)$. That is, (4.7) gives information essentially about a_1 .

For our extreme case, we also note that $p(a_1, a_2 | \delta_1^2 , \delta_2^2 ; \underset{\sim}{y}) \cong h_{ij}(a_1, a_2)$, and indeed $h_{ij}(a_1, a_2)$ is located at (b_1, b_2) , say, where

$$(4.8) \qquad b_s = \sqrt{2}\ \Gamma(\tfrac{n}{2})\ z_{s(ij)} \Big/ \left[\Gamma(\tfrac{n-1}{2})(1 - \varepsilon_{ij})^{\frac{1}{2}} \right] .$$

An illustrative example putting the above results to use is given in Guttman and Khatri (1974), and for reasons of space is omitted here.

4b. Mean Shifts related: $a_1 = -a_2 = a$

In this sub-section, we assume that 2 observations may be spurious of the type described in the previous sub-section, but assume that the spuriousness is such that one of the observations has mean that is shifted by " a " standard deviations $[(\sigma/\delta_1)$, the standard deviation of this spurious source] above μ , while the other has mean that is shifted by " a " standard deviations $[(\sigma/\delta_2)$, the standard deviation of this spuriousness] below μ . Specifically, we assume that if y_i and y_j are spurious, then $y_i \sim N(\mu + a\sigma/\delta_1 ; \sigma^2/\delta_1^2)$ and $y_j \sim N(\mu - a\sigma/\delta_2 ; \sigma^2/\delta_2^2)$. We of course assume that all $n(n-1)$ pairs of observations are equally likely to be spurious, if spuriousness is to occur, so that the likelihood is as in (4.1) with $a_1 = -a_2 = a$. If we use the prior $p(\mu , \sigma^2 , a) \propto (\sigma^2)^{-1}$, then, by following the arguments of Appendix IIIa, with $a_1 = -a_2 = a$, and/or by putting $a_1 = -a_2 = a$ in (4.3) - (4.3i), Guttman and Khatri (1974) prove (Appendix IIIb of that paper) that the posterior of a is given by

$$(4.9) \qquad p(a | \delta_1^2 , \delta_2^2 ; \underset{\sim}{y}) = \sum_{i \neq j}\sum \omega_{ij}\ h_{ij}(a)$$

where the weights ω_{ij} are given by

(4.9a)

$$\omega_{ij} = (u_{1(ij)})^{-(n-1)/2} \Big/ \sum_{i \neq j}\sum (u_{1(ij)})^{-(n-1)/2} , \quad \sum_{i \neq j}\sum \omega_{ij} = 1 ,$$

and the function h_{ij} is defined by

(4.9b)

$$h_{ij}(a) = \sum_{t=0}^{\infty} \frac{\Gamma(\frac{n-1+t}{2})\ 2^{t/2}(1-\varepsilon_{1(ij)}^2)^{(n-1)/2}}{t!\ \Gamma(\frac{n-1}{2})\ \sqrt{2\pi}} (\varepsilon_{1(ij)}a)^t \ .$$

$$\cdot\ c^{(t+1)/2} \exp\{-ca^2/2\} \ ,$$

and where

(4.9c) $$c = [2n-4+(\delta_1+\delta_2)^2]/[n-2+\delta_1^2+\delta_2^2]$$

$$p_{ij} = \frac{[(n-2)(\delta_2-\delta_1)\bar{y}^{(ij)} + \delta_1(n-2+\delta_2^2+\delta_1\delta_2)y_i - \delta_2(n-2+\delta_1^2+\delta_1\delta_2)y_j]}{2n-4+(\delta_1+\delta_2)^2}$$

$$\varepsilon_{1(ij)} = \sqrt{c}\ p_{ij} / [A^{(ij)} + B_{ij}]^{\frac{1}{2}} \ , \ -1 < \varepsilon_{1(ij)} < 1 \ , \text{ and}$$

$$u_{1(ij)} = A^{(ij)} + B_{ij} - c\,p_{ij}^2 \ .$$

[Recall that $A^{(ij)}$ and $\bar{y}^{(ij)}$ are defined in (4.3b), while B_{ij} is defined in (4.3f).] From (4.9) we are able to show that

(4.10)

$$E(a|\delta_1^2, \delta_2^2; \underset{\sim}{y}) = \frac{\sqrt{2}\ \Gamma(\frac{n}{2})}{\Gamma(\frac{n-1}{2})} \sum_{i\neq j}\sum \omega_{ij}\, \varepsilon_{1(ij)} \Big/ \sqrt{c(1-\varepsilon_{1(ij)}^2)}$$

and

(4.10a)

$$\mathrm{Var}(a|\delta_1^2, \delta_2^2; \underset{\sim}{y}) = \frac{1}{c}\left\{1 + (n-1)\sum_{i\neq 1}\sum \omega_{ij}\left(\frac{\varepsilon_{1(ij)}^2}{1-\varepsilon_{1(ij)}^2}\right)\right\}$$

$$- [E(a|\delta_1^2, \delta_2^2; \underset{\sim}{y})]^2 \ .$$

Further, it is also shown in Appendix III of Guttman and Khatri (1974), that

(4.11) $$P(a > 0|\delta_1^2, \delta_2^2; \underset{\sim}{y}) = \frac{1}{2}\sum_{i\neq j}\sum \omega_{ij}\, b(i,j)$$

with

(4.11a)

$$b(i,j) = \sum_{m=0}^{\infty} \frac{\Gamma(\frac{n}{2} + m) 2^{2m} (1 - \varepsilon^2_{1(ij)})^{(n-1)/2} \varepsilon^{2m+1}_{1(ij)} \Gamma(m+1)}{\Gamma(2m+2) \Gamma(\frac{n-1}{2}) \sqrt{\pi}} .$$

An illustrative example putting the results of this subsection to work is given in Section 4b of Guttman and Khatri (1974).

Acknowledgements

The authors would like to thank Malcolm Cairns for his patient and skillful programming necessary for the completion of this paper.

References

[1] Box, G.E.P., and G.C. Tiao, (1968), "A Bayesian Approach to Some Outlier Problems," Biometrika, 55, pp. 119-129.

[2] Cramer, H. (1946), "Mathematical Methods of Statistics," Princeton University Press (p. 133).

[3] De Finetti, Bruno, (1961), "The Bayesian Approach to the Rejection of Outliers," Fourth Berkeley Symposium, pp. 199-210, University of California Press, Berkeley, California.

[4] Dempster, A. P. and B. Rosner, (1971), "Detection of Outliers," Research Report 5-11, Department of Statistics, Harvard University, Cambridge, Mass.

[5] Guttman, Irwin (1973), "Care and Handling of Univariate or Multivariate Outliers in Detecting Spuriosity -- A Bayesian Approach," Tecnometrics, Vol. 15, pp. 723-738.

[6] Guttman, Irwin, and C. G. Khatri, (1974), "A Bayesian Approach to Detection of Spuriosity," Technical Report, Department of Mathematics, University of Toronto.

[7] Kendall, M. G., and Alan Stuart, (1958), "The Advanced Theory of Statistics," Vol. 1, C. Griffin and Co., London, England, (pp. 155-156).

Irwin Guttman
Department of Mathematics
University of Toronto
Toronto, Ontario
Canada

C. G. Khatri
Department of Statistics
Gujrat University
Ahmedasad, India

Applied Statistics, R.P. Gupta, (editor)

EDUCATIONAL MEASUREMENT IN THE UNIVERSITY: QUANTIFYING TEACHING QUALITY AND VARIATIONS IN GRADING STANDARDS

by Paul B. Huber*

Abstract

This paper predicts grades of graduate students in economics on the basis of certain student characteristics, and on the basis of current-year and past-year instructor effects. Current-year instructor effects are interpreted as reflecting differing grading standards among instructors. Previous-year instructor effects -- the effect on the current-year grade of a given student in a class with a given instructor of having been taught by a particular (different) instructor in the previous year -- are interpreted as a measure of teaching quality. A weighted multiple regression model is employed, which could readily be extended, for example to be applied to undergraduates and departments in order to measure departmental differences in standards and/or teaching effectiveness.

Two aspects of the estimation may exhibit some statistical interest: (a) differing grading standards are treated as errors in the grading variables, these errors being removed by an iterative procedure; (b) in contrast to the usual individual tests of significance of estimates of dummy variable coefficients, collective tests are applied.

Estimated previous year instructor effects are not correlated with current year effects. Both types of effects yield estimated distributions which modestly but significantly differ from random distribution of estimated dummy variable coefficients. For both effects, the maximum difference between instructors is about 20 points on a 100 point grading scale in which 65 is the minimum passing grade.

BACKGROUND

The evaluation of learning performance and of teaching performance is obviously central to the educational process. In North American universities, the usual practice is for the individual instructor to assess the learning performance of his (her) students and to assign grades which reflect these assessments. The evaluation of teaching performance at the post-secondary level is sometimes

*I am grateful for the assistance of colleagues in the Economics and Mathematics Departments at Dalhousie. Responsibility for errors is mine.

undertaken by students; other attempts to assess teaching quality are sketchy and incoherent. The weaknesses of these evaluative procedures have long been recognized, [e.g., by 11] a recognition reflected for example in the development from the early 1900's of objective tests and normalized scoring as a way of eliminating bias in test administration and grading. Yet these procedures persist while much of the large literature on educational evaluation is devoted to interminable methodological discussions and qualitative approaches.[1]

In this paper the usual grading practices are postulated. If all students were associated with all instructors, or if the distribution of students by instructor were random with respect to student characteristics and the distribution of instructors by student random with respect to instructor characteristics, random differences among instructors in evaluative standards would balance out with regard to measurement of learning performance of students. Under these conditions high grades by instructors would reflect either effective teaching or easy assessment standards, which without employing externally-normed examinations would still be difficult to separate. But such desirable distributions of students and instructors do not generally prevail, even though one or the other is often assumed when assessing teaching -- or alternatively learning -- performance.[2] Hence it would be desirable to have some way of measuring differences in the grading standards of instructors in order to measure gross student learning performance without bias. Net student learning could then be judged by comparing successive performance through time, and significant variations among instructors in net performance of students might be attributed to differences in teaching quality.

Considerable effort has been devoted to the prediction of academic performance. More than ten years ago researchers at the Educational Testing Service used what might be termed a multivariate "input-output" approach which essentially treats academic (learning) performance as a linear function of student and environmental inputs, one of which is teaching. Multiple regression models which elaborate and apply this approach have recently been surveyed by Guthrie [13 pp. 25-54] and Jencks [8].

Elsewhere [7] I have presented a model of this general type which differs from earlier work in two respects: (a) it attempts to adjust the measure of output for variations in instructor grading standards, and (b) it incorporates as arguments teaching inputs in the period before the current output period. In the present paper, attention is centred on unbiased statistical estimation of this model and on testing the power of the model to measure significant differences among instructors with respect to variations in teaching quality and with regard to variations in grading standards. The model is applied to data, the results generated being somewhat stronger than might have been expected.

THE STATISTICAL MODEL AND ITS ESTIMATION

In order to focus on measurement of teaching quality and of variations in grading, let us consider the model in its simplest form. Let gross learning output, the dependent variable, be represented by G_{ijt} -- the numerical grade (or the numerical equivalent of the letter grade) of the i_{th} student from the j_{th} instructor in the t_{th} year.[3] We restrict student input characteristics to $\bar{G}_{i,t-1}$ -- the weighted average grade of the i_{th} student in the $t-1_{th}$ year.[4] The effects of current instructors are estimated by the vector, d_k, of coefficients of the binary variable $N_k(k=1, K+1)$, defined as follows:

$$N_k = 0 \; ; \quad k \neq j$$

$$N_k = 1 \; ; \quad k = j \; .$$

The effects of instructors from the previous year are estimated by the vector, e_m of coefficients of the dummy variable $T_m(m=1,M+1)$, defined as follows: let $c_{mi,t-1}$ be the extent of involvement (in numbers of credits) of the m_{th} instructor teaching the i_{th} student in the $t-1_{th}$ year. Then

$$c_{mi,t-1} = (0.0, \; 0.5, \; 1.0, \; 1.5 \; \ldots); \text{ and}$$

$$T_m = 0 \; ; \quad m = j^5$$

$$T_m = c_{mi,t-1} \; ; \quad m \neq j$$

Thus the model may be expressed in linear form:

$$G_{ijt} = a + b\,\bar{G}_{i,t-1} + \sum_{k=1}^{K} d_k N_k + \sum_{m=1}^{M} e_m T_m + u_{ijt} \quad . \tag{1}$$

Clearly k and m refer to instructors teaching in the t_{th} year and in the $t-1_{th}$ years respectively, there being respectively $K+1$ instructors teaching more than one student in the course of the period covered, M+1 of which taught any t_{th} year student in the $t-1_{th}$ year; u_{ijt} is the error term which is assumed to be normally distributed with mean 0 and variance σ^2 .

In this model e_m measures the net effect on current gross learning performance (G_{ijt}) of having been exposed to a particular (different) instructor in the previous year; d_k measures the effect on the current assessed grade of the particular instructor making the assessment of G_{ijt} . In the former case it has been explicitly assumed that the corresponding instructor input is the presence of the instructor. This seems an appropriate assumption for the current instructor input as well. Observations are weighted in accordance with the credit applicable to the j_{th} instructor's class: for full credit classes, the weight is unity; for half-credit classes the weight equals 0.5 . Therefore the estimation is carried out using weighted multiple linear regression.[6]

The net variation in assessments which are not predicted by $\bar{G}_{i,t-1}$ reflect specification errors and random variation in measuring G_{ijt} as well as the effects of d_k and e_m . *A priori*, the effects of e_m involve variations in the amounts of learning by students. If, in fact, the estimated d_k also measure improvements in learning by students as well as systematic differences among instructors with respect to standards of assessment, then -- provided that true learning is not merely evanescent -- e_m and d_k must be positively correlated for $m=k$, since both measure learning produced in students by a given instructor. Hence we may construct a null hypothesis

$$(2) \qquad H_0 : r_{e_m, d_k} = 0 \ .$$

If this hypothesis cannot be rejected -- even at a low level of certainty -- then we may suppose that current instructor effects solely measure systematic variations among instructors with respect to judging (gross) student performance. Another way of putting this is to say that G_{ijt} is an erroneous measure of student performance to the extent that d_k differs from zero. But in this event $\bar{G}_{i,t-1}$ will also be erroneously measured since there is every reason to assume that students will not make a random choice among instructors who grade differently. Hence if d_k are treated as reflecting systematic errors of measurement, the model turns out to be one which involves errors in variables as well as errors in the equation. As is well known, biased and inconsistent estimators result. Define d_k^*

$$d_k^* = \hat{d}_k - \sum_k \hat{d}_k/K = \hat{d}_k - \bar{d} \tag{3}$$

in order to centre the estimated current year instructor effects around zero. If for any given observation we then subtract d_j^* from both sides of equation (1) we obtain (neglecting terms equal to zero)

$$G_{ijt} - \hat{d}_j + \bar{d} = a + b\,\bar{G}_{i,t-1} + \bar{d} + \sum_m e_m T_m + u_{ijt}\ . \tag{4}$$

This will leave the average grade essentially unchanged. And it clearly adjusts G_{ijt} to eliminate the systematic assessment error of the j_{th} instructor. But consider the effect of eliminating this error from $\bar{G}_{i,t-1}$. We would obtain a term

$$b(\bar{G}_{i,t-1} - \sum_m \hat{d}_m c_{mi,t-1} / \sum_m c_{mi,t-1} + \sum_k d_k/K) \tag{5}$$

which only by accident will equal $b\,\bar{G}_{i,t-1}$. Hence $\hat{b}$ will likely be biased as will $\hat{e}_m$.

Assuming still that hypothesis (2) cannot be rejected, we may eliminate the error in assessment (as measured by $\hat{d}_k$) from G_{ijt} and $\bar{G}_{i,t-1}$ by replacing these terms in equation (1) by the corrected terms from equations (4) and from (5) and regressing (1) again. If repeated, this iterative procedure in the limit will have the effect of reducing the estimated $\hat{d}_k$ to zero, thus eliminating any bias in the estimation of b or other coefficients. By summing the d_k^* over all iterations we will also obtain in the limit unbiased estimates of the current-year instructor effects, $\hat{D}_k$ -- under the assumption that these effects measure only systematic assessment biases of instructors:

$$\hat{D}_k = \sum_g {}_g d_k^* \text{ ; where } g \text{ refers to the } g_{th} \text{ iteration.} \tag{6}$$

In fact only three iterations were calculated since the rate of convergence is rapid -- estimated $\hat{d}_k$ being reduced by about 10^{-1} on each iteration.

THE DATA

The sample consists of the grades and characteristics of all students regularly enrolled in the graduate programmes of the Department of Economics at Dalhousie University who received grades in classes taken in the years 1965 through 1972. All grades of all such students received while they were formally enrolled as graduate students are included, whether or not the classes involved were at the graduate or undergraduate level and whether or not they were in the Department of Economics. There are 586 grades earned by 99 students in classes given by 42 instructors in this sample.[7]

RESULTS

Although a number of specifications were regressed, essentially two are reported here, the first suppressing the intercept, a ; the second including it. Since all the variables are dummies except $\bar{G}_{i,t-1}$, it is reasonable to postulate linear homogeneity between this and the dependent variable and hence suppress the intercept. With no intercept, the coefficient of $\bar{G}_{i,t-1}$ reflects the tendency of the current year grades of students to exceed (or be smaller than) $\bar{G}_{i,t-1}$.

The values from estimation of these specifications are presented in Table I. The first three columns refer to the model without intercept (Model 1), the last three to the model which includes the intercept (Model 2). Columns one, two, four and five present estimated coefficients derived from the third iteration. For purposes of comparison, estimates of e_m without iteration are provided in columns three and six. Overall, the two models "explain" respectively 33 percent and 39 percent (unadjusted) of the variance in G_{ijt} using essentially three variables.[8] This does not compare unfavourably with results of other attempts to predict academic performance with multiple regression models: Michelson [13, pp. 120-168] reported coefficients of determination from .07 to .48 using 10 or more input variables, Creager [3] values of around .50 using a minimum of 12 inputs selected stepwise from a pool of more than a hundred possible variables.

Obviously many of the estimated instructor effects turn out not to differ significantly from zero. Since the estimated values of D_k and e_m both form distributions it is inappropriate to apply individual tests of significance; inevitably some estimates lie near the middle of distribution and even in a chance distribution, one out of twenty values may be expected to be "significant" at the 95% level. Instead collective tests may be applied to the $\hat{D}_k$ and the $\hat{e}_m$ distributions to test the hypotheses that these distributions of estimates might have been generated by chance. The standard approach [5, pp. 218-231; or 14] is to apply an F test to the hypothesis that the variance "explained" by adding a set of dummy variables to the regression exceeds the variance remaining unexplained. Let us use the notation $SSR(I_g)$, $SSRR(I_g)$, $SSU(I_g)$ and $SST(I_g)$ to refer respectively to the sums of squares "explained" by regression, the sums of squares "explained" by regression with dummy variable coefficients restricted to equal zero, the sums of squares unexplained by unrestricted regression, and the total sum of squares -- all in the gth iteration. Further, let DFU be the degrees of freedom remaining after unrestricted estimation.[9] Then we may express the null hypotheses with respect to estimates of D_k and e_m as follows:[10]

$$(7) \qquad H_{0,\hat{D}_k} : \frac{(SST(I_1) - SST(I_3))/K}{SSU(I_3)/DFU} \leq 1.0$$

$$(8) \qquad H_{0,\hat{e}_m} : \frac{(SSR(I_3) - SSRR(I_3))/M}{SSU(I_3)/DFU} \leq 1.0$$

Note that in (7) the reduction in the sums of squares attributable to inclusion of $\hat{D}_k$ is measured by the extent to which the total sums of squares of G_{ijt} is reduced by iteration.

Because of the iterative estimation procedure this standard test is expensive computationally. An alternative approach is to investigate the distributions of "t" values of the dummy variable estimates, where "t" is the ratio of estimated coefficient (as a deviation from the mean of such estimates) divided by its estimated standard error.

$$(9) \qquad \hat{t} = \frac{\hat{V}-\bar{V}}{\hat{S}_V};$$ where $\hat{V}$ is an estimated coefficient from either $\hat{e}_m$ or $\hat{D}_k$

It the t's are independently distributed -- which is assumed -- then the population of t's relating to $\hat{e}_m$ or to $\hat{D}_k$ will be normal with zero mean and unit variance. The sum of squares of a random sample from such a population is χ^2 distributed with n-1 degrees of freedom [6, pp. 216-217]. Hence we may test the possibility of random occurence of the distribution of $\hat{V}$'s by setting up the following null hypotheses:

$$(10) \qquad H_{0,\hat{D}_k} : \sum_K \hat{t}_{D_k}^{\,2} \leq K$$

$$(11) \qquad H_{0,\hat{e}_m} : \sum_M \hat{t}_{e_m}^{\,2} \leq M$$

Let us test hypothesis (2) at the 30% level using both tails. With 25 instructors for whom there are estimates of current-year and previous-year effects, r_{e_m,D_k} will be normally distributed with a variance of $1/(25-3)$. Hence the critical value of $r_{30\%}$ is .0809 . The values obtained were respectively .045 and .078 for the two models. Hence hypothesis (2) that D_k and e_m are uncorrelated cannot be rejected even at a low significance level.

The critical F values for one-tailed tests of hypotheses (7) and (8) at the 99% level are $F_{30,530} = 1.73$, $F_{30,529} = 1.73$ for (7), and for (8) $F_{25,530} = 1.82$, $F_{25,529} = 1.82$ for models 1 and 2 respectively. The values obtained are indicated in Table I. All indicate rejection of these hypotheses, although the distribution of e_m is not significant at the 99.9% level. The critical values of χ^2 for one-tailed test of hypotheses (10) and (11) at the 99% level are 49.6 and 43.0 respectively. Again the values obtained lead to the rejection of the hypotheses, although with respect to e_m the hy othesis could not be rejected at the 99.9% level.

In conclusion, the method outlined in this paper appears capable of isolating and quantifying the current-year effects and the previous-year effects of instructors on the grades received by their students. In practical terms these effects appear large enough to warrant further refinements of the model, and possibly adjustments in scholarship granting practices and other reward systems in the universities. It may be tempting however to overstate the results from a model such as this. Note that the performance measure used is internal to the education system: specifically it consists of those cognitive elements recognized and rewarded in graduate training in economics. Moreover the input of instructors from several departments reflects only their efforts in teaching graduate students from one department. Obviously there are other dimensions of student learning and of professorial input which are therefore neglected.

Doubtless the model could be extended in several dimensions, given a larger data base and an adequate regression programme. For example lagged effects of two or three periods could be introduced to try to discover rates of attenuation of learning, or variations in grading standards through time of individual instructors could be explored in an analogous manner. And attempts could be made to improve the measurement of some of the student and instructor inputs.

TABLE I

REGRESSION ESTIMATES OF INSTRUCTOR EFFECTS

	Model 1			Model 2		
a Intercept		-			36.156(5.666)**	
b		1.000(0.27)			.508(.073)**	
INSTRUCTOR	$\hat{D}_k(I_3)$	$\hat{e}_m(I_3)$	$\hat{e}_m(I_1)$	$\hat{D}_k(I_3)$	$\hat{e}_m(I_3)$	$\hat{e}_m(I_1)$
A	-2.331	1.286	2.022	-2.341	1.176	1.637
B	7.717+	-8.527**	-10.854**	5.168	-7.196*	-8.891**
C	-1.164	3.831	3.948	-1.902	3.538	3.648
D	-1.645	-.559	-.232	-1.730	-.021	.124
E	-8.266**	-2.598+	.916	-9.725**	-3.134*	-1.977
F	2.063	1.031	.163	1.365	-.592	-1.086
G	.255	.878	.072	-1.242	.070	.010
H	-6.919*	-2.488	-.522	-7.786**	-1.779	-.780
I	-1.701	-.152	.652	-1.815	-1.583	-1.252
J	4.570	5.602*	4.133+	3.859	4.102+	3.924+
K	-1.785	5.424	5.320	-2.610	2.601	2.547
L	-7.037**	-2.828	-.730	-7.662**	-3.760*	-2.672
M	-2.945	-.289	.646	-3.185	1.921	2.488
N	-11.437**	.711	3.333	-10.957**	4.068	5.554+
O	.895	-.923	-.296	1.084	.640	.446
P	3.291	-1.941	-2.414+	4.065+	-.082	-.606
Q	-.435	3.897	4.582	-1.039	-1.348	-.274
R	5.270+	-.127	-1.355	4.303	-2.918	-3.570
S	4.860*	1.116	-.587	5.994*	1.693	-.626
T	2.077	10.196**	9.840**	2.884	8.925**	8.689**
U	-1.962	4.773	5.139+	-1.857	6.359*	6.654*
V	10.135**	-4.434	-8.641**	9.450**	-1.872	-3.848
W	-.103	-3.134	-4.129	.538	-2.504	-3.043
X	-4.712	-5.310	-2.680	-5.167	-10.585+	-9.206
Y	-1.343	-	-	-1.703	-	-
Z	-6.343	-	-	-7.659+	-	-
AA	1.996	-3.827	-4.764	1.397	1.583	.861
BB	1.631	-	-	.929	-	-
CC	1.583	-	-	3.556	-	-
DD	9.311*	-	-	9.287*	-	-
F	5.669**	2.171**		7.201**	2.130**	
Σt^2	69.111**	47.695**		77.580**	48.796**	

Significant at + 90% level : * 95% level : ** 99% level

FOOTNOTES

1. *c.f.* in recent years in the Unites States, [11][4][12]. It is symptomatic that in assessment of "teaching effectiveness" in schools only in the last few years [e.g., 13] has measurement been on the basis of learning outputs -- however measured -- rather than on the basis of teaching inputs. In 1948, nine-tenths of the 140 studies surveyed by Barr [2] assessed effectiveness *solely* in terms of the latter.

2. *c.f.* [9], especially Ch. 2 "Academic Performance: The Meaning of the Concept and Problems of its Measurement."

3. Dalhousie shifted from a numerical grading system for graduate students to a letter system in 1972. To relate the two grading systems, the Economics Department used a standard conversion scale which has been used here.

4. Weighting is by the number of credit hours per class. If the t_{th} year was the first at Dalhousie, $\bar{G}_{i,t-1}$ is the anticipated average grade of the student in graduate work on entrance to the Dalhousie programme. This was calculated for North American students as the average in social science and mathematics classes in the year prior to entrance; for other students a judgemental estimate was made.

5. This restriction avoids the bias which might be involved if students become attached to a particular instructor and merely learn to adjust to his (her) idiosyncracies.

6. There were no jointly-instructed classes offered in this period. If there had been and it were assumed that current instructor effects on grades are independent and additive, it would have been appropriate for each such class to set the values of N_k for all participating instructors positive in proportion to participation in the grading process, the sum of N_k for any observation equalling unity. Alternatively, without such assumptions, joint instructorship can be treated as creating artificial composite instructors, each such composite being separately identified by its own current and previous year dummies.

7. In the particular application of this model, the "reference" $K + 1_{th}$ and $M + 1_{th}$ instructor is a composite of twelve instructors -- mostly from outside the Economics Department,

each of whom taught one graduate student during the period from which the data is taken. With respect to the M instructors, it was necessary to eliminate one visiting professor because the students who went on from his half class were the same as those going on from a full year class taught by instructor X . As a consequence the coefficient of instructor X includes the effects of both men.

8. Adjusted percentages are 26% and 33% respectively. These values are derived by subtracting the ratio of unexplained sum of squares after iteration to the initial total sum of squares from unity (and adjusting as indicated for degrees of freedom).

9. This equals the number of observations minus the number of non-dummy explanatory variables minus M minus K if there is no intercept; and it is reduced by one if there is an intercept.

10. Note that $SSRR(I_3)$ is estimated by omitting T_m at each iteration, not just on the third iteration.

REFERENCES

[1] Astin, Alexander W., (1963), "Differential College Effects on the Motivation of Talented Students to Obtain the Ph.D.", Journal of Educational Psychology 54, 63-71.

[2] Barr, A. S., (1948), "The Measurement and Prediction of Teaching Efficiency: A Summary of Investigations," Journal of Experimental Education 16, 203-283.

[3] Creager, John A., (1971), "Academic Achievement and Institutional Environments: Two Research Strategies," Journal of Experimental Education 40, 9-23.

[4] The Evaluation of Teaching, (Washington: Pi Lambda Theta, 1967).

[5] Goldberger, Arthur S., (1964), Econometric Theory, New York, Wiley.

[6] Hoel, Paul G., (1954), Introduction to Mathematical Statistics, 2nd ed., New York, Wiley.

[7] Huber, Paul B., "An 'Internal' Measure of the Quality of University Teaching" (unpublished manuscript).

[8] Jencks, Christopher et. al., (1972), Inequality: A Reassessment of the Effect of Family and Schooling in America, New York, Basic Books.

[9] Lavin, David E., (1965), The Prediction of Academic Performance, New York, Russell Sage.

[10] Suchman, Edward A., (1967), Evaluative Research, New York, Russell Sage.

[11] Trabue, Marion R., (1924), Measuring Results in Education, New York, American Book Company.

[12] Tyler, Ralph W., (1969), ed. Educational Evaluation: New Roles, New Means, The Sixty-eighth Yearbook of the National Society for the Study of Education, Part II, Chicago, University of Chicago Press.

[13] U.S. Department of Health, Education and Welfare, (1970), Do Teachers Make a Difference? A Report on Recent Research on Pupil Achievement, Washington, USGPO.

[14] Ward, Joe H., Jr., (1962), "Multiple Linera Regression Models," Ch. 10 in Harold Borko, ed., Computer Applications in the Behavioral Sciences, Englewood Cliffs, Prentice Hall.

Paul B. Huber
Department of Economics
Dalhousie University
Halifax, Nova Scotia
Canada

Applied Statistics, R.P. Gupta, (editor)

CLUSTER ANALYSIS AS A MULTIPLE COMPARISON METHOD

I.T. Jolliffe

Multiple comparison methods are widely used in biological and other sciences to decide which of a set of sample means differ 'significantly' from one another. An alternative to the well-known methods of Scheffé, Tukey, Duncan etc. is to use cluster analysis to put the means (or, equivalently, the corresponding populations) into groups, and then to decide that two means differ 'significantly' if and only if they belong to different groups. It is possible to construct methods based on cluster analysis which are closely linked to conventional multiple comparison methods. This is shown for the Newman-Keuls multiple comparison method. It is suggested that the methods based on cluster analysis may have some advantages over conventional methods.

1. Introduction

Consider the multiple comparison situation, where samples of size n are taken from each of p populations, corresponding to p treatments in an experiment, and the value of a random variable, X, is observed for each sample member. Suppose that for the ith population

$$X \sim N(\mu_i, \sigma^2) \qquad i = 1,2,\ldots,p \quad ,$$

that $\bar{x}_1 \leq \bar{x}_2 \leq \ldots \leq \bar{x}_p$, are the sample means, and that s^2 is an independent estimate of σ^2 with ν degrees of freedom. The assumptions of normality and of equal variances are necessary for all the well-known multiple comparison methods, although the assumption of equal sample sizes is not essential for some e.g. Scheffé's (1953) method.

Multiple comparison methods may be used to find simultaneous confidence intervals for differences between population means, or for linear combinations of population means. Alternatively, they may be used to test hypotheses of the form

$$H_{k,\ell} : \mu_{k+1} = \mu_{k+2} = \ldots = \mu_{k+\ell} \quad ,$$

$$\text{for} \quad k = 0,1,\ldots,(p-\ell); \; \ell = 2,3,\ldots,p \quad .$$

If all such hypotheses are tested, and if populations $k+1, k+2, \ldots, k+\ell$ are assigned to the same cluster when $H_{k,\ell}$ is accepted, then a set of overlapping clusters is obtained e.g. if $p = 4$ and $H_{0,2}$, $H_{1,3}$ are accepted but $H_{0,4}$, $H_{0,3}$ are rejected, then there are two overlapping clusters, $\{1,2\}$, $\{2,3,4\}$.

It may be required, as suggested by Calinski (1969), that the populations be put into non-overlapping clusters, and one possible way to do this is to combine all overlapping clusters which have common elements. This will, in general, give a few large, relatively non-homogeneous, clusters; in the example above with $p = 4$ all populations are put into the same group.

Another approach to the problem of finding non-overlapping clusters, which may divide the populations into smaller, more homogeneous groups, is to define a distance or similarity measure between pairs of populations and then to apply one of the numerous published methods of cluster analysis to the matrix of distances or similarities. For example, Calinski (1969) defines the distance, d_{ij}, between the ith and j^{th} populations in the most obvious way i.e.

$$d_{ij} = |\bar{x}_i - \bar{x}_j| \quad ,$$

and uses the clustering method suggested by Edwards and Cavalli-Sforza (1965), which attempts to minimize 'within-groups' sums of squares.

By defining suitable similarity measures it is possible to construct clustering methods which are, in a sense, extensions of conventional multiple comparison methods, and this is demonstrated below for the Newman-Keuls (NK) multiple range test (Newman, 1939, Keuls, 1952).

A suitable similarity measure, based on probability, is introduced in Section 3, and this measure is derived from consideration of multiple range tests, which are discussed in Section 2. In Section 4, it is shown that the use of this similarity measure, in conjunction with single-linkage cluster analysis, leads to clusters which could also be obtained from NK multiple range tests. Finally, in Section 5, an example is given which illustrates some interesting properties of this clustering method when used in the multiple comparison situation.

2. Multiple Range Tests.

Suppose the multiple comparison problem is defined as in Section

1 and it is required to test all hypotheses of the form $H_{k,\ell}$, $k = 0,1,\ldots,(p-\ell)$; $\ell = 2,3,\ldots,p$.

Multiple range procedures consist of $(p-1)$ stages and, at the qth stage, hypotheses of the form

$$H_{k,(p-q+1)} , \quad k = 0,1,\ldots,(q-1) \quad \text{are tested.}$$

However, it is usually unnecessary to test all such hypotheses since acceptance of $H_{k,\ell}$ will imply automatic acceptance of $H_{k',\ell'}$ when $k' \geq k$ and $k' + \ell' \leq k + \ell$ i.e. if $H_{k,\ell} : \mu_{k+1} = \mu_{k+2} = \ldots = \mu_{k+\ell}$ is accepted then so is $H_{k',\ell'} : \mu_{k'+1} = \mu_{k'+2} = \ldots = \mu_{k'+\ell'}$ whenever $\{\mu_{k'+1} , \mu_{k'+2},\ldots,\mu_{k'+\ell'}\}$ is a subset of $\{\mu_{k+1}, \mu_{k+2}, \ldots , \mu_{k+\ell}\}$. Because of this the procedure may stop before the $(p-1)$th stage is reached.

The results of such a procedure may be simply represented by writing down labels for the p populations in ascending order of their sample means, and underlining populations $\{k+1,k+2,\ldots,k+\ell\}$ if and only if hypothesis $H_{k,\ell}$ is accepted. This is illustrated by a simple example for which $p = 6$.

1 2 3 4 5 6

This diagram would be constructed if the procedure gave the following results at each stage.

<u>Stage 1</u> $H_{0,6}$ tested and rejected.

<u>Stage 2</u> $H_{0,5}$, $H_{1,5}$ tested and rejected.

<u>Stage 3</u> $H_{0,4}$, $H_{1,4}$, $H_{2,4}$ tested;
$H_{1,4}$, $H_{2,4}$ rejected, $H_{0,4}$ accepted, so underline populations $\{1,2,3,4\}$.

<u>Stage 4</u> $H_{0,3}$, $H_{1,3}$ automatically accepted;
$H_{2,3}$, $H_{3,3}$ tested; $H_{2,3}$ rejected,
$H_{3,3}$ accepted, so underline populations $\{4,5,6\}$.

The procedure stops at this stage since all hypotheses which could be tested at Stage 5 i.e. $H_{k,2}$, $k = 0,1,\ldots,4$, have already been automatically accepted.

In this example two overlapping clusters, $\{1,2,3,4\}$ and $\{4,5,6\}$ have been formed.

The NK multiple range test, at level α, rejects $H_{k,\ell}$ at the appropriate stage of the above procedure if and only if

$$\bar{x}_{k+\ell} - \bar{x}_{k+1} > sn^{-\frac{1}{2}} q^{\alpha}_{\ell,\nu} ,$$

where $\bar{x}_i$, s, n and ν are defined as in Section 1, and $q^{\alpha}_{\ell,\nu}$ is the upper $100\alpha\%$ point of the Studentised range distribution with parameters ℓ and ν. To justify this test, suppose that the hypothesis $H^*_{k,\ell}$ is defined to be true when $H_{k,\ell}$ is true and

$$\left.\begin{array}{lll} H_{k,(\ell+1)} \text{ is false} & & k = 0 , \ell < p \\ H_{k,(\ell+1)}, H_{(k-1),(\ell+1)} & \text{are false} & 0 < k < p - \ell \\ H_{(k-1),(\ell+1)} \text{ is false} & & k = p - \ell, \ell < p \end{array}\right\} \quad (1)$$

Note that $H_{k,\ell}$ is only tested if condition (1) is accepted at the previous stage in the multiple range procedure, so acceptance/rejection of $H_{k,\ell}$ at the $(p-\ell+1)$th stage of the procedure is equivalent to acceptance/rejection of $H^*_{k,\ell}$. If $H^*_{k,\ell}$ holds, and $\bar{X}_i$, S are random variables corresponding to $\bar{x}_i$, s, then

$$Q_{\ell,\nu} = n^{\frac{1}{2}}(\bar{X}_{k+\ell} - \bar{X}_{k+1})/S \quad (2)$$

has the Studentised range distribution with parameters ℓ, ν, which provides the justification for the above test.

The other well-known multiple range test, due to Duncan (1955), differs from the above only in the fact that the significance level, α, depends on ℓ, whereas in the Newman-Keuls test α is independent of ℓ. For the moment attention will be restricted to the NK test, but Duncan's test will be mentioned again in Section 5.

3. A Similarity Measure Based on Probability.

Consider any pair, (i,j), from the p populations, and suppose, without loss of generality, that

$$i = j + \ell - 1 , \text{ where } \ell \geq 2 .$$

Then the probability, α_{ij}, that

$$D^*_{ij} = n^{\frac{1}{2}}|\bar{X}_i - \bar{X}_j|/S$$

exceeds the observed value

$$d_{ij}^* = n^{\frac{1}{2}}|\bar{x}_i - \bar{x}_j|/s = n^{\frac{1}{2}}d_{ij}/s \quad ,$$

given that $H^*_{(j-1),\ell}$ holds, can easily be found. This follows since, from (2), D_{ij}^* has the Studentised range distribution when $H^*_{(j-1),\ell}$ holds, and tables of this distribution are available, (Pearson and Hartley, 1943, Harter et. al., 1959).

Now it is reasonable to use α_{ij} as a measure of similarity, since, for a fixed pair of populations i and j, α_{ij} increases as d_{ij}, the absolute difference between the sample means, decreases, $\alpha_{ij} = 1$ when $d_{ij} = 0$, and $\alpha_{ij} \to 0$ as $d_{ij} \to \infty$. Notice, however, that because α_{ij} depends on the number of other sample means which lie between $\bar{x}_i$ and $\bar{x}_j$, $d_{gh} \leq d_{ij}$ does not necessarily imply $\alpha_{ij} \leq \alpha_{gh}$.

Mulitple range tests, rather than other multiple comparison methods, were used above as a starting point for constructing a similarity measure. The reason for this is that methods such as Scheffé's (1953) are designed mainly to produce numerous simultaneous confidence intervals, rather than for testing hypotheses like $H_{k,\ell}$, and in these methods all comparisons of pairs of means are dealt with in the same way i.e. independently of the value of $|i - j|$. Thus any similarity measure derived from these methods would just be a monotone transformation of d_{ij}.

4. Single Linkage Cluster Analysis and the Newman-Keuls Multiple Range Test.

Single linkage cluster analysis (SLCA) is only one of the great many methods which have been suggested for grouping observations. It is, however, one of the best known, one of the easiest to use, and it has many desirable properties (See, for example, Sibson, 1972). Furthermore, when used in multiple comparison problems, with the similarity measure defined above, it gives the same clusters as could be obtained from the NK multiple comparison method by combining overlapping clusters which have common elements.

This is easily seen by noting that if two populations are underlined in the NK method, at significance level α, then the similarity between them will be greater than α. If two populations are in the same NK cluster, then they are linked by a series of overlapping underlinings, which means that the two populations may be linked via a number of intermediate populations, with all links corresponding to similarities greater than α. Thus the two populations will be in the same single-linkage cluster, at similarity level α.

The main advantage that SLCA has over a straightforward application of the NK method is that it gives a more complete picture. The results of the NK method will normally be given at one or two significance levels only, whereas a dendrogram may be drawn for SLAC which gives, at a glance, clusters at all possible significance levels. [See Fig. 1, for example].

It is also of interest to note that simple probability statements may be made about the populations, conditional on whether the populations belong to the same, or different clusters. For example, if populations i and j are in different clusters at level α, then we know that

$$P[D_{ij}^* > d_{ij}^* | H^*_{(j-1),\ell}] < \alpha ,$$

where D_{ij}^*, d_{ij}^* and $H^*_{(j-1),\ell}$ are as defined previously. Such statements are noteworthy because of the complete absence of probability statements of any kind in most forms of cluster analysis, though there are exceptions such as Ling (1973).

The fact that probability statements can be made shows that cluster analysis is being used here as rather more than a purely descriptive method.

5. Example.

The example which follows was used by Duncan (1965) to illustrate and compare many different multiple comparison methods. In the present section the cluster analysis method suggested above will be tried on the same data, which come from a bread baking experiment and consist of measurements of loaf volume for 17 varieties of wheat.

The results of the NK multiple range test at the 5 per cent significance level may be represented as

M P D C Q L H G N B F I K J E A O

where M,P,...,A,O are labels for the 17 wheat varieties and, with obvious notation,

$$\bar{x}_M \leq \bar{x}_P \leq \cdots \leq \bar{x}_A \leq \bar{x}_O \quad .$$

These results may also be represented as a set of ten overlapping clusters,

{M}, {P,D}, {D,C}, {C,Q,L,H,G,N}, {Q,L,H,G,N,B}, {H,G,N,B,F,I}, {N,B,F,I,K}, {B,F,I,K,J,E}, {I,K,J,E,A} and {E,A,O} or as a set of two non-overlapping clusters, {M} {P,D,C,Q,L,H,G,N,B,F,I,K,J,E,A,O}.

The results of other well-known multiple comparison methods may be represented in a similar way, and all will give only one or two non-overlapping clusters.

Alternatively, Fig. 1 gives the dendrogram for these bread baking data when SLCA, with similarity coefficient α_{ij} , is used. This dendrogram illustrates several interesting points.

FIGURE 1

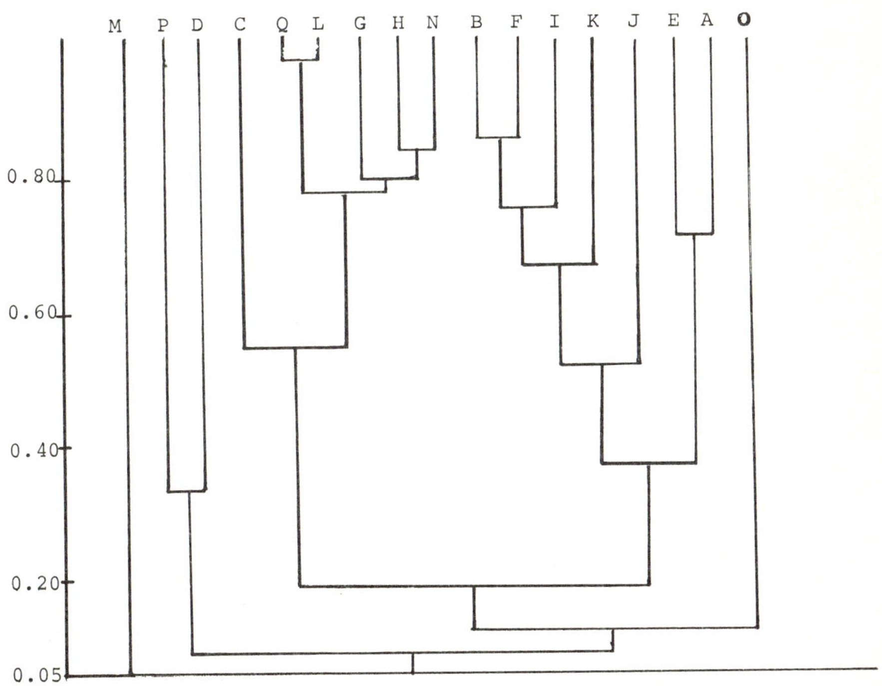

SLCA dendrogram for bread baking data.

FIGURE 2

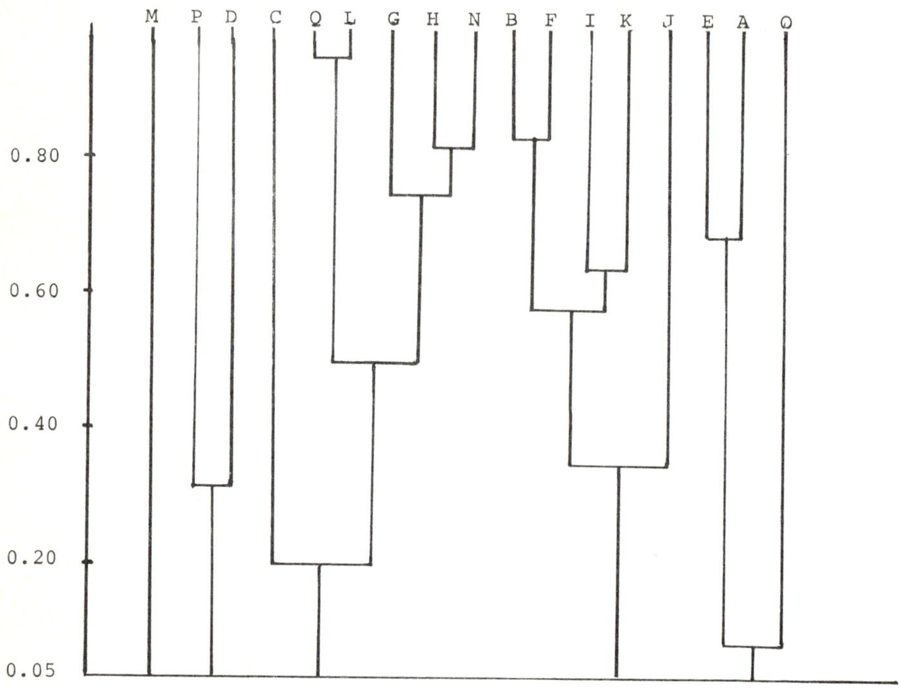

CLCA dendrogram for bread baking data.

As stated above, the clusters formed at $\alpha = 0.05$ are precisely the same as the non-overlapping clusters given by the NK multiple range test at the $\alpha = 0.05$ significance level. It is possible to use the dendrogram to find the NK clusters corresponding to any specified significance level α, but it is probably more useful to look at the entire lower part of the dendrogram. For example, for these data a striking feature of the dendrogram is the separation of those populations to the left of N from those to the right of B. Most conventional multiple comparison methods, applied at standard significance levels, fail to detect this (even Tukey's (1949) method which looks for large gaps).

One undesirable property of the clustering method, which is caused by the choice of similarity measure, is that for large values of α clusters may have gaps in them, in the sense that populations i and j may be assigned to one cluster and population k to another, even though $\bar{x}_j \leq \bar{x}_k \leq \bar{x}_i$. This is illustrated by populations H , G and N in the present example. The method could be modified to avoid this difficulty, but this additional complication is probably not worthwhile since, in general, only small values of α will be of interest (say $\alpha < 0.5$), and clusters with gaps can only occur at levels of α greater than this.

To conclude, three variations of the above clustering method will be mentioned. First, we could use the same similarity measure, but a different method of cluster analysis. One alternative, which also allows simple probability statements to be made about the populations, is to use complete linkage cluster analysis (CLCA). CLCA, however, does not have such a simple connection with multiple range tests as SLCA. Fig. 2 shows the dendrogram for CLCA, and it is seen that the structure is very similar to Fig. 1, the dendrogram for SLCA, except that populations join each other at lower levels of α than for SLCA.

Secondly, we could use Duncan's multiple range test, instead of Newman-Keuls, to construct a measure of similarity. This might be useful since Duncan's test is the more widely used. However, there are no tables in existence from which the similarities between populations could be easily found. Single-linkage clusters at a few selected levels of α , but not the whole dendrogram, may be found by using tables of critical points for Duncan's test e.g. Harter (1960). For $\alpha = 0.05$ the results of Duncan's test may be represented as

```
M   P   D   C   Q   L   H   G   N   B   F   I   K   J   E   A   O
    _____   _________________       _________________   _________
        _____   _________________           _____________
                            _____________       _____________
```

It can be seen that single-linkage clusters, {M} ,{P,D,...,A,O} are exactly the same as for the NK method, but the overlapping clusters are smaller for Duncan's method. A similar pattern emerges for $\alpha = 0.10$; the single-linkage clusters are the same for both Duncan and NK, but the overlapping clusters for Duncan's method are smaller. Thus, it seems unlikely that the full SLCA for Duncan's method will differ very much from that for NK.

Finally, it is worth noting that the use of d_{ij} instead of α_{ij} in the above analysis does not greatly alter the structure of the dendrograms. With either d_{ij} or α_{ij} the main use of cluster analysis in the multiple comparison situation is descriptive, but if α_{ij} is used, there is the advantage of a close link with well-known multiple range procedures, and the possibility of making probability statements like those in Section 4.

References

[1] Calinski, T. (1969). On the application of cluster analysis to experimental results. Bull. Int. Statist. Inst., 42, 101-103.

[2] Duncan, D.B. (1955). Multiple range and multiple F tests. Biometrics, 11, 1-42.

[3] Duncan, D.B. (1965). A Bayesian approach to multiple comparisons. Technometrics, 7, 171-222.

[4] Edwards, A.W.F. and Cavalli-Sforza, L.L. (1965). A method for cluster analysis. Biometrics, 21, 362-375.

[5] Harter, H.L. (1960). Critical values for Duncan's new multiple range test. Biometrics, 16, 671-685.

[6] Harter, H.L., Clemm, D.S. and Guthrie, E.H. (1959). The probability integrals of the range and of the studentized range, Vol. II - probability integral and percentage points of the studentised range; critical values for Duncan's new multiple range test. Wright Air Development Center Technical Report 58-484.

[7] Keuls, M. (1952). The use of the 'studentised range' in connection with an analysis of variance. Euphytica, 1, 112-122.

[8] Ling, R.F. (1973). A probability model of cluster analysis. J. Am. Statist. Ass., 68, 159-164.

[9] Newman, D. (1939). The distribution of the range in samples from a normal population, expressed in terms of an independent estimate of standard deviation. Biometrika, 31, 20-30.

[10] Pearson, E.S. and Hartley, H.O. (1943). Tables of the probability integral of the studentised range. Biometrika, 33, 89-99.

[11] Scheffé, H. (1953). A method for judging all constrasts in the analysis of variance. Biometrika, 40, 87-104.

[12] Sibson, R. (1972). Order invariant methods for data analysis. J. R. Statist. Soc. B, 34, 311-349.

[13] Tukey, J.W. (1949). Comparing individual means in the analysis of variance. Biometrics, 5, 99-114.

I. T. Jolliffe
Department of Mathematics
Dalhousie University
Halifax, Nova Scotia
Canada

Applied Statistics, R.P. Gupta, (editor)

SOME RESULTS FOR THE REDUCED FORM MODEL

by D.G. Kabe

Although the reduced form model or the classical normal multivariate analysis of variance (MANOVA) model is extensively explored in Econometric Theory, the generalized MANOVA (G MANOVA) model appears to have been neglected. The purpose of the present paper is to show an application of the G MANOVA model to the MANOVA model. This application is useful in optimal aggregation theory.

1. Introduction and summary

Let the reduced form or the MANOVA model be defined by

$$Y = \beta X + E, \tag{1}$$

where Y is $p\times N$, known X is $q\times N$ and of rank $q<N$, β is $p\times q$, and E, in the usual sense, is pN variate normal with mean zero and covariance $\Sigma\otimes I$. Further, let β be subjected to the linear restrictions

$$F\beta = 0 \quad , \quad \beta H = 0 \quad , \tag{2}$$

where F is a known $(p-q)\times p$ matrix of rank $(p-q)<p$, and H is a known $q\times(q-m)$ matrix of rank $(q-m)$. Then we may choose a $q\times p$ matrix B' orthogonal to F, and a $q\times m$ matrix D' orthogonal to H, and write β as

$$\beta = B\xi D \quad , \tag{3}$$

where ξ is a $q\times m$ matrix, and the elements of ξ are unrestricted, see Fisher [2]. It follows that (1) may now be written as

$$Y = B\xi DX + E = B\xi A + E \quad , \tag{4}$$

where $A = DX$ is a known $m\times N$ matrix of rank $m<N$.
The model $Y = B\xi A + E$ is known in Multivariate Statistical Analysis as the G MANOVA model of Potthoff and Roy [5]. Gleser and Olkin [3] have noted down a number of properties of the G MANOVA model which are similar to the properties of the MANOVA model. Note that B in model (4) is $p\times q$ and of rank $q<p$. If $p=q$, then the G MANOVA model reduces to the MANOVA model. Gleser and Olkin [3] derive all their results in terms of canonical transformation of the

original variates, and before using their results one must transform them back in terms of original variates, or else every time one uses their results the observation matrix Y must be transformed to its canonical form.

Since the estimation and testing of the linear parametric functions of β and the prediction of future independent observations on Y of model (1) are important factors in optimal aggregation theory, Fisher [2], we express and derive some results of Gleser and Olkin [3] in terms of the original variates and apply them to the MANOVA model (1) in our context.

In this section we list down the results we need and in section 3 we derive them. The (full information) maximum likelihood estimators of the parameters ξ and Σ are

$$\hat{\xi} = (B'S^{-1}B)^{-1}B'S^{-1}W, \; W = YA'(AA')^{-1} , \tag{5}$$

$$\hat{\Sigma} = S + (W-B\hat{\xi})AA'(W-B\hat{\xi}), \; S = YY' - WAA'W' , \tag{6}$$

and $\hat{\xi}$ and S are sufficient for ξ and Σ.

A likelihood ratio test (LRT) criterion for testing the composite linear hypothesis that $C\xi V = \Omega$, C is $c\times q$, V is $m\times v$, Ω known against general ξ is

$$\lambda_1 = |(C\hat{\xi}V-\Omega)' \; (C(B'S^{-1}B)^{-1}C')^{-1}(C\hat{\xi}V-\Omega)|$$
$$|\phi|^{-1}|I + \phi^{-1}(\Lambda - B\hat{\xi}V)'S^{-1} \; (\Lambda - B\hat{\xi}V)|^{-1} , \tag{7}$$

where $\Lambda = WV$, $\phi = V'(AA')^{-1}V$. It may be shown that

$$\lambda_1 = |Z| \; |I - Z|^{-1} , \tag{8}$$

where the $v\times v$ matrix Z has the density

$$g(Z) = K|I-Z|^{\frac{1}{2}(N-p-m+q-c-1)}$$
$$|Z|^{\frac{1}{2}(c-v-1)} , \tag{9}$$

where K is used in this paper, as a generic letter, to denote the normalizing constants of density functions. The results (7), (8), (9), are given by Gleser and Olkin ([3], p. 280, equation (4.8)) in terms of canonical variates. Some particular cases (7), (8), and (9) are as follows.

In case V is an $m\times m$ identity matrix, then (7) reduces to

$$\lambda_2 = |AA'|\ |(C\hat{\xi}-C\xi)'\ (C(B'S^{-1}B)^{-1}C')^{-1}(C\hat{\xi}-C\xi)|$$
$$|I + AA'(W-B\hat{\xi})'\ S^{-1}(W-B\hat{\xi})|^{-1}\ , \tag{10}$$

The LRT criterion λ_2 may be written as

$$\lambda_2 = |Z|\ |I-Z|^{-1}\ , \tag{11}$$

where now $m\times m$ Z has the density

$$g(Z) = K|I-Z|^{\frac{1}{2}(N-p-m+q-c-1)}|Z|^{\frac{1}{2}(c-m-1)}\ . \tag{12}$$

The criterion λ_2 is useful for testing that $C\xi$ has a specified value against a general ξ , Σ unknown.

Next by taking $c=q$, and C to be a $q\times q$ identity matrix, we find a LRT criterion for testing that ξV has a specified value, against general ξ , Σ unknown. This criterion is

$$\lambda_3 = |(\hat{\xi}V-\xi V)'B'S^{-1}B(\hat{\xi}V-\xi V)|\ |\phi|^{-1}$$
$$|I+\phi^{-1}(\Lambda-B\hat{\xi}V)'S^{-1}\ (\Lambda-B\hat{\xi}V)|^{-1}\ , \tag{13}$$

where λ_3 has the density of $U_{N-p-m+v,\ v,\ q,}$ see Anderson ([1], p. 192).

Again in (13), we take V to be an identity matrix, and we have that

$$\lambda_4 = |AA'|\ |(\hat{\xi}-\xi_0)'\ B'S^{-1}B(\hat{\xi}-\xi_0)|$$
$$|I + AA'(W-B\xi_0)'S^{-1}(W-B\xi_0)|^{-1}\ , \tag{14}$$

where λ_4 is useful for testing $\xi = \xi_0$, ξ_0 specified, and has the density of $U_{N-p,\ m,\ q}$.

Now let G be any $N\times N$ positive semidefinite symmetric matrix of rank $t<N$, then the unique Moore-Penrose inverse of G is $G^- = R'\Delta^{-1}R = R'(RGR')^{-1}R$, where R is the $t\times N$ matrix of the first t latent vectors of G corresponding to the first t nonzero roots, and Δ is the diagonal matrix of these nonzero roots. Now if X is a $q\times N$ matrix of arbitrary rank, then

$$X^- = (X'X)^- X', \; X^- \text{ is } N\times q, \tag{15}$$

is the unique Moore-Penrose inverse of X. It follows from (3) that

$$\xi = B^-\beta D^- , \tag{16}$$

is a particular solution of (3). Thus we observe from (3) that testing linear parametric functions of β, subject to (2) is the same as testing linear parametric functions of ξ, ξ unrestricted. Testing that $\beta = \beta_0$, β_0 specified, subject to (2), is the same as testing $\xi = \xi_0$, ξ_0 specified. Thus the criteria $\lambda_1,\lambda_2,\lambda_3,\lambda_4$ are useful for testing linear parametric functions of β.

Further, Gleser and Olkin ([3], p. 284, equation (5.6)) show that the density of $\hat{\xi}$ and an $m\times m$ arbitrary positive definite symmetric matrix L is

$$g(\hat{\xi},L) = K\phi(L)$$

$$\exp\{-\tfrac{1}{2}\mathrm{tr}(B'\Sigma^{-1}B)(\hat{\xi}-\xi)AA'(I-AA'L)(\hat{\xi}-\xi)' , \tag{17}$$

where

$$\phi(L) = |I-AA'L|^{\frac{1}{2}(N-p-m+2q-1)} \, |L|^{\frac{1}{2}(p-q-m-1)} . \tag{18}$$

Let now $y = B\xi A_0 + e$, given A_0 $m\times 1$, and e normal mean zero covariance Σ, be a future observation, then $B\hat{\xi}A_0$ is its predictor. The error of prediction $(y-B\hat{\xi}A_0)$, conditional on L, is obviously normal mean zero and covariance $(1+A_0'((AA')^{-1}-L)^{-1}A_0)\Sigma$. Thus given (1) and (2), a future observation y corresponding to given X_0, $q\times 1$, can be predicted by $B\hat{\xi}A_0$, where $A_0 = DX_0$.

Now we proceed with the derivation of the distribution of the criterion λ_1. The derivation given here is in terms of original variates, unlike Gleser and Olkin [3] who resort to canonical transformations of the original variates.

Some results found useful in the sequel are stated in the next section and the distribution theory is given in section 3. We assume that all integrals occurring in this paper are evaluated over the appropriate ranges of the variables of the integration. Sometimes the same symbol denotes different quantities, but its meaning is made explicit in the context.

2. Some useful results

Let Y be $p\times N$, $-\infty<Y<\infty$, D a given $q\times N$ matrix of rank $q<N$, Δ an $N\times N$ positive definite symmetric matrix, $N\geq p+q$. Then Kabe [4] proves that

$$\int_{Y\Delta Y' = G,\ DY' = V'} f(Y\Delta Y')dY = K|D\Delta^{-1}D'|^{-\frac{1}{2}p}|\Delta|^{-\frac{1}{2}p}f(G)\,|G-V(D\Delta^{-1}D')^{-1}V'|^{\frac{1}{2}(N-p-q-1)}, \quad (19)$$

where G is $p\times p$, V is $p\times q$. When f is a suitable density function, then the right hand side of (19) represents the joint density of G and V.

Let Λ be any $p\times v$ matrix, ϕ^{-1} a $v\times v$ positive definite symmetric matrix, S a $p\times p$ positive definite symmetric matrix, then we have the following system of matrix identities.

$$(S+\Lambda\phi^{-1}\Lambda')^{-1} = S^{-1} - S^{-1}\Lambda(\phi+\Lambda' S^{-1}\Lambda)^{-1}\Lambda' S^{-1} . \quad (20)$$

On setting $\rho = S + \Lambda\phi^{-1}\Lambda'$, we may prove that

$$\Lambda' S^{-1}\Lambda = H(I-\phi^{-1}H)^{-1}, \quad H = \Lambda'\rho^{-1}\Lambda, \quad (21)$$

and for any $p\times q$ matrix B we may observe that

$$B' S^{-1}\Lambda = B'\rho^{-1}(I-\phi^{-1}H)^{-1} . \quad (22)$$

Further, it is easily deduced that

$$B' S^{-1}B = B'\rho^{-1}B + B'\rho^{-1}\Lambda(\phi-H)^{-1}\Lambda'\rho^{-1}B , \quad (23)$$

for which it follows that

$$(B' S^{-1}B)^{-1} = (B'\rho^{-1}B)^{-1} - \theta(\phi-T)^{-1}\theta' , \quad (24)$$

where $\theta = (B'\rho^{-1}B)^{-1}B'\rho^{-1}\Lambda$, $M = \theta' B'\rho^{-1}B\theta$, $H - M = T$.

Again by using (22) and (23), we find that

$$(B' S^{-1}B)^{-1} B' S^{-1}\Lambda = \theta(I-\phi^{-1}T)^{-1} , \quad (25)$$

and that

$$\Lambda' S^{-1} B(B'S^{-1}B)^{-1} B'S^{-1}\Lambda = (I-\phi^{-1}T)^{-1} M(I-\phi^{-1}H) \quad . \tag{26}$$

Note that θ is $q\times v$, H. M. T. and $v\times v$. Although we have used only the identity (26) in the sequel, we hvae given the identities (20), ..., (25), because the derivation of (26) depends on them.

Now we proceed with the distribution theory.

3. Distribution Theory

On integrating the density of Y with respect to (w.r.t.) Y, by using (19) over the region

$$YY' = G, \ (AA')^{-1}AY' = W \ , \tag{27}$$

We find the density of G and W, and on setting $S = G - WAA'W'$, $\Lambda = WV$, we obtain the joint density of S and Λ to be

$$g(S,\Lambda) = K|S|^{\frac{1}{2}(N-p-m-1)}$$
$$\exp\{-\tfrac{1}{2}\mathrm{tr}\ \Sigma^{-1}[S+(\Lambda-B\xi V)\phi^{-1}(\Lambda-B\xi V)']\}. \tag{28}$$

The density of $\rho = S + (\Lambda-B\xi V)\ \phi^{-1}(\Lambda-B\xi V)'$, and Λ is

$$g(\Lambda,\xi) = K \exp\{-\tfrac{1}{2}\mathrm{tr}\ \Sigma^{-1}\rho\}|\rho|^{\frac{1}{2}(N-p-m-1)} \ .$$
$$|\phi - (\Lambda-B\xi V)'\rho^{-1}(\Lambda-B\xi V)|^{\frac{1}{2}(N-p-m-1)} \ . \tag{29}$$

We now integrate (29) w.r.t. Λ over the region

$$(\Lambda-B\xi V)'\rho^{-1}(\Lambda-B\xi V) = H \ , \ (B'\rho^{-1}B)^{-1}B'\rho^{-1}(\Lambda-B\xi V) = \theta \tag{30}$$

and then integrate the result w.r.t. θ over the region

$$\theta' B'\rho^{-1}B\theta = M \ , \ C\theta = \delta(I-\phi^{-1}T) \ , \tag{31}$$

and we find that

$$g(\rho,H,M,\delta) = K \exp\{-\tfrac{1}{2}\mathrm{tr}\ \Sigma^{-1}\rho\}|\rho|^{\frac{1}{2}(N-p-1)}$$
$$|\phi-H|^{\frac{1}{2}(N-p-m-1)}|H-M|^{\frac{1}{2}(p-v-q-1)}|C(B'\rho^{-1}C'|^{-\frac{1}{2}v}$$
$$|I-\phi^{-1}T|^{C}$$
$$|M-(I-\phi^{-1}T)'\delta'(C(B'\rho^{-1}B)^{-1}C')^{-1}\delta(I-\phi^{-1}T)|^{\frac{1}{2}(q-v-c-1)} . \tag{32}$$

Note that $\delta = (C\hat{\xi}V - C\xi V)$. Further, setting

$$U = \delta'(C(B'\rho^{-1}B)^{-1}C')^{-1}\delta \ , \tag{33}$$

and integrating (32) w.r.t. δ we obtain the joint density of ρ , H-M = T, M , and U. However, the density of ρ is independent of T , M. U whose density is

$$g(T,M,U) = K|\phi-M-T|^{\frac{1}{2}(N-p-m-1)}|T|^{\frac{1}{2}(p-v-q-1)}$$
$$|I-\phi^{-1}T|^{c}\ |M-(I-\phi^{-1}T)U(I-\phi^{-1}T)'|^{\frac{1}{2}(q-v-c-1)}$$
$$|U|^{\frac{1}{2}(c-v-1)} \ , \tag{34}$$

where we assume that $c\geq v$, $q\geq v+c$, $p\geq v+q$, $N\geq p+m$. The criterion λ_1 may now be written as

$$\lambda_1 = |I-\phi^{-1}T|\ |U|\ |\phi-(I-\phi^{-1}T)U|^{-1} = |Z|\ |I-Z|^{-1} \ , \tag{35}$$

where $Z = (\phi-T)^{\frac{1}{2}}\phi^{-1}U\phi^{-1}(\phi-T)^{\frac{1}{2}}$. Thus the density T , M , and Z can be found from (34) and on integrating this density w.r.t. T and M we find the density of Z as given by (9).

References

Anderson, T.W. (1958), An introduction to multivariate statistical analysis. John Wiley, New York.

Fisher, W.D. (1962), Optimal Aggregation in multiequation prediction models. Econometrica, 30, 744-768.

Gleser, L.J. and Olkin, I. (1970), Linear models in multivariate analysis. pp. 267-292, in Essays in Probability and Statistics. John Wiley, New York.

Kabe, D.G. (1965), Generalization of Sverdrup's Lemma and its applications to multivariate distribution theory. Ann. Math. Statist. 36 , 671-676.

Potthoff, R.F. and Roy, S.N. (1964), A G MANOVA model useful especially for growth curve problems. Biometrika 51, 315-328.

D. G. Kabe
Mathematics Department
St. Mary's University
Halifax, N. S.
Canada

Applied Statistics, R.P. Gupta, (editor)

TRIMMED MEANS AND THE METHOD OF MAXIMUM LIKELIHOOD WHEN SPURIOUS OBSERVATIONS ARE PRESENT

B. K. KALE[1]

Summary

Consider a situation in which a sample size n contains m spurious observations $(x_{i_1}, x_{i_2}, \ldots, x_{i_m})$ distributed as $G(x)$ and the remaining $(n-m)$ distributed as $F(x)$. Any of the $\binom{n}{m}$ possible subsets is equally likely to be the set of spurious observations. Assuming that $\frac{dG}{dF} = \psi(x)$ is strictly increasing in x, it is shown that $(x_{(n-m+1)}, \ldots, x_{(n)})$ has maximum probability of being the set of spurious observations. This result is then used to derive maximum likelihood estimators of the parameters when

$$\frac{dG}{dx} = \beta(\theta_1)e^{\theta_1 x_1} \quad \text{and} \quad \frac{dF}{dx} = \beta(\theta_2)e^{\theta_2 x_2} \qquad \theta_2 > \theta_1 .$$

It is shown that for normal and exponential situation these estimators are appropriately trimmed means, as opposed to Winsorized means recommended earlier on the basis of premium protection approach.

KEY WORDS: Spurious Observation; Stochastically Larger Populations; Order Statistics; Method of Maximum Likelihood; Trimmed Means.

1. Introduction

The problem of analyzing data involving outliers and/or spurious observations has received considerable attention in the past few years. In general, the inference procedures here are based on linear combination of the order statistics of the sample, in particular, either trimmed or winsorized means. Anscombe [1] suggested the premium protection approach and this has been followed by Guttman and Smith ([2], [3]), Kale and Sinha [6], and Veale and Kale [9], among others.

In general, we are given a set of n observations $(x_1, x_2, \ldots, x_n)$ which ideally comes from a population P with probability density function (p.d.f.) $f(x;\underset{\sim}{\theta})$ and the objective is to make inferences about $\psi(\underset{\sim}{\theta})$. The suspicion is that some of the sample observations $(x_{i_1}, x_{i_2}, \ldots, x_{i_m})$ are not from population P but come from

[1] Supported by the research grant from the National Research Council of Canada.

another population, Q , with p.d.f. given by $f(x;\xi)$, and have been mixed with the given sample either through an error of measurement or through the non-homogeneity of the experimental material which is as yet undetected. Kale and Sinha [6] and Veale and Kale [9] considered this problem for life testing experiments with $m = 1$ and $f(x;\underset{\sim}{\theta}) = \frac{1}{\sigma}\exp(-\frac{x}{\sigma})$ and $f(x;\underset{\sim}{\xi}) = \frac{k}{\sigma}\exp\{-\frac{xk}{\sigma}\}$, $x \geq 0$, $\sigma > 0$, $0 < k \leq 1$. Here they recommended procedures based on winsorized sum $S_{n-1} = \sum_{i=1}^{n-2} x_{(i)} + 2x_{(n-1)}$. This recommendation was based on the premium protection approach. In the present paper we will show that the method of maximum likelihood leads to estimators which are trimmed means rather than winsorized means. The result is proved for the general case when $f(x;\theta)$ belongs to single parameter exponential family.

We will make the following assumptions:

(A) Prior to the experiment there is no way to identify the set of m spurious observations $(m \geq 1)$ and that every subset $(x_{i_1}, x_{i_2}, \ldots, x_{i_m})$ of $(x_1, x_2, \ldots, x_n)$ has the equal prior probability $\frac{1}{\binom{n}{m}}$ of being the set of spurious observations

(B) If $F(x)$ and $G(x)$ are the d.f.'s corresponding to P and Q respectively, then $\frac{dG}{dF} = \psi(x)$ where ψ is strictly increasing in x.

The above assumption (B) implies that the population Q is stochastically larger than P and the large observations are suspected of being the spurious observations. If we assume that $\frac{dG}{dF} = \psi(x)$ is strictly decreasing in x , then the small observations are suspected of being the spurious observations. We will concentrate here on the case where $\psi(x)$ is strictly increasing. The results for the other situation can be worked out in a similar fashion.

Under these assumptions, we show that the $(x_{(n-m+1)}, \ldots, x_{(n)})$, where $(x_{(1)}, x_{(2)}, \ldots, x_{(n)})$ is the order statistic of the sample, has maximum probability of corresponding to the set of spurious observations. This result is a direct generalization of an earlier result of Mount and Kale [8] for $m = 1$. We utilize this result to obtain the Maximum Likelihood Estimators (MLE) of the parameters when populations are characterized by single parameter exponential family. The case of life testing experiments is considered in some detail.

2. Selecting the Set of Possible Spurious Observations

First we will consider the case of $m = 2$ under these assumptions. Let $\phi(r_1,r_2)$ denote the probability that $(x_{(r_1)},x_{(r_2)})$ corresponds to the set of spurious observations distributed as G. Then an elementary argument shows that:

$$\phi(r_1,r_2) = \frac{(n-2)!2!}{(r_1-1)!(r_2-r_1-1)!(n-r_2)!} \iint_{R_2} [F(x)]^{r_1-1} [F(y)-F(x)]^{r_2-r_1-1} \cdot [1 - F(y)]^{n-r_2} dG(x)dG(y) . \quad (1)$$

We now prove that $\phi(r_1,r_2)$ is strictly increasing in each of the variables r_1 , r_2 , $1 \leq r_1 < r_2 \leq n$, and $\max_{r_1<r_2}\phi(r_1,r_2) = \phi(n-1,n)$.

Proof: By making transformation $u = F(x)$, $v = F(y)$, we have

$$\phi(r_1,r_2) = \frac{(n-2)!2!}{(r_1-1)!(r_2-r_1-1)!(n-r_2)!} \int\int_{0\le u<v<1} u^{r_1-1} (v-u)^{r_2-r_1-1} (1-v)^{n-r_2} \cdot \psi[F^{-1}(u)]\psi[F^{-1}(v)]dudv. \quad (2)$$

Fix r_2 ; then show that $\phi(r_1,r_2)$ is strictly increasing in r_1 for $1 \leq r_1 < r_2$, i.e., $\phi(1,r_2) < \phi(2,r_2) <\ldots< \phi(r_2-1,r_2)$.

First consider v fixed in (2) and let

$$I(r_1,v) = \frac{\Gamma(r_2)}{\Gamma(r_1)\Gamma(r_2-r_1)} \frac{1}{u^{r_2-1}} \int_0^v u^{r_1-1} (v-u)^{r_2-r_1-1} \psi[F^{-1}(u)]du; \quad (3)$$

then

$$\phi(r_1,r_2) = \frac{2}{n(n-1)} \frac{\Gamma(n+1)}{\Gamma(r_2)\Gamma(n-r_2+1)} \int_0^1 I(r_1,v)v^{r_2-1} (1-v)^{n-r_2}\psi[F^{-1}(v)]dv. \quad (4)$$

$I(r_1,v) = E[\psi^{-1}(F(u))|v]$ where the conditional distribution of u given v is beta over the interval $(0,v)$ with parameters (r_1,r_2-r_1) with density denoted by $h_1(u|r_1, r_2-r_1, v)$. We observe that

$$\frac{h_1(u|r_1+1 , r_2-r_1 , v)}{h_1(u|r_1 , r_2-r_1 , v)} \propto \frac{u}{v-u}$$

which is strictly increasing in u . Further, $\psi[F^{-1}(u)]$ is assumed strictly increasing. We now apply the results of Lehmann ([7], Lemma 2 , p. 74) for conditional distributions of u , given v , and conclude that $I(r_1,v)$ is strictly increasing in r_1 for $1 \leq r_1 < r_2$ or

$$I(1,v) < I(2,v) < \ldots < I(r_2-1\ ,\ v)\ . \tag{5}$$

Next, consider the expression (4). Noting that v is marginally beta over the interval $(0,1)$, $\psi[F^{-1}(v)] > 0$ a.e. (Lebesgue) and from (5), it immediately follows that for fixed r_2, we have

$$\phi(1,r_2) < \phi(2,r_2) < \ldots < \phi(r_2-1\ ,\ r_2)\ .$$

Similarly, we can show that $\phi(r_1,r_2)$ for fixed r_1 is strictly increasing in r_2. Thus, consider

$$J(r_2,u) = \frac{\Gamma(n-r_1+1)}{(1-u)^{n-r_1}\Gamma(r_2-r_1)\Gamma(n-r_2+1)} \int_u^1 (v-u)^{r_2-r_1-1}(1-v)^{n-r_2} \cdot \psi[F^{-1}(v)]dv\ ; \tag{6}$$

then

$$\phi(r_1,r_2) = \frac{2}{n(n-1)} \frac{\Gamma(n+1)}{\Gamma(r_1)\Gamma(n-r_1+1)} \int_0^1 J(r_2,u)u^{r_1-1}(1-u)^{n-r_1} \cdot \psi[F^{-1}(u)]du\ . \tag{7}$$

Again we have $J(r_2,u) = E[\psi[F^{-1}(v)]|u]$ where the conditional distribution of v, given u, is beta over the interval $(u,1)$ with parameters $(r_2-r_1\ ,\ n-r_2+1)$, with p.d.f. given by $h_2(v|r_2-r_1\ ,\ n-r_2+1\ ,\ u)$. We observe that

$$\frac{h_2(v|r_2+1-r_1\ ,\ n-(r_2+1)+1\ ,\ u)}{h_2(v|r_2-r_1\ ,\ n-r_2+1\ ,\ u)} \propto \frac{v-u}{1-v} \qquad u < v < 1$$

which is monotone increasing in v. Further, $\psi[F^{-1}(v)]$ being strictly increasing again applying Lehmann's result quoted earlier to conditional distributions of v, given u, we conclude that

$$J(r_2+1\ ,\ u) < J(r_2+2\ ,\ u) < \ldots < J(n,u)\ . \tag{8}$$

Next consider the expression (7). Noting that u is marginally beta over $(0,1)$ with parameters $(r_1,\ n-r_1+1)$, and $\psi[F^{-1}(u)] > 0$ a.e. (Lebesgue) and from (8), it follows that

$$\phi(r_1\ ,\ r_1+1) < \phi(r_1\ ,\ r_1+2) < \ldots < \phi(r_1\ ,\ n)\ . \tag{9}$$

Using (5) and (9), it now immediately follows that

$$\max_{1\le r_1<r_2\le n} \phi(r_1,r_2) = \phi(n-1\ ,\ n)\ .$$

The proof for the general m is carried out in a similar fashion. Let $\phi(r_1,r_2,\ldots,r_m)$ denote the probability that $(x_{(r_1)},\ldots,x_{(r_m)})$ correspond to the set of spurious observations. Then we have the following expression for $\phi(r_1,r_2,\ldots,r_m)$:

$$\phi(r_1,r_2,\ldots,r_m) = \frac{(n-m)!m!}{r_1!(r_2-r_1-1)!(r_3-r_2-1)!\ldots(n-r_m)!} \int_{0<w_1<w_2<\ldots<w_m<1}$$

$$\cdot\, w_1^{r_1-1}(w_2-w_1)^{r_2-r_1-1}(w_3-w_2)^{r_3-r_2-1}\ldots.(1-w_m)^{n-r_m}$$

$$\cdot\, \psi[F^{-1}(w_1)]\psi[F^{-1}(w_2)]\ldots.\psi[F^{-1}(w_m)]dw_1dw_2\ldots.dw_m\,. \quad (10)$$

Now fixing $(r_2,r_3,\ldots,r_m)$, we can show that $\phi(r_1,r_2,\ldots,r_m)$ is monotone increasing function of r_1 for $1 \le r_1 < r_2$. The basic tool is the Lehmann's result quoted earlier. Fixing $(r_1,r_3,\ldots,r_m)$, we can show that $\phi(r_1,r_2,\ldots,r_m)$ is monotone increasing in r_2 for $r_1 < r_2 < r_3$. In general, fixing $(r_1,r_2,\ldots,r_{i-1},r_{i+1},\ldots,r_m)$, we can show that $\phi(r_1,r_2,\ldots,r_m)$ is strictly increasing in r_i for $r_{i-1} < r_i < r_{i+1}$. Thus, Max $\phi(r_1,r_2,\ldots,r_m) = \phi(n-m+1,n-m+2,\ldots,n)$ and $(x_{(n-m+1)}, x_{(n-m+2)},\ldots,x_{(n)})$ has maximum probability of corresponding to the set of spurious observations.

3. Maximum Likelihood Estimators of the Parameters

Now consider the case where $(n-m)$ observations are drawn from a population P which admits a p.d.f. $f(x;\theta_1)$ and the remaining m observations are drawn from a population Q with p.d.f. $f(x;\theta_2)$, where both densities belong to exponential class of densities. Without loss of generality we can assume that

$$f(x\ \theta_1) = \beta(\theta_1)\exp\{\theta_1 x\} \quad\text{and}\quad f(x;\theta_2) = \beta(\theta_2)\exp\{\theta_2 x\}$$

with $\theta_2 > \theta_1$. The likelihood of the sample now can be written down as:

$$L(\underset{\sim}{x}|\theta_1,\theta_2,I) = \frac{1}{\binom{n}{m}}[\beta(\theta_2)]^m\exp\{\theta_2 S_I\}[\beta(\theta_1)]^{n-m}\exp\{\theta_1(T-S_I)\}$$

where

$$T = \sum_{i=1}^{n} x_i \quad , \quad I = (x_{i_1}, x_{i_2},\ldots, x_{i_m}) \quad , \quad S_I = \sum_{x_i \varepsilon I} x_i$$

the set of m spurious observations. The parameter space is $\theta_1 < \theta_2$ and $I \varepsilon \mathcal{T}$ where $\mathcal{T}$ = the collection of all combinations of m observations out of n .

$\underset{\substack{\theta_1<\theta_2 \\ I\epsilon T}}{\text{Max}}\ L(\underset{\sim}{x}|\theta_1,\theta_2,I)$ can be obtained by stepwise procedure, first fixing θ_1,θ_2 and maximizing for $I \epsilon T$. When $\theta_1 < \theta_2$, the conditions of the previous section are satisfied and $\underset{I\epsilon T}{\text{Max}}\ L(\underset{\sim}{x}|\theta_1,\theta_2,I)$ with $\theta_1 < \theta_2$ fixed, occurs at $I = \hat{I}$ where $\hat{I}$ corresponds to the set of observations given by $(x_{(n-m+1)},x_{(n-m+2)},\ldots,x_{(n)})$. Since this maximum does not depend upon the fixed values of θ_1, θ_2,

$$\underset{\substack{\theta_1<\theta_2 \\ I\epsilon T}}{\text{Max}}\ L(\underset{\sim}{x}|\theta_1,\theta_2,I) = \underset{\theta_1<\theta_2}{\text{Max}}\ L(\underset{\sim}{x}|\theta_1,\theta_2,\hat{I}) .$$

Let

$$T_2 = S_{\hat{I}} = \sum_{i=n-m+1}^{n} x_{(i)} \quad \text{and} \quad T_1 = T - S_{\hat{I}} = \sum_{i=1}^{n-m} x_{(i)} .$$

Then we have

$$L(\underset{\sim}{x}|\theta_1,\theta_2,\hat{I}) \propto [\beta(\theta_1)]^{n-m} \exp\{\theta_1 T_1\}[\beta(\theta_2)]^m \exp\{\theta_2 T_2\} .$$

The maximization of $L(\underset{\sim}{x}|\theta_1,\theta_2,\hat{I})$ is simple and leads to

$$\hat{\theta}_1 = \psi^{-1}(T_1/n-m) \quad \text{and} \quad \hat{\theta}_2 = \psi^{-1}(T_2/m)$$

where

$$\psi(\theta) = \frac{-d \log \beta(\theta)}{d\theta}$$

and is known to be strictly monotone and has a unique inverse ψ^{-1}. We also note that $\frac{T_1}{n-m} < \frac{T_2}{m}$ with probability one and no complications arise out of the constraint $\theta_1 < \theta_2$. To prove that $\frac{T_1}{n-m} < \frac{T_2}{m}$, let

$$C = (n-m) \sum_{i=n-m+1}^{n} x_{(i)} - m \sum_{i=1}^{n-m} x_{(i)} ;$$

then

$$C \geq (n-m)m\, x_{(n-m+1)} - m \sum_{i=1}^{n-m} x_{(i)}$$

or

$$\frac{C}{m} \geq \sum_{i=m}^{n-m} \{x_{(n-m+1)} - x_{(i)}\} \geq 0 .$$

In particular, if we are dealing with the case of life testing experiments with $f(x;\sigma_1) = \frac{1}{\sigma_1} e^{-x/\sigma_1}$, $x \geq 0$, $\sigma_1 > 0$, and $f(x;\sigma_2) = \frac{1}{\sigma_2} \exp\{-\frac{x}{\sigma_2}\}$, $x \geq 0$, $\sigma_2 > \sigma_1$, then $(x_{(1)},x_{(2)},\ldots,x_{(n-m)}$

and $(x_{(n-m+1)},\ldots,x_{(n)})$ has maximum probability of being sets of observations from populations indexed by σ_1 and σ_2 respectively and

$$\hat{\sigma}_1 = \sum_{i=1}^{n-m} x_{(i)}/(n-m) \quad \text{and} \quad \hat{\sigma}_2 = \sum_{i=n-m+1}^{n} x_{(i)}/m$$

are the MLEs of σ_1 and σ_2 respectively.

Similarly, if we are dealing with normal populations indexed by μ_1 and μ_2 with $\mu_2 > \mu_1$, then $\hat{\mu}_1 = \sum_{i=1}^{n-m+1} x_{(i)}/(n-m)$ $\hat{\mu}_2 = \sum_{i=n-m+1}^{n} x_{(i)}/m$. Here, of course, it is assumed that σ^2 is known and common to both populations.

We note that the MLEs here correspond to the trimmed means rather than winsorized ones suggested by Kale and Sinha [6] based on the premium protection approach. It is quite natural to compare $\hat{\sigma}_1 = \frac{1}{n-m}\sum_{i=1}^{n-m+1} x_{(i)}$ obtained here with $T_{n-m} = \frac{1}{n-m+1}\{\sum_{i=1}^{n-m} x_{(i)} + m/x_{(n-m)}\}$ suggested earlier by Kale and Sinha [6]. This will be considered in the next section for $m = 1$, for which the distribution theory is tractable and known.

4. Application to Life Testing Experiments

Let $\frac{\sigma_2}{\sigma_1} = \frac{1}{k}$, where $0 < k \leq 1$. Since we can compare the behaviour of $\frac{\hat{\sigma}_1}{\sigma_1}$ with that of T_{n-1}/σ_1 without loss of generality, we can assume $\sigma_1 = 1$. Using the transformation

$$Z_i = (n-i+1)(x_{(i)} - x_{(i-1)}), \quad i = 1,2,\ldots,n \quad \text{and} \quad x_{(0)} = 0$$

we have

$$T_{n-1} = \frac{1}{n}\sum_{i=1}^{n-1} Z_i \quad \text{and} \quad \hat{\sigma}_1 = \frac{1}{n-1}\sum_{i=1}^{n-1} \frac{n-i}{n-i+1} Z_i .$$

It is already known that (Joshi [5])

$$\text{MSE}(T_{n-1}|k) = \frac{1}{n} + \frac{2\theta^2}{n}[1 - u_{n-1}\sum_{i=1}^{n} \frac{1}{n-i+k}\}$$

where

$$\theta = \frac{1-k}{k} \quad \text{and} \quad u_{n-1} = \prod_{r=1}^{n} \frac{n-r+k}{n-r+1} .$$

We note that

$$\text{MSE}(T_{n-1}|k=1) = \frac{1}{n} \quad \text{and} \quad \text{MSE}(T_{n-1}|k=0) = \frac{1}{n^2}\sum_{i=1}^{n-1}\left(\frac{n-i+1}{n-i}\right)^2 .$$

If we consider

$$\hat{\sigma}_1 = \frac{1}{n-1} \sum_{i=1}^{n-1} \frac{n-i}{n-i+1} Z_i ,$$

then $MSE(\hat{\sigma}_1 | k=1)$ would certainly be larger than $MSE(T_{n-1} | k=1)$. But $MSE(\hat{\sigma}_1 | k=0)$ can be shown to be equal to $\frac{1}{n-1}$ which is smaller than $MSE(T_{n-1} | k=0)$. Thus, in terms of premium protection terminology, we buy more protection (reduction in MSE near $k = 0$) by paying more premium (increase in MSE near $k = 1$). These results follow immediately from the fact that under $k = 0$, $\{Z_i\}_1^{n-1}$ are independent exponential r.v. with the parameter $E(Z_i) = \frac{n-i+1}{n-i}$, $i=1,2,\ldots,n-1$. For this result we refer to Veale and Kale [9]. The exact expression for $MSE(\hat{\sigma}_1 | k)$ can be obtained by using Joshi's [5] expression for $E(Z_i | k)$ and $E(Z_i^2 | k)$.

Concluding Remarks and Comments

In most of the applications for samples of moderate size n , the value of m would be small, say between 1 to 3 . Reference may be made here to the discussion on this point in a paper by Guttman [4]. For large m and n , the problem could be viewed from asymptotic (large sample theory) with the further condition that $\lim \frac{m}{n} = \delta < 1$. However, in our view, such problems are more in the domain of classification/discrimination analysis rather than inference problems involving outlier/spurious observations. One of the most important and possibly the more difficult problem is the determination of m , the number of spurious observations, which we have assumed to be known. Techniques for determination of m can be developed on the basis of "residual" analysis for specific distributions like exponential (scale parameter) and normal distribution (location parameter).

One may also consider the problem of <u>at most</u> m outliers by allowing different distributions $G_1, G_2, \ldots, G_m$ and these being parameterized in such a way that some of the G_i's could be identical to F for some set of parameter values. This is probably a more realistic situation than the one considered in this paper. These results are being worked out and hopefully would be completed in the near future.

References

[1] Anscombe, J. F. (1960): Rejection of Outliers, Technometrics, 2, 123-147.

[2] Guttman, I. and Smith, D. E. (1969): Invistigation of rules for dealing with outliers in small samples from normal distribution: I . Estimation of the mean, Technometrics, 11, 527-550.

[3] Guttman and Smith, D. E. (1971): Investigation of rules for dealing with outliers in small samples from normal distribution: II. Estimation of the variance, Technometrics, 13, 101-111.

[4] Guttman, I. (1973): Care and handling of univariate and multivariate outliers in detecting spuriousity - a Bayesian approach, Technometrics, 15, 723-738.

[5] Joshi, P. C. (1972): Efficient estimation of the mean of an exponential distribution when an outlier is present, Technometrics, 14, 137-143.

[6] Kale, B. K. and Sinha, S. K. (1971): Estimation of expected life in the presence of an outlier observation, Technometrics, 13, 755-759.

[7] Lehmann, E. L. (1959): Testing of Statistical Hypotheses, John Wiley and Sons, New York.

[8] Mount, K. S. and Kale, B. K. (1972): On selecting a spurious observation, Canad. Math. Bulletin, 16, 75-78.

[9] Veale, J. R. and Kale, B. K. (1972): Tests of hypotheses for expected life in the presence of a spurious observation, Utilitas Math., 2, 9-23.

B. K. Kale
Department of Statistics
The University of Manitoba
Winnipeg, Manitoba
Canada R3T 2N2

Applied Statistics, R.P. Gupta, (editor)

ON THE LIKELIHOOD RATIO TEST FOR COVARIANCE MATRIX IN GROWTH CURVE MODEL[1]

by C. G. Khatri & M. S. Srivastava

Summary

Let the columns of X be independently normally distributed with covariance matrix Σ and, let $E(X) = B\xi A$, the growth curve model. This paper considers the problem of testing the hypothesis that the covariance matrix $\Sigma = \Sigma_0$, Σ_0 specified against the general alternatives. The exact hth moment of the modified likelihood ratio criteria (LRC) is derived under the alternative hypothesis. Using this result, an asymptotic expansion up to $0(n^{-3})$ is derived for the distribution of the modified LRC under the alternatives close to the hypothesis.

1. Introduction

In the linear growth curve model of Potthoff and Roy (1964), the columns of $p \times N$ random matrix X are independently and normally distributed with nonsingular covariance matrix Σ and $E(X) = B\xi A$ where B and A are known matrices of order $p \times q$ and $m \times N$ respectively. Symbolically we will write $X \sim N_{p,N}(B\xi A, \Sigma, I_N)$. Without loss of generality, we may assume that B and A are of full rank, $q(\leq p)$ and $n = N-m(\geq p)$ respectively; the reader is referred to Khatri (1966a and 1973) for the case when B and A are not of full rank. The problem with which we are concerned in this paper is that of testing the hypothesis $H_0 : \Sigma = \Sigma_0$, where Σ_0 is specified against the alternative $A \neq H_0$. We derive in Section 2 the likelihood ratio test for this problem. The exact hth moment of the modified test statistic is derived in Section 3 under the alternative A. Using this result, an asymptotic expansion up to $0(n^{-3})$ is derived in Section 4 for the statistic under the alternatives close to the hypothesis.

---

[1]Research supported by the National Research Council of Canada, No. A-4060.

Key Words and Phrases: Growth curve model, Covariance matrix, Modified likelihood ratio test, hth moment, Asymptotic distribution.

2. Likelihood Ratio Test

The likelihood function denoted by L is given by

$$L = (2\pi)^{-\frac{1}{2}pN} |\Sigma|^{-\frac{1}{2}N} \operatorname{etr}\{-\tfrac{1}{2}\Sigma^{-1}(X-B\xi A)(X-B\xi A)'\} \tag{2.1}$$

where etr stands for the exponential of the trace of a matrix. The maximum likelihood estimates of $\hat{\Sigma}$ and $\hat{\xi}$ are respectively given by

$$N\hat{\Sigma} = (X - B\hat{\xi}A)(X - B\hat{\xi}A)' \tag{2.2}$$

and

$$B'\hat{\Sigma}^{-1}XA' = (B'\hat{\Sigma}^{-1}B)\hat{\xi}(AA') . \tag{2.3}$$

Let

$$S = X[I_N - A'(AA')^{-1}A]X' \tag{2.4}$$

and

$$S_1 = XX' - S \tag{2.5}$$

where I_n denotes the $n\times n$ identity matrix. Then from (2.2) and (2.3), we get

$$NB'S^{-1} = B'\hat{\Sigma}^{-1} \quad \text{and} \quad B'S^{-1}XA' = (B'S^{-1}B)\hat{\xi}AA' . \tag{2.6}$$

Let

$$T = B(B'S^{-1}B)^{-1}B' \quad \text{and} \quad T_0 = B(B'\Sigma_0^{-1}B)^{-1}B' . \tag{2.7}$$

It is easy to see from (2.6) that

$$N\hat{\Sigma} = S + (I - TS^{-1})S_1(I - S^{-1}T) . \tag{2.8}$$

Hence the maximum of the likelihood is given by

$$(2\pi)^{-\frac{1}{2}pN} |\hat{\Sigma}|^{-\frac{1}{2}N} \exp(-\tfrac{1}{2}Np) . \tag{2.9}$$

Also, the maximum value of the likelihood under H_0 is given by

$$(2\pi)^{-\frac{1}{2}pN} |\Sigma_0|^{-\frac{1}{2}N} \operatorname{etr}\{-\tfrac{1}{2}\Sigma_0^{-1}[S + (I - T_0\Sigma_0^{-1})S_1]\} . \tag{2.10}$$

This gives the likelihood ratio statistic as

$$(2.11) \qquad \lambda^* = |\Sigma_0^{-1}\hat{\Sigma}|^{\frac{1}{2}N} \operatorname{etr}\{\tfrac{1}{2}[NI_p - \Sigma_0^{-1}(S + (I - T_0\Sigma_0^{-1})S_1)]\}$$

and the likelihood ratio test (LRT) procedure is given by

$$(2.12) \qquad \text{reject } H_0 \text{ if } \lambda^* \leq c_\alpha$$

where c_α is to be determined from $P(\lambda^* \leq c_\alpha | H_0) = \alpha$. α is the specified level of significance. However, the test based on λ^* will not be unbiased as shown by Sugiura nd Nagao (1968) in the special case of $q = p$ and $m = 1$. Thus, we will modify the LRT procedure λ^* to λ by replacing N with n , the degrees of freedom. Tnus the modified likelihood ratio test statistic is given by

$$(2.13) \qquad \lambda = |(N/n)\Sigma_0^{-1}\hat{\Sigma}|^{\frac{1}{2}n} \operatorname{etr} \tfrac{1}{2}[nI_p - \Sigma_0^{-1}(S + (I - T_0\Sigma_0^{-1})S_1)]; \qquad n = N-m .$$

3. h-th Moment of λ

If P is a positive definite matrix, then there exists a unique positive definite matrix Q such that $P = Q^2$ and we shall denote $Q = P^{\frac{1}{2}}$, a square root of P .

Let us write

$$(3.1) \quad \begin{cases} B_0 \equiv \Sigma_0^{-\frac{1}{2}} B(B'\Sigma_0^{-1}B)^{-\frac{1}{2}} , \quad B_1' \equiv (B_0, B_2) \text{ an orthogonal matrix,} \\[2ex] \Gamma \equiv (A'(AA')^{-\frac{1}{2}}, A_2) \text{ an orthogonal matrix, } \Sigma_1 \equiv \Sigma_0^{-\frac{1}{2}}\Sigma\Sigma_0^{-\frac{1}{2}} , \\[2ex] \Lambda \equiv \begin{matrix} q \\ p-q \end{matrix}\begin{pmatrix} \Lambda_{11} & \Lambda_{12} \\ \Lambda_{12}' & \Lambda_{22} \end{pmatrix} = \begin{pmatrix} B_0' \\ B_2' \end{pmatrix} \Sigma_1(B_0 B_2) , \quad \Lambda_{1\cdot 2} = \Lambda_{11} - \Lambda_{12}\Lambda_{22}^{-1}\Lambda_{12}' , \\ \qquad\qquad q \quad p-q \\[2ex] \begin{matrix} q \\ p-q \end{matrix}\left(\begin{matrix} Z_1 \\ Z_2 \end{matrix} \,\vdots\, Z\right) = B_1\Sigma_0^{-\frac{1}{2}}X\Gamma , \quad W \equiv \begin{matrix} q \\ p-q \end{matrix}\begin{pmatrix} W_{11} & W_{12} \\ W_{12}' & W_{22} \end{pmatrix} = ZZ' \text{ and } n \geq p , \end{cases}$$

where $n = N - m$. Then, it is easy to verify that

(3.2) $$\begin{pmatrix} Z_1 \\ Z_2 \end{pmatrix} \text{ and } Z \text{ are independently distributed}$$

and

(3.3) $$Z_2 \sim N_{p-q,m}(0,\ \Lambda_{22},\ I_m) \quad \text{and} \quad Z \sim N_{p,n}(0,\ \Lambda,\ I_n)\ .$$

Note that

$$W = ZZ' = B_1\Sigma_0^{-\frac{1}{2}}S\Sigma_0^{-\frac{1}{2}}B_1' \quad , \quad Z_2Z_2' = B_2'\Sigma_0^{-\frac{1}{2}}S_1\Sigma_0^{-\frac{1}{2}}B_2$$

and by Lemma 1 of Khatri (1966a)

$$S^{-1} - S^{-1}B(B'S^{-1}B)^{-1}B'S^{-1} = \Sigma_0^{-\frac{1}{2}}B_2W_{22}^{-1}B_2'\Sigma_0^{-\frac{1}{2}}$$

and

$$\Sigma_0^{-1} - \Sigma_0^{-1}B(B'\Sigma_0^{-1}B)^{-1}B'\Sigma_0^{-1} = \Sigma_0^{-\frac{1}{2}}B_2B_2'\Sigma_0^{-\frac{1}{2}}\ .$$

Hence, if $W_{1\cdot 2} = W_{11} - W_{12}W_{22}^{-1}W_{12}'$, we have

(3.4) $$|N\Sigma_0^{-1}\hat{\Sigma}| = |\Sigma_0^{-1}S| \cdot |I + (S^{-1} - S^{-1}B(B'S^{-1}B)^{-1}B'S^{-1})S_1|$$

$$= |W|\ |I + W_{22}^{-1}Z_2Z_2'| = |W_{1\cdot 2}|\ |W_{22} + Z_2Z_2'|$$

and

(3.5) $$\operatorname{tr} \Sigma_0^{-1}(S + (I - T_0\Sigma_0^{-1})S_1) = \operatorname{tr} W + \operatorname{tr} Z_2Z_2'\ .$$

From (3.1) and (3.3), we notice that W is distributed as Wishart with n degrees freedom and $E(W) = n\Lambda$, and this will be denoted by $W \sim W_p(n,\Lambda)$. From this we have

(3.6) $$\begin{cases} W_{1\cdot 2} \text{ and } (W_{12}',W_{22}) \text{ are independently distributed,} \\ W_{1\cdot 2} \sim W_q(n-p+q,\Lambda_{1\cdot 2}),\ W_{22} \sim W_{p-q}(n,\Lambda_{22}) \quad \text{and} \\ W_{12} \text{ given } W_{22} \sim N_{q,p-q}(\beta W_{22},\Lambda_{1\cdot 2},W_{22})\ ,\ \beta = \Lambda_{12}\Lambda_{22}^{-1} \end{cases}$$

Now, using (2.13), (3.4), (3.5) and (3.6), we get

$$(3.7)\quad E\lambda^h = (e/n)^{\frac{1}{2}nph}\{E|W_{1\cdot 2}|^{\frac{1}{2}nh}\ \mathrm{etr}(-\tfrac{1}{2}hW_{1\cdot 2})\}$$

$$\{E|W_{22}+Z_2Z_2'|^{\frac{1}{2}nh}\ \mathrm{etr}(-\tfrac{h}{2}(W_{22}+Z_2Z_2'))\mathrm{etr}(-\tfrac{h}{2}\ W_{12}W_{22}^{-1}W_{12}')\}$$

$$= (e/n)^{\frac{1}{2}nph}\ g_2(h)g_1(h)\ ,\ (\text{say}),$$

where the expressions inside the curly brackets { } on the right side of (3.7) are respectively denoted by $g_2(h)$ and $g_1(h)$.

It is easy to see that

$$(3.8)\qquad g_2(h) = \frac{2^{nhq/2}\ \Gamma_q(\frac{n+nh-p+q}{2})\ |\Lambda_{1\cdot 2}|^{nh/2}}{\Gamma_q(\frac{n-p+q}{2})\ |I_q + h\Lambda_{1\cdot 2}|^{\frac{1}{2}(n+nh-p+q)}}$$

Further, define,

$$(3.9)\qquad f(W_{22}) \equiv E\{\mathrm{etr}(-\tfrac{h}{2}\ W_{12}W_{22}^{-1}W_{12}')|W_{22}\}$$

$$= [(2\pi)^{-q(p-q)/2}|\Lambda_{1\cdot 2}|^{-(p-q)/2}||W_{22}|^{-q/2}]\ \cdot$$

$$\cdot \int (\mathrm{etr} - \tfrac{1}{2}T)dW_{12}$$

where

$$\mathrm{tr}\ T = \mathrm{tr}(\Lambda_{1\cdot 2}^{-1}+hI_q)W_{12}W_{22}^{-1}W_{12}' - 2\mathrm{tr}\ \Lambda_{1\cdot 2}^{-1}\beta W_{12}' + \mathrm{tr}\ \Lambda_{1\cdot 2}^{-1}\beta\ W_{22}\beta'$$

$$= \mathrm{tr}(\Lambda_{1\cdot 2}^{-1}+hI_q)(W_{12}-\nu)W_{22}^{-1}(W_{12}-\nu)' + \mathrm{tr}\ \Lambda_{1\cdot 2}^{-1}\ \beta W_{22}\beta'$$

$$- \mathrm{tr}(\Lambda_{1\cdot 2}^{-1}+hI_q)\nu W_{22}^{-1}\nu'\quad ,$$

and

$$\nu = (\Lambda_{1\cdot 2}^{-1} + hI_q)^{-1}\ \Lambda_{1\cdot 2}^{-1}\ \beta W_{22}\ .$$

From this, we get

$$(3.10)\quad f(W_{22}) = |I_q+h\Lambda_{1\cdot 2}|^{-(p-q)/2}(\mathrm{etr}-\tfrac{1}{2}h\beta'(I_q+h\Lambda_{1\cdot 2})^{-1}\beta W_{22})\ .$$

Hence, $g_1(h)$ defined in (3.6) is

(3.11) $$g_1(h) \equiv E|W_{22}+Z_2Z_2'|^{nh/2} f(W_{22}) \mathrm{etr}(-\tfrac{1}{2}hW_{22} - \tfrac{h}{2} Z_2Z_2') .$$

Then

(3.12) $$g_1(h) = (E|V+YY'|^{nh/2}) \Big/ \{|\Sigma_4|^{m/2}|\Sigma_3|^{n/2}|\Lambda_{22}|^{N/2} \cdot |I_q+h\Lambda_{1\cdot 2}|^{(p-q)/2}\}$$

where V and Y are independently distributed,

(3.13) $$V \sim W_{p-q}(n, \Sigma_3^{-1}) \quad \text{and} \quad Y \sim N_{p-q,m}(0, \Sigma_4^{-1}, I_m) ,$$

with

(3.14) $$\Sigma_3 = \Sigma_4 + h\Lambda_{22}^{-1}\Lambda_{12}'(I + h\Lambda_{1\cdot 2})^{-1}\Lambda_{12}\Lambda_{22}^{-1} \quad \text{and}$$

$$\Sigma_4 = \Lambda_{22}^{-1} + hI_{p-q} .$$

Making the transformation $V_1 = V + YY'$ and $Y_1 = V_1^{-\frac{1}{2}} Y$ in

(3.15) $$a(h) \equiv E|V + YY'|^{nh/2}$$

and breaking the density function of V_1 and Y_1 appropriately we can write

(3.16) $$a(h) = |\Sigma_4\Sigma_3^{-1}|^{m/2} E\{|V_1|^{nh/2} \mathrm{etr}(\tfrac{1}{2} V_1^{\frac{1}{2}}(\Sigma_3-\Sigma_4)V_1^{\frac{1}{2}}Y_1Y_1')\}$$

where V_1 and Y_1 are independently distributed,

(3.17) $$\begin{cases} Y_1 \sim \Gamma_{p-q}(\tfrac{1}{2} N)\{\pi^{m(p-q)/2}\Gamma_{p-q}(n/2)\}^{-1}|I_{p-q}-Y_1Y_1'|^{\frac{1}{2}(n-p+q-1)} \\ \text{and} \\ V_1 \sim W_{p-q}(N, \Sigma_3^{-1}) . \end{cases}$$

To evaluate (3.16), we integrate first with respect to Y_1. That is, we need to find

(3.18) $$b(B) \equiv E\, \mathrm{etr}(BY_1Y_1') .$$

If $m \leq p - q$, then

$$\Gamma_{p-q}(N/2) \Big/ \Gamma_{p-q}(n/2) = \Gamma_m(N/2) \Big/ \Gamma_m(\tfrac{N-p+q}{2})$$

and using Lemma 8 of Khatri (1966b), we get

$$b(B) = \{B_m(\tfrac{N-p+q}{2}, \tfrac{p-q}{2})\}^{-1} \sum_k \sum_\kappa \int_{W>0} |W|^{\frac{1}{2}(p-q-m-1)} |I_m - W|^{\frac{1}{2}(N-p+q-m-1)} \cdot$$

$$\cdot [C_\kappa(W) C_\kappa(B) \Big/ k!\, C_\kappa(I_{p-q})] dW$$

$$= \sum_{k=0}^{\infty} \sum_\kappa (\tfrac{1}{2}(p-q))_\kappa \{(\tfrac{1}{2} N)_\kappa\}^{-1} C_\kappa(I_m) C_\kappa(B) \Big/ k!\, C_\kappa(I_{p-q})$$

using the Beta integral of Constantine (1963, Theorem 3, page 1275), $(a)_\kappa = \Gamma_m(a,\kappa) \Big/ \Gamma_m(a)$, $B_m(a,b) = \Gamma_m(a)\, \Gamma_m(b) \Big/ \Gamma_m(a+b)$, and $\Gamma_m(a,\kappa) = \pi^{m(m-1)/4} \sum_{i=1}^{m} \Gamma(a + k_i - \frac{i-1}{2})$. Noting a result of James (1964), namely

$$C_\kappa(I_m) \Big/ C_\kappa(I_{p-q}) = (\tfrac{1}{2} m)_\kappa \Big/ (\tfrac{1}{2}(p-q))_\kappa ,$$

we get

$$b(B) = \sum_{k=0}^{\infty} \sum_\kappa (\tfrac{1}{2} m)_\kappa \{(\tfrac{1}{2} N)_\kappa k!\}^{-1} C_\kappa(B) . \tag{3.18a}$$

Use (3.18a) in (3.16) with $C_\kappa(B) = C_\kappa(\frac{1}{2} V_1^{\frac{1}{2}}(\Sigma_3 - \Sigma_4)V_1^{\frac{1}{2}}) = C_\kappa(\frac{1}{2}(\Sigma_3 - \Sigma_4)V_1)$ and then integrate over V_1 using a result of Constantine (1963). The final answer is given by

$$a(h) = |\Sigma_4|^{\frac{1}{2}m} |\Sigma_3|^{-\frac{1}{2}(m+nh)} 2^{\frac{1}{2}nh(p-q)} \Gamma_{p-q}(\tfrac{n(1+h)+m}{2}) \{\Gamma_{p-q}(\tfrac{1}{2}N)\}^{-1} \cdot \tag{3.19}$$

$$\cdot\ {}_2F_1(\tfrac{n(1+h)+m}{2}, \tfrac{1}{2} m ; \tfrac{1}{N} ; I - \Sigma_4 \Sigma_3^{-1}) .$$

Similarly, if $m \geq p - q$, we use Anderson's lemma 13.3.1 (1958, p. 319) to get the density of $Y_1 Y_1'$ and then integrate over $Y_1 Y_1'$ and then over V_1 using the results of Constantine (1963), we arrive at the same final expression as in (3.19).

Note that (3.12), (3.15) and (3.19) give us

$$(3.20)\qquad g_1(h) = 2^{\frac{1}{2}nh(p-q)}|\Sigma_3|^{-\frac{1}{2}n(1+h)-\frac{1}{2}m}|\Lambda_{22}|^{-\frac{1}{2}N}|I_q+h\Lambda_{1\cdot 2}|^{-\frac{1}{2}(p-q)} \;.$$

$$\cdot\; \Gamma_{p-q}\left(\frac{n(1+h)+m}{2}\right)\{\Gamma_{p-q}(\tfrac{N}{2})\}^{-1} \;.$$

$$\cdot\; {}_2F_1\left(\frac{n(1+h)+m}{2} ,\ \frac{1}{2}m ;\ \frac{1}{2}N ;\ I - \Sigma_4\Sigma_3^{-1}\right) .$$

Hence, using (3.8) and (3.20) in (3.7), we get the hth moment of λ as

$$(3.21)\qquad E\lambda^h = \left\{(2e/n)^{\frac{1}{2}nph}\,\frac{\Gamma_q\left(\frac{n(1+h)-p+q}{2}\right)\Gamma_{p-q}\left(\frac{n(1+h)+m}{2}\right)}{(1+h)^{\frac{1}{2}f}\Gamma_q\left(\frac{n-p+q}{2}\right)\Gamma_{p-q}\left(\frac{n+m}{2}\right)}\right\} .$$

$$\cdot\; \left\{(1+h)^{\frac{1}{2}f}|\Lambda_{1\cdot 2}|^{\frac{1}{2}nh}|\Lambda_{22}|^{-\frac{1}{2}(n+m)}|\Sigma_3|^{-\frac{1}{2}n(1+h)-\frac{1}{2}m} \;.\right.$$

$$\left.\cdot\; |I + h\Lambda_{1\cdot 2}|^{-\frac{1}{2}n(1+h)}\right\}$$

$$\cdot\; \left\{{}_2F_1\left(\frac{n(1+h)+m}{2} ,\ \frac{1}{2}m ;\ \frac{1}{2}N ;\ I - \Sigma_4\Sigma_3^{-1}\right)\right\}$$

$$\equiv a_1(h)a_2(h)a_3(h)$$

where the expressions in the curly brackets { } on the right side of (3.21) are respectively $a_1(h)$, $a_2(h)$ and $a_3(h)$, and

$$(3.22)\qquad f = pn(1+h) + m(p-q) \;.$$

Notice that (3.21) is valid for any complex h for which the real part of h is positive.

Notice that $\Lambda_{1\cdot 2}$, Λ_{22} are defined in (3.1) and Σ_4 and Σ_3 are defined in (3.14). We shall write these parameters in terms of B_0 , B_2 and Σ_1 . Using lemma 1 of Khatri (1966a),

$$(3.23)\qquad \Lambda_{1\cdot 2}=(B_0'\Sigma_1^{-1}B_0)^{-1},\ \Lambda_{22} = B_2'\Sigma_1B_2,\ \Sigma_4 = hI+(B_2'\Sigma_1B_2)^{-1} ,$$

and

$$(3.24)\qquad \Sigma_3 = \{B_2'(hI + \Sigma_1^{-1})^{-1}B_2\}^{-1}$$

for if

$$D \equiv \begin{pmatrix} D_{11} & D_{12} \\ D_{12}' & D_{22} \end{pmatrix} \equiv hI + \begin{pmatrix} B_0' \\ B_2' \end{pmatrix} \Sigma_1^{-1}(B_0, B_2) = hI + \Lambda^{-1} = hI + \begin{pmatrix} \Lambda_{11} & \Lambda_{12} \\ \Lambda_{12}' & \Lambda_{22} \end{pmatrix}^{-1} ,$$

then

$$D_{22} = hI + \Lambda_{22}^{-1} + \Lambda_{22}^{-1}\Lambda_{12}'\Lambda_{1\cdot 2}^{-1}\Lambda_{12}\Lambda_{22}^{-1} , \quad D_{12} = -\Lambda_{1\cdot 2}^{-1}\Lambda_{12}\Lambda_{22}^{-1} , \quad D_{11} = Ih + \Lambda_{1\cdot 2}^{-1} ,$$

and

$$\Sigma_3 = hI + \Lambda_{22}^{-1} + h\Lambda_{22}^{-1}\Lambda_{12}'(I + h\Lambda_{1\cdot 2})^{-1}\Lambda_{12}\Lambda_{22}^{-1}$$

$$= D_{22} - D_{12}'D_{11}^{-1}D_{12} .$$

Further,

$$(3.25) \qquad \Sigma_3 - \Sigma_4 = h(B_2'\Sigma_1^{-1}B_0)(B_0'\Sigma_1^{-1}B_0 + hI)^{-1}(B_0'\Sigma_1^{-1}B_0)^{-1}(B_2'\Sigma_1^{-1}B_0)'$$

$$= h\Lambda_{22}^{-1}\Lambda_{12}'(I_q + h\Lambda_{1\cdot 2})^{-1}\Lambda_{12}\Lambda_{22}^{-1} .$$

Notice also that

$$(3.26) \qquad |\Sigma_3| = |D|/|D_{11}| = |hI_p + \Sigma_1^{-1}|/|hI_q + B_0'\Sigma_1^{-1}B_0| .$$

4. Asymptotic Expansion of the Distribution

In this section we derive an asymptotic expansion up to $0(n^{-3})$ for $P(-2\ell n\ \lambda \leq z)$ under the condition that

$$(4.1) \qquad \Sigma_1^{-1} = I - (2/n)P ,$$

where P is a fixed matrix, as $n \to \infty$. Thus

$$(4.2) \qquad (h\ I + \Sigma_1^{-1})^{-1} = ((1+h)I - (2/n)P)^{-1}$$

$$= (1+h)^{-1}[I + (2/n(1+h))P + 0(n^{-2})] ,$$

$$(4.3) \qquad \Sigma_3^{-1} = B_2'(h\ I + \Sigma_1^{-1})^{-1}B_2$$

$$= (1+h)^{-1}I + (2/n)(1+h)^{-2}B_2'PB_2 + 0(n^{-2}) ,$$

$$\begin{aligned}\Sigma_3-\Sigma_4 &= hB_2'(I-\tfrac{2}{n}P)B_0[B_0'(1-\tfrac{2}{n}P)B_0+hI]^{-1}(B_0'(I-\tfrac{2}{n}P)B_0)^{-1}B_0'(I-\tfrac{2}{n}P)B_2 \\ &= \frac{4}{n^2}h(B_2'PB_0)[(1+h)I-\frac{2}{n}B_0'PB_0]^{-1}(I+\frac{2}{n}B_0'PB_0+0(n^{-2}))B_0'PB_2 \\ &= (4h/n^2(1+h))(B_2'PB_0)(B_0'PB_2)+0(n^{-3}) \ ;\end{aligned}$$

since $B_0'B_2 = 0$, and so

$$(\Sigma_3-\Sigma_4)\Sigma_3^{-1} = (4h/n^2(1+h)^2)(B_2'PB_0)(B_0'PB_2)+0(n^{-3}) \ . \tag{4.4}$$

Hence

$$\begin{aligned}a_3(h) &\equiv {}_2F_1(\frac{n(1+h)+m}{2}, \frac{1}{2}m ; \frac{1}{2}N ; (\Sigma_3-\Sigma_4)\Sigma_3^{-1}) \\ &= 1+\frac{2mh}{n^2(1+h)}\operatorname{tr}(B_2'PB_0)(B_0'PB_2)+0(n^{-3}) \ .\end{aligned} \tag{4.5}$$

Similarly, using

$$\ell n|I+n^{-1}Q| = n^{-1}\operatorname{tr}Q - \frac{1}{2}n^{-2}\operatorname{tr}Q^2+\frac{1}{3}n^{-3}\operatorname{tr}Q^3 - \dots ,$$

we find that

$$\begin{aligned}\ell n\, a_2(h) &= \frac{1}{2}N\,\ell n|I-2n^{-1}P| - \frac{1}{2}(N+nh)\,\ell n|I-(2/n(1+h))P| \\ &\quad -\frac{1}{2}m\,\ell n|I-2n^{-1}B_o'PB_o| + \frac{1}{2}m\,\ell n|I-(2/n(1+h))B_o'PB_o| \\ &= n^{-1}b(1-(1+h)^{-1})+n^{-2}c(1-(1+h)^{-2})+0(n^{-3})\end{aligned} \tag{4.6}$$

where

$$\left\{\begin{aligned} b &= m(\operatorname{tr}B_o'PB_o - \operatorname{tr}P) - \operatorname{tr}P^2 \\ \text{and}\quad & \\ c &= m(\operatorname{tr}(B_o'PB_o)^2 - \operatorname{tr}P^2) - (4/3)\operatorname{tr}P^3 \ .\end{aligned}\right. \tag{4.7}$$

Hence

$$\begin{aligned}a_2(h) &= 1+n^{-1}b(1-(1+h)^{-1})+n^{-2}c(1-(1+h)^{-2}) \\ &\quad +\frac{1}{2}n^{-2}b^2(1-(1+h)^{-1})^2+0(n^{-3}) \ .\end{aligned} \tag{4.8}$$

Also, following as in Anderson (1958, p. 203), we find that

(4.9)
$$\ell n\ a_1(h) = -\frac{1}{2} r\ \ell n(1+h) + n^{-1}((1+h)^{-1} - 1)d_0 + n^{-2}((1+h)^{-2} - 1)d_1 + 0(n^{-3}) ,$$

where

$$r = \frac{1}{2} p(p+1) + q(p-q)$$

$$d_0 = \frac{1}{24} p(2p^2+3p-1) + \frac{1}{4}(p-q)(m^2-m(p-q+1) + q(p+1))$$

(4.10)
$$d_1 = \frac{1}{48} p(p-1)(p+2)p+1) + \frac{1}{24}(p-q)(-2m^3 + 3m^2(p-q+1) - m(p-q-1)(2p-2q+5) + 4(q-m) + 2q(p-q)^2 + 3q(p-q)(q+1) + q(q-1)(2q+5)) .$$

Hence

(4.11)
$$a_1(h) = (1+h)^{-\frac{1}{2}r}[1 + n^{-1}((1+h)^{-1}- 1)d_0 + n^{-2}((1+h)^{-2}-1)d_1 + \frac{n^{-2}((1+h)^{-1}-1)^2 d_0^2}{2} + 0(n^{-3})] .$$

Consequently

(4.12)
$$E(\lambda^h) = a_1(h)a_2(h)a_3(h) = (1+h)^{-r/2}[1 + n^{-1}\ \beta(1-(1+h)^{-1}) + n^{-2}(\gamma_o + \gamma_1(1+h)^{-1})(1-(1+h)^{-1}) + 0(n^{-3})]$$

where

$$\beta = b - d_o\ ,\ e = 2m\ tr\ (B_2'PB_o)(B_o'PB_2) ,$$

(4.12a)
$$r_o = e - bd_o - d_1 + c + \frac{1}{2} d_o^2 + \frac{1}{2} b^2\ , \text{ and}$$

$$r_1 = bd_o - d_1 + c - \frac{1}{2} d_o^2 - \frac{1}{2} b^2 .$$

Now, inverting (4.12), we get the asymptotic distribution of $- 2\ \ell n\ \lambda$ as

(4.13)
$$P(-2\ \ell n\ \lambda \le x) = P(\chi_r^2 \le x) + n^{-1}\beta[P(\chi_r^2 \le x) - P(\chi_{r+2}^2 \le x)] + n^{-2}[r_o\{P(\chi_r^2 \le x) - P(\chi_{r+2}^2 \le x)\} + r_1\{P(\chi_{r+2}^2 \le x) - P(\chi_{r+4}^2 \le x)\}] + 0(n^{-3})$$

where χ_f^2 is a Chi-square variate with f degrees of freedom.

When $p = q$, we get the same result as the corresponding result given by Khatri and Srivastava (1970). There are various types of asymptotic expansion given by the authors when $p = q$, but here we consider only one type for illustration. Other types can be developed in the same way.

References

[1] Anderson, T. W. (1958): Introduction to Multivariate Statistical Analysis, Wiley, New York.

[2] James, A. T. (1964): Distribution of matrix variates and latent roots derived from normal samples. Ann. Math. Statist. 35, 475-501.

[3] Khatri, C. G. (1966a): A note on a MANOVA model applied to problems in growth curve. Ann. Inst. Statist. Math. 18, 75-86.

[4] Khatri, C. G. (1966b): On certain distribution problems based on positive definite quadratic functions in normal vectors. Ann. Math. Statist. 37, 468-479.

[5] Khatri, C. G. (1973): Testing some covariance structures under a growth curve model. J. Multivariate Analysis 3 (To appear).

[6] Khatri, C. G. and Srivastava, M. S. (1970): Asymptotic Expansion of the Non-null distributions of the covariance matrices II. Unpublished.

[7] Potthoff, R. F. and Roy, S. N. (1964): A generalized MANOVA model useful especially for growth curve problems. Biometrika 51, 315-328.

[8] Sugiura, N. and Nagao, H. (1968): Unbiasedness of some test criteria for the equality of one or two covariance matrices. Ann. Math. Statist. 39, 1686-1692.

C. G. Khatri
Department of Statistics
Gujrat University
Ahmedabad, India

M. S. Srivastava
Department of Mathematics
University of Toronto
Toronto, Ontario
Canada

Applied Statistics, R.P. Gupta, (editor)

STRUCTURALISM

REMARKS ON SPHERICAL AND HYPERSPHERICAL STRUCTURES

Andre G. Laurent

Introduction.

The expression "structure" is used here within the frame of reference of Structuralism, a paradigmatic and unifying -- but somewhat elusive -- concept and system of thought, nowadays of paramount importance in the fields of Linguistics, Anthropology, Psychology, and many other fields. In the abstract, a structure is self-sufficient and does not require any reference to objects or principles that are foreign to its nature. In the concrete, a formalization of a system of invariances with paradigmatic value to describe and unify a variety of situations, conceived as realizations of the system, is a structure. A masterful introduction to Structuralism and its philosophical aspects can be found in Piaget [14]. In [10] one has tried to outline some aspects and uses of the notion of structure in Statistics and to define -- if not a theory -- at least an attitude that puts emphasis on the consideration of statistics rather than of *a priori* models, as the foundation of Statistics, and whose utopian goal is an extension of the statistical discipline: a framework within which to unify, reconcile, and blend, information theoretic, utility theoretic, group theoretic, integral geometry, and statistical concepts, with a view to reduce so-called random mechanisms to "structures" and their "integral geometries." With the present writer, this utopia originated in a 1951 lecture that remained unpublished but was recently mentioned by G. A. Barnard [2]. There is evidence in the literature that this utopia is shared by others.

In this paper we outline a few definitions and results of a theoretical nature and present some remarks on the spherical and hyperspherical structures as the latter illustrate definitions and concepts in a way that is especially concrete and -- technically -- simple, and as many basic results are familiar to all (*e.g.*, basic Jacobians are well known since their gathering by Olkin [3], [13]; and their abstract significance through group theory and differential forms considerations has been made clear by James [4]).

Structural concepts.

Let $\mathcal{X} = (X^1,\ldots,X^p) = (X_1,\ldots,X_n)'$ be a random matrix of dimension $n \times p$, $n \geq p$, generally thought of as a column of n p-dimensional random vectors; let $d\mathcal{X}$ be the differential element of Lebesgue measure in the np-dimensional (sample) space E_{np}. Let $\{\mathcal{J}\}$ be a partition of the sample space (or perhaps of a parameter free restriction of the sample space) into bounded equivalence classes which are characterized and determined by -- or, according to the viewpoint, which induce -- a set of rank $r < n$ of statistics $T = \{T_i(X_1,\ldots,X_n)\}$, $i = 1$ to r. Let us assume that $T = t$ is as smooth as will be needed and such that one can define statistics $U = \{u_j\}$, $j = 1$ to $n - r$, so that $\mathcal{X} \to (T,V)$ be $1 - 1$; possibly $U = \{X_j\}$, $j = r+1,\ldots,n$. At all points of continuity of Lebesgue measure with respect to T, the mass of the shell $t < T \leq t + dt$, is given by $K(t)dt$.

$$K(t) = (d/dt) \int\cdots\int_{c<T\leq t} d\mathcal{X},$$

written

$$K(t) = \int\cdots\int_{T=t} d\mathcal{X}$$

for convenience, where (d/dt) denotes $(\partial^r/\partial t_1\ldots\partial t_r)$. $K(t)$ is possibly obtained using the Jacobian $\frac{D(\mathcal{X})}{D(T,V)}$ and integrating out V. We call $K(t)$ the "integral geometry area" of Σ_t, defined by $T = t$. Then $\frac{D(\mathcal{X})}{K(T)D(T,V)}\,dV$, $T = t$, is a probability measure on $T = t$, that can be written as $\frac{|G_V|^{1/2}}{|\text{grad } T|}\,dV$, $T = t$, where G_V is the metric tensor of Σ_t, with parametric representation $\mathcal{X}_t = \mathcal{X}_t(V)$, and where $|\text{grad } T|$ denotes the square root of the Grammian $((\text{grad } T_i, \text{grad } T_j))$. It is the V coordinate representation of the coordinate free probability measure $\frac{d\sigma_t}{K(t)} = \frac{dS_t}{K(t)|\text{grad } T|}$ on Σ_t where dS_t denotes Lebesgue measure in the tangent space to Σ_t (indeed, $d\mathcal{X} = dT \cdot dS_t/|\text{grad } T|$, $T = t$). We call this probability measure the "integral geometry probability measure" on $T = t$; $K(t)$ is the measure of $T = t$ using the measure $d\sigma_t = dS_t/|\text{grad } T|$, which we call integral geometry measure or Liouville measure on $T = t$ (as it would be the Liouville measure on the "phase space" $T = t$). The latter, induced by Lebesgue measure inside the shell,

$t < T \leq t + dt$ generalizes the notion of Lebesgue measure on Σ_t when Σ_t is not isolated but is considered as an equivalence class of the partition $\{T\}$, and it is invariant in the group of motions on Σ_t. When $U = \underset{\sim}{\xi}^* = (X_{r+1},\ldots,X_n)$, the corresponding representation of the integral geometry probability is called the T-Thompson distribution of $\underset{\sim}{\xi}^*$. There exists a system of "coordinates" $\tilde{\Phi} = \tilde{\Phi}(\mathfrak{X},t)$ on $T = t$ for which the representation is independent from t, i.e., $d\sigma_t/K(t) = M(\tilde{\Phi})d\tilde{\Phi}$, $\tilde{\Phi} = (\tilde{\Phi}_1,\ldots,\tilde{\Phi}_{n-r})$. When one wants to keep some symmetry in the formulae one may define Φ by $\Phi = (\Phi_1,\ldots,\Phi_n)$, restricted by $\underset{\sim}{H}(\Phi) = 0$ or $H[\Phi] = \underset{\sim}{h}$ of rank r. Then the mapping $\mathfrak{X} \to (T,\phi)$ is replaced by the mapping $\mathfrak{X} \to (T,\phi)$, the latter restricted by $H[\Phi] = \underset{\sim}{h}$ -- a mapping from E_{np} to a subset $H[\Phi] = \underset{\sim}{h}$ of a $np + rp$ degrees of freedom sample space, where $H[\Phi] = \underset{\sim}{h}$ is a cylinder. The partition $\{\phi\}$ defines a 1 - 1 correspondence between a given equivalence class $T = t_0$ of the partition $\{T\}$ and an equivalence class $T = t$, for every t, and $H[\Phi] = \underset{\sim}{h}$ is a relation of consistency implied by the correspondence between $T = t_o$ and $T = t$ (i.e., $T = t$ can be rewritten as $H[\Phi] = h$). Symbolically, $d\sigma_t/|\text{grad } T| = M(\Phi)d\Phi$, $T = t$. We call $\Phi = \Phi(\mathfrak{X},t)$ a conditional T-Thompsonization of $\mathfrak{X}$. Starting with the partition $\{T\}$, establishing a mapping $T = t_0 \to T = t$, for every t, will provide a conditional T-Thompsonization of $\mathfrak{X}$. $\mathfrak{X}$ can be viewed as $\mathfrak{X} = T \otimes \Phi$, "sum" of T and Φ and an approximation of T with an "error" or residual Φ. This is one point where our interpretation of $\mathfrak{X}$ is germane to Fraser's representation of $\mathfrak{X}$ as $\mathfrak{X} = \theta \otimes \varepsilon$, where θ is a parameter and ε is structural (in Fraser's sense). Whatever interpretation, one has $d\mathfrak{X} = K(t)dT\ M(\Phi)d\Phi$, a basic factorization.

Useful factorizations of $d\mathfrak{X}$ -- without Jacobians.

A special case of interest is when $T = (X_1,\ldots,X_k,T_{k+1},\ldots,T_r) = (\underset{\sim}{\xi},T^*)$, say, and several interesting but rather straightforward remarks can be made about the most expedient techniques to obtain the corresponding factorization of $d\mathfrak{X}$ and as to whether it is advantageous to consider first the factorization $d\mathfrak{X} = d\underset{\sim}{\xi} \cdot d\underset{\sim}{\xi}^*$, $\mathfrak{X} = (\underset{\sim}{\xi},\underset{\sim}{\xi}^*)$, (it depends whether T^* is "separable" or not).

Another case is when $\{T_1,\ldots,T_k\} = D\mathfrak{X}$, instead of $T_i = X_i$, $i = 1$ to k, where D is a $k \times n$ matrix of rank k, $k < n$; a case that is hardly a generalization of the preceding one but is useful in the applications. In this case, it is useful to factor $d\mathfrak{X}$ into the Lebesgue measure of the D row space and that of its orthogonal component D^* so as to obtain (since the metric tensor of the D row space expressed in function of $D\mathcal{X}$ is $|DD'|^{-p}$),

$$d\mathcal{X} = |DD'|^{-p/2}\, d(D\mathcal{X}) \cdot dZ_{D^*} \quad . \tag{1}$$

In further manipulations one will remember that $Z = \begin{pmatrix} Z_D \\ Z_{D^*} \end{pmatrix} = 0\mathcal{X}$, $0 = \begin{pmatrix} 0_D \\ 0_{D^*} \end{pmatrix}$ orthogonal, $0_D = L_D^{-1}D$, with $LL' = DD'$, L triangular with positive diagonal, as Z_D is the projection of $\mathcal{X}$ on the D row space).

Let us consider now the case of $T = L$, a triangular matrix with positive diagonal such that $\mathcal{X}'\mathcal{X} = L'L$. L is the matrix of rectangular coordinates of $\mathcal{X}$ made famous by the Indian school [11]. Then

$$d\mathcal{X} = \prod_{i=1}^{p} \ell_{ii}^{n-i}\, dL \prod_{i=1}^{p} d\Sigma_{n-i+1} = K(L)dL \prod_{i=1}^{p} d\Sigma_{n=i+1}/\Sigma_{n-i+1} \;, \tag{2}$$

where $d\Sigma_k$ stands for the elementary measure on the unit sphere in E_k. Hsu and others showed early that $K(L)dL$, $L = \ell$, was the mass of the shell $\ell < L \leq \ell + d\ell$, after proving by algebraic methods the unicity of the factorization of $\mathcal{X}'\mathcal{X}$ as $L'L$, and, therefore, gave implicitly the factorization above. Another proof [7] uses a Gram-Schmidt orthogonalization process to construct an orthonormal basis $(Y^1,\ldots,Y^p)$ in the space spanned by $(X^1,\ldots,X^p)$ and uses polar coordinates in the spaces $E_n,\ldots,E_{n-i+1},\ldots,E_{n-p+1}$, where E_{n-i+1} is the orthogonal complement of the space spanned by $(Y^1,\ldots,Y^{i-1})$. Then,

$$d\mathcal{X} = \prod_{i=1}^{p} dX^i = \prod_{i=1}^{p} \ell_{ii}^{n-i} d\ell_{ii} d\Sigma_{n-i+1} \prod_{j=1}^{i-1} d\ell_{ij} = \prod_{i=1}^{p} K(\ell_{ii})d\ell_{ii} \frac{d\sigma_{ii}}{|\text{grad } \ell_{ii}|}$$

$$\prod_{i=1}^{p} \prod_{j=1}^{i-1} d\ell_{ij} = K(L)dL\, d\mathcal{Y} \Big/ \prod_{i=1}^{p} \Sigma_{n-i+1} \;, \tag{3}$$

where $K(\ell_{ii}) = \ell_{n-i+1}\;{}^{n-i}_{ii}$, $\mathcal{Y} = (Y^1,\ldots,Y^p)$; and $\mathcal{X} = \mathcal{Y}L$ is obtained by purely geometric means and exhibits $K(L)$, $L = \ell$, as $\prod_{i=1}^{p} K(\ell_{ii})$, and the integral geometry probability measure $d\mathcal{Y}$ as the product $\prod_{i=1}^{p} d\Sigma_{n-i+1}/\Sigma_{n-i+1}$ of the integral geometry probability measures on $X'^{i}X^{i} = \ell_{ii}$.

Let us now examine the case $T = \mathcal{X}'\mathcal{X}$. Then $\mathcal{X}'\mathcal{X} \to L$ is a 1-1 mapping, defining the same partition $\mathcal{J}$ on the sample space, i.e., $L \sim \mathcal{X}'\mathcal{X}$, modulo $\{\mathcal{J}\}$, (if $\mathcal{X}'\mathcal{X} = \beta'\beta$, β not triangular, then β is not $\sim \mathcal{X}'\mathcal{X}$). On the basis of purely geometrical considerations, one has

$$\frac{D(\mathcal{X}'\mathcal{X})}{D(L)} = 2^{p} \prod_{i=1}^{p} \ell_{ii}^{p-i+1} , \tag{4}$$

so that

$$d\mathcal{X} = 2^{-p} |\mathcal{X}'\mathcal{X}|^{(n-p+1)/2} \prod_{i=1}^{p} \Sigma_{n-i+1} \, d\mathcal{X}'\mathcal{X} \, d\mathcal{Y}, \tag{5}$$

with $\mathcal{Y} = \mathcal{X}L^{-1}$. Also $\mathcal{Y} = H\Gamma$, where Γ is orthogonal if $(H^{1},\ldots,H^{p}) = H$ is an orthonormal basis for the space spanned by $(X^{1},\ldots,X^{p})$ and Γ are the coordinates of $\mathcal{Y}$ with respect to H ; H , depending on $p(n-p)$ variates, defines the p dimensional space spanned by $\mathcal{Y}$. Then $K(\mathcal{X}'\mathcal{X}) = 2^{-p}|\mathcal{X}'\mathcal{X}|^{(n-p-1)/2} \prod_{i=1}^{p} \Sigma_{n-i+1}$ and the integral geometry probability on the partition $\mathcal{J}$ is

$$\frac{d\sigma_{\mathcal{J}}}{K(\mathcal{X}'\mathcal{X})} = \prod_{i=1}^{p} d\Sigma_{n-i+1} = d\mathcal{Y} = d(\mathcal{X}L^{-1}) , \tag{6}$$

where $\mathcal{Y} = \mathcal{X}L^{-1}$, restricted by $\mathcal{Y}'\mathcal{Y} = I$ is the $\mathcal{X}'\mathcal{X}$ or, equivalently, the L-Thompsonization of $\mathcal{X}$. The partition $\{\mathcal{Y}\}$ establishes a 1-1 correspondence $\mathcal{Y} = \mathcal{X}_{\ell}\ell^{-1} = \mathcal{X}_{I}$ between the equivalence class $\mathcal{X}_{I}'\mathcal{X}_{I} = I$ and every equivalence class $\mathcal{X}_{\ell}'\mathcal{X}_{\ell} = \ell'\ell$.

Consider further the case when $T = (\mathcal{X}'\mathcal{X}, D\mathcal{X}) = (\mathcal{X}'\mathcal{X}, V)$, where D is a $k \times p$ nonsingular matrix, $k \leq p$; then, for $n-1 \geq p$,

$$dZ_{D*} = dP_{D*}\mathcal{X} = 2^{-p}|\mathcal{X}'P_{D*}\mathcal{X}|^{(n-k-p-1)/2} \prod_{i=1}^{p} d\Sigma_{n-k-i+1} ,$$

where $P_{D*} = I - D'(DD')^{-1}D$ is the projection matrix on $D*$, and therefore,

$$d\mathcal{X} = 2^{-p}|\mathcal{X}'(I-D'(DD')^{-1}D)\mathcal{X}|^{(n-k-p-1)/2} \, d\mathcal{X}'\mathcal{X} \, dD\mathcal{X} \prod_{i=1}^{p} d\Sigma_{n-1-i+1} \tag{7}$$

$$= 2^{-p}|\mathcal{X}'\mathcal{X} - V'(DD')^{-1}V|^{(n-k-p-1)/2} \, d\mathcal{X}'\mathcal{X} \, dV \prod_{i=1}^{p} d\Sigma_{n-k-i+1} \,. \tag{8}$$

Kabe [5] exhibited the first factor (up to a constant) as the mass of the shell associated with $(\mathcal{X}'\mathcal{X}, D\mathcal{X})$ and, therefore, gave implicitly the factorization above. Kabe's proof was nongeometric and less direct than this one, and was resting on a result, logically equivalent to (5), given by T. W. Anderson in [1] without the use of rectangular coordinates as in [11] nor of an orthogonalization process as in [7]. Kabe had the insight to realize the considerable importance of this shell mass and used it to derive most of multivariate analysis classical results, [5] and [6].

Absolutely sufficient statistics, maximum entropy, and integral transforms.

We call a statistic T, or the corresponding partition $\mathcal{F}$ (induced by $T = t$, all t's, on the sample space or on a parameter free restriction of the sample space) absolutely sufficient, AS, iff $K(t)$ is finite. We call a distribution of $\mathcal{X}$ T-Heaviside if the probability density function of is uniform on all sets $\{c < T \leq t\}$, for $t \leq \theta$, and null for $t > \theta$. T is AS iff the T-Heaviside distribution exists. The family of distributions of $\mathcal{X}$, whose pdf has structure $f(\mathcal{X}) = g(T)$ is called the T-family of distributions of $\mathcal{X}$; it admits T as a sufficient statistic and the pdf of T is $h(t) = K(t)g(t)dt$. A special case is the T-Heaviside distribution. We will admit in the family the T-Dirac pdf, degenerate distribution with all the probability mass concentrated on $T = t$. Then the degenerate conditional pdf of $\mathcal{X}$, given $T = t$, is the T-Dirac distribution, to be relabeled as the T-Thompson distribution of $\mathcal{X}$. T, sufficient for the whole T-family, is called AS with respect to the family since its conditional distribution does not depend on which member of the family is considered. Let $\phi_{Dir.T}(U)$ be the characteristic function of the T-Dirac distribution, then the characteristic function of $\mathcal{X}$ is $\phi_{\mathcal{X}}(U) = E_T[\phi_{Dir.T}(U)]$, hence a multivariate integral transform calculus relates $\phi_{\mathcal{X}}(U)$ and $g(t)$, through the kernel

$K(t)\phi_{Dir\ T}(U) = W(t,U)$, namely,

$$\phi\ (U) = \int\ldots\int g(t)K(t) \cdot \phi_{Dir.T}(U)dt\ . \tag{9}$$

The conditional superficial probability measure on $T = t$, given $T = t$, induced by the pdf of $\mathcal{X}$, is the integral probability measure $d\sigma_t/K(t)$ on $T = t$, previously defined, and the conditional pdf of $(X_1,\ldots,X_{n-r})'$, submatrix of $\mathcal{X}$, is a representation of that measure and also the conditional pdf of the projection of $\mathcal{X}$ on the corresponding space $E_{(n-r)p}$. The conditional pdf of a submatrix $(X_1,\ldots,X_k)'$, given $T = t$, is the projection of the T-Dirac measure on the corresponding space E_{kp} and is called a T-Thompson distribution of order k of $\mathcal{X}$. With t replaced by T and as a function of T, it is a minimum variance unbiased estimator of the joint pdf of $(X_1,\ldots,X_k)'$ and, in case $\mathcal{X}$ is a random sample of rows, it converges almost surely toward the pdf of $(X_1,\ldots,X_k)'$. $\{T = t\}$ defines a constant likelihood partition, i.e., the pdf of $\mathcal{X}$ is uniform in the shell $t < T \leq t + dt$ and, therefore, has there local maximum entropy (within a broad sense of the word entropy). One can therefore say that the integral probability measure on $T = t$ is maximum entropy induced, when $T = t$ is not isolated but considered as the equivalence class of a partition (the entropy concept involved relates to the partition not to its equivalent classes, a notion to make more precise). The conditional pdf of $\Phi_t = \Phi(\mathcal{X},t)$ is independent from T ; therefore $\Phi = \Phi(\mathcal{X},T)$ is parameter free distributed and an ancillary statistic; Φ is called the T-Thompsonization of $\mathcal{X}$ and the distribution of $\mathcal{X}$ is a mixture of the integral geometry distribution on $T = t$ with $h(t)$ as a mixer. More technical entropy theoretic statements can be made on the integral geometry measure. Let $I_\mu(\mathcal{P})$ denote the entropy (in the sense of Kolmogorov) of the partition $\mathcal{P}$ of the sample space Ω, relative to a measure μ on a, large enough for our purpose, Borel field of subsets of Ω. A partition $\{T\}$ is μ-sufficient, if, for all $\{V\}$ -- $\{V\}$ not a refinement of $\{T\}$ -- one has $I_\mu(\{T\} \cap \{V\}) = I_\mu(\{T\}) + I_\mu(\{V\}|\{T\})$ with $I_\mu(\{V\}|\{T\})$ independent from μ's unspecified parameters, all $\{V\}$'s ; $\{T\}$ is AS with respect to a family $\{\mu\}$, if $I_\mu(\{U\}|\{T\})$ is independent from the member μ in $\{\mu\}$. If T is AS, then $\{T\}$ is AS and there exists $\{\Phi\}$ such $I_\mu(\{\phi\}|\{T\}) = I_\mu(\{\phi\}) = I(\{\phi\})$, independent from μ in $\{\mu\}$, and

$I_\mu(\{T\} \cap \{\phi\}) = I_\mu(\{T\}) + I(\{\Phi\})$, all μ's in $\{\mu\}$; that is, $\{T\} \perp \{\Phi\}$, where $\perp$ is taken in the sense of partition independence. As $I_\mu(\{V\}|\{T\}) \leq I_\mu(\{V\})$, if T is AS there exists a partition $\{\Phi\}$ with maximum entropy among all $\{U\}$ given $\{T\}$; this entropy is $i_\lambda(\Phi)$, where, notationally, λ refers to a measure on the quotient set $\{\Phi\}$ and, in this instance, is the integral probability measure on $T = t$. Furthermore, if ν is a probability measure on the quotient set $\{\Phi\}$, then $i_\nu(\Phi) \leq i_\lambda(\Phi)$; therefore the integral probability measure on $T = t$ is maximum entropy within the frame of reference of partition entropy.

The joint pdf $f_k(\underset{\sim}{x}_1,\ldots,\underset{\sim}{x}_k)$ of the submatrix $(X_1,\ldots,X_k)$, projection of $\mathcal{X}$ on the corresponding E_{kp} , is a mixture of the T-Thompson distribution of order k of $\mathcal{X}$ by the pdf $h(t)$ of T. One has, therefore

$$f_k(\underset{\sim}{x}_1,\ldots,\underset{\sim}{x}_k) = \int\cdots\int W_{n,k}(\underset{\sim}{x}_1,\ldots,\underset{\sim}{x}_k\ ,\ t_n)g_n(t_n)dt_n\ , \qquad (10)$$

where $W_{n,k}g_n$ is the joint pdf of $(X_1,\ldots,X_k)$ and T_n , which defines an operational calculus relating $f_n(x) = g_n$ and its projection $f_k = \mathbb{P}_{n,k}f_n$, and calls for the inversion, under proper restrictions, of the transform (10) and the corresponding definition of an operator $\mathbb{P}^{-1}_{n,k}$ of anti-projection allowing for the T_n-generalization, to n rows matrices, of pdf of k rows matrices. If $\{T_i\}$ is a <u>projective sequence</u> of $T_i = T_i(x_1,\ldots,x_i)$, by which we mean that $f_i(\underset{\sim}{x}_1,\ldots,\underset{\sim}{x}_i) = g_i(t_i)$, at least for some i , then

$$f_i = g_i(t_i) = \int\cdots\int G_{n,i}(t_i,t_n)g_n(t_n)dt_n = \mathbb{P}_{n,i}f_n \qquad (11)$$

involves a simpler multivariate operational calculus than that defined by (10).

We call T_n <u>symmetrically separable,</u> SS , if the family $f(\mathcal{X}) = g(T)$ contains a member for which the rows of $\mathcal{X}$ are a random sample of n observations of a random vector X . Should one restrict one's attention to those members of the family that belong to the exponential family, then the characterization of T_n as SS would be trivial as T_n should have structure $T_n(\mathcal{X}) = \sum_{i=1}^{n} t(X_i)$. In this case, if the dimensionalities k , $n-k$, and p are comfortable with the following method, the T_n-Thompson distribution of order k of $(X_1,\ldots,X_k)'$ is easily obtained by writing the joint pdf of T_{n-k}, sufficient for $(X_{k+1},\ldots,X_n)'$, and of $(X_1,\ldots,X_k)'$; then deriving the joint pdf $(X_1,\ldots,X_k)$ and T_n by means of the change of variables $T_n = T(T_{n-k},X_1,\ldots,X_k)$; and, finally, dividing by the pdf of T_n. The projection operator is easily obtained from the consideration of the joint pdf of $(X_1,\ldots,X_k)'$ and T_n.

Note that if T is not AS it can be completed by a statistic S so that (T,S) be AS and determine an AS partition $\mathcal{J}$; then there exists a (non unique) statistic L such that $(T,S) \sim (T,L)$, modulo $\mathcal{J}$, and $Ld\mathfrak{X}$ define a finite measure on the sample space. Then T is said to be <u>L-absolutely sufficient</u>, that is, T is AS with respect to $Ld\mathfrak{X}$ in the sense that $f(\mathfrak{X}) = g(T)L(\mathfrak{X})$. For example $(\bar{X}^1,\ldots,\bar{X}^p)$ cannot be AS without a restriction on the sample space but $(\bar{X}^1,\ldots,\bar{X}^p,\mathfrak{X}'\mathfrak{X})$ is AS, and the Wishart distribution, with parameter $\Sigma = I$, is, among others, a measure $L(\mathfrak{X})dx$ with respect to which $(\bar{X}^1,\ldots,\bar{X}^p)$ is AS.

A sequence $\{T_i\}$, $n_0 < i \leq n \leq \infty$, is said to be <u>strongly projective</u> if $\{f_i\}$, with $f_i(x_1,\ldots,x_i = g_i(t_i)$, is strongly projective, that is, if $f_{i-1} = \mathbb{P}_{i,i-1}f_i$, all i's.

Spherical structures and fractional calculus.

In case $p = 1$ and $T = X'X$, the family with structure $f(X) = g(X'X)$ is called the spherical family. Its properties are well known. The partition $\{X'X = R^2\}$, together with the integral probability distribution associated with it (the uniform probability distribution on the sphere -- in reduced form $d\,\Sigma_n / \Sigma_n$) and the Thompsonization of X as X/R, define what we call a spherical structure. Important points about the spherical structure are that the family $g(R^2)$ is symmetrically separable (into a product of normal pdf's), that the sequence $\{g_n(R_n^2)\}$ is strongly projective, and that the conditional pdf of $(X_1,\ldots,X_k)$, given $R_n^2/n = t$, as n tends to infinity, tends to that of a sample of k normal variables with variance t : the projection of a uniform measure on a sphere in Hilbert space, and the maximum entropy pdf of $X_1,\ldots,X_k$ under the constraints $E[X_i^2] = t$. The latter exhibits the normal model as generated, not by a random mechanism involving central limit theorem type considerations, but by the spherical structure itself. Similar asymptotic properties hold in the case of non-spherical structures. By a "structure" we mean (see [10]) an AS partition together with the integral geometry probability associated with it. Some structures generate, asymptotically, parametric probability models, characterized by maximum entropy properties under constraints that thus characterize also the partitions considered. Such constraints define the respective structures of the probability models so obtained, and the random mechanisms which these models are assumed to represent can be accounted for, under a hypothesis of maximum symmetry and homogeneity, by the constraints themselves--a geometrization of random phenomena. For example, the class of spherical distributions, which include the distribution of samples of normally distributed variables, is a class whose construction and interpretation can be entirely based on, and are reducible to, considerations of maximum uniformity on the spherical structure. Similarly, the class of simplicial distributions (see [10]), for which $f(\underset{\sim}{x}) = g(x_1 + \dots + x_n)$, with AS partition $\{T = X_1 + \dots + X_n = t\}$, and among which the negative exponential model characterizes separability, rests on the integral geometry of the simplex. We want, here, to make only a few remarks about the integral transform calculus involved by the projection operator associated with the spherical structure.

Let $\phi_n(\underset{\sim}{V}_n) = E[\exp(iX\underset{\sim}{V}_n')] = \Phi_n(\rho_n^2)$, where $\underset{\sim}{V}_n = (V_1,\dots,V_n)$ and $\rho_n^2 = \underset{\sim}{V}_n\underset{\sim}{V}_n'$, be the characteristic function of a spherical distribution in E_n with argument X. Then the characteristic function of the projection of X on E_k is $\phi_k(\underset{\sim}{V}_k) = \Phi_k(\rho_k^2) = \Phi(\rho_k^2)$ and, as is well known, is invariant under projection when considered as a function of the generic argument ρ^2, namely, $\phi_j(\underset{\sim}{V}_j) = I_j\Phi(\rho^2)$, where I_j is an indicator of the dimensionality j to be associated with the argument.

Let $F_nf_n = (2\pi)^{-n/2}\phi_n(\underset{\sim}{V}_n)$ denote the Fourier transform of f_n, then it is well known that

$$F_nf_n = \rho_n^{1-n/2}\int_0^\infty r_n^{n/2}\, J_{n/2-1}(\rho_n r_n)g_n(r_n^2)dr_n \,, \tag{12}$$

which, as a univariate transform of $g_n(r_n^2)$, will be denoted

$$F_nf_n = I_n\,\mathbb{F}_{n/2}\, g_n(r_n^2) \,. \tag{13}$$

The operator $\mathbb{F}_{n/2}$ can be generalized to $\mathbb{F}_{\lambda/2}$, λ not an integer, so as to generalize the dimensionality of multivariate Fourier transforms and characteristic functions to non integral values.

Also the joint pdf of the projection of X on E_k, and of $X'X$ is

$$f(\xi,r_n^2) = (\pi^{(n-k)/2}/\Gamma[(n-k)/2])(r_n^2 - r_k^2)^{(n-k)/2-1}g_n(r_n^2) , \qquad (14)$$

as obtained by application of the Sverdrup lemma [6] or other methods [7]. The pdf $f_k(\xi) = g_k(r_k^2)$ of the projection X on E_k is then

$$f_k(\xi) = g_k(r_k^2) = \pi^{(n-k)/2} {}_{r_k^2}D_\infty^{-(n-k)/2} g_n(r_n^2) , \qquad (15)$$

that is, $g_k(r_k^2)$ is the fractional derivative of order $-(n-k)/2$ (Weyl's transform of order $(n-k)/2$) of $g_n(r_n^2)$. The projection $f_k = \mathbb{P}_{n,k}f_n$ of f_n is, therefore, operated by fractional integration and the antiprojection $f_n = \mathbb{P}_{k,n}f_k$, denoted $\mathbb{P}^*_{(k-n)/2}f_k$, by fractional differentiation. Given a spherical pdf in E_k, its antiprojections, if they exist, generalize that pdf to those of spherical distributions in higher dimensional spaces. In particular, the antiprojection by 2 dimensions of a Heaviside spherical distribution results in a Dirac spherical distribution. By the use of the antiprojection operator, a symmetric univariate pdf with characteristic function $\Phi(\rho)$ may be generalized into a spherical distribution in E_n. Whether such a generalized pdf exists depends on whether or not $I_n\Phi(\rho)^2$ is an n dimensional characteristic function $\Phi_n(\rho_n^2)$. The operator $\mathbb{P}^*_{(n-1)/2}$ can be generalized to $\mathbb{P}^*_\lambda$, λ not a half integer, by means of the fractional integration of order λ, to define projection by a non integral number of dimensions. One has shown in [11] that the rule of compositions of the operator of fractional differentiation D^ν and the Fourier operator $\mathbb{F}_\lambda$, as applied to functions that are pdf or characteristic functions of spherical distributions and subject to their existences as such, are

$$\mathbb{F}_\alpha \mathbb{F}_\alpha = I_\alpha \qquad (16)$$

$$D^\alpha D^{-\alpha} = 1 \qquad (17)$$

$$D^\alpha D^\beta = D^\beta D^\alpha = D^{\alpha+\beta} \qquad (18)$$

$$\mathbb{F}_\alpha I_{2\alpha} \mathbb{F}_\beta I_{2\beta} = D^{\alpha-\beta} \qquad (19)$$

$$\mathbb{F}_{\alpha} I_{2\alpha} D^{\beta} = \mathbb{F}_{\alpha-\beta} I_{2(\alpha-\beta)} \quad (20)$$

$$D^{\alpha} \mathbb{F}_{\beta} I_{2\beta} = \mathbb{F}_{\alpha+\beta} I_{2(\alpha+\beta)} ; \quad (21)$$

the equalities are exact or defined up to a multiplicative constant in π, that can be taken care of at the end of the computations. This calculus is most useful for different kinds of generalizations of spherical models and for the study of problems related to these models.

The use of the words "projection" and "antiprojection," to stand for integration and differentiation, makes this calculus intuitively clear. A different approach may be found in Thomas [15].

Hyperspherical distributions.

1. Random bases and integral geometry

In case $T = \mathcal{X}'\mathcal{X}$, the family $f(\mathcal{X}) = g(\mathcal{X}'\mathcal{X})$, which is naturally induced by the partition $\{\mathcal{X}'\mathcal{X}\}$, is called the hyperspherical family. This family was introduced in [7], 1959, and was discussed since in several contributed papers and abstracts by the same author. All the formulae in the following sections can be found already in [7]; some were used in the applications to bombing and firing problems [9], and some (in several abstracts) for the study of the multivariate linear model; Thomas [15] comments on [7].

$\mathcal{X}'\mathcal{X}$ is AS for the family and is invariant under an orthogonal transformation in E_n. $\mathcal{Y} = \mathcal{X}L^{-1}$, L triangular with positive diagonal, $L'L = \mathcal{X}'\mathcal{X}$, is ancillary. The family is strongly projective and symmetrically separable. Separation occurs for that member of the family which is the distribution of a random sample, hence, that of n observations of a normally distributed random vector X. It may be surmised that, in many applications, the hypothesis of normality is made for the distribution of a set of n random vectors when only a hypothesis of rotational invariance, that is, or hypersphericity, is called for because of some tacit assumption made on the independence of the observations. It is, therefore, important to study the general properties implied by hypersphericity. As $\mathcal{X}'\mathcal{X}$ is the generalization of $X'X$, considered in the univariate case, $\mathcal{Y}$ generalizes the notion of normed vector and the distribution of $\mathcal{Y}$ generalizes the concept of uniform distribution on the unit sphere. Similarly, the notion of a unit vector "picked at random" (that is, distributed uniformly on the unit sphere -- an integral geometry concept, as the distribution is invariant under a rotation) is generalized to that of an orthonormal basis "picked at random," or a "random orthonormal basis," or a basis "at random" for short. Let $\mathcal{Z} = (Z^1,\ldots,Z^n)$ be an orthonormal basis in E_n ; we shall say that $\mathcal{Z}$ is at random if its probability distribution is invariant under all orthogonal transformations in E_n or, equivalently, under a change of orthogonal basis in E_n ; hence, this distribution is an integral geometry measure with respect to the orthogonal group. There exists at most one probability measure on $\{\mathcal{Z}\}$, $\mathcal{Z}'\mathcal{Z} = I$, that is invariant under that group. Consider then $\mathcal{X}$ hyperspherical and $\mathcal{Y} = \mathcal{X}_L L^{-1}$; the distribution of $\mathcal{Y}$ is invariant under the orthogonal group and $\mathcal{Y}$ is, therefore, a random orthonormal basis. From the results of the preceding sections on the factorization of the differential element $d\mathcal{X}$ it is seen that a p dimensional random orthonormal basis can be constructed in E_n by taking any hyperspherically distributed matrix $\mathcal{X}$ of dimensions $p \times n$ and normalizing it as $\mathcal{Y} = \mathcal{X}L^{-1}$, and that $\mathcal{Y}$ may be obtained by a Schmidt orthogonalization process. Of special convenience, because of separability, will be to use as $\mathcal{X}$ a sample of normal rows. Further, as $d\mathcal{Y} = \prod_{i=1}^{p} d\,\Sigma_{n-i+1}$, the factorization of $d\mathcal{X}$ shows the probability distribution of $\mathcal{Y}$ as a product, $\prod_{i=1}^{p} d\,\Sigma_{n-i+1} / \Sigma_{n-i+1}$, of uniform distributions on the $n-i+1$ dimensional unit spheres. This provides a parametric representation of the distribution and indicates that a p dimensional random orthonormal basis in E_n can be obtained by taking Y^1 at random in E_n , Y^2 at random in the space E_{n-1} orthogonal to Y^1 , and, more generally, Y^i at random in the space E_{n-i+1} orthogonal to $(Y^1,\ldots,Y^{i-1})$. Note that a random subspace E_p of E_n can be obtained by taking $\mathcal{Y}$ at random and considering all linear combinations of its columns. Analytically, $\mathcal{Y} = H\Gamma$, $H'H = I$, Γ orthogonal; $H = \mathcal{Y}\Gamma^{-1}$, depending on $p(n-p)$ variates, characterizes E_p and its distribution is induced by $\mathcal{Y}$ and Γ taken consistently "at random."

In case $n = p$, $\mathcal{Y}$ is an orthogonal matrix, and an orthonormal basis for the space E_n . One can then use Cayley's parametric representation of $\mathcal{Y}$. If $\mathcal{Y}$ is nonexceptional, that is, if $|I + \mathcal{Y}| \neq 0$, then there exists S skew symmetric, non exceptional, such that $\mathcal{Y} = (I-S)/(I+S)$, $S = (I-\mathcal{Y})/(I+\mathcal{Y})$. If $\mathcal{Y}$ is exceptional, characterized by $|\mathcal{Y}+I| = 0$, $|J_i\mathcal{Y}+I| = 0$, $i = 1$ to $r-1$, $|J_i\mathcal{Y}+I| \neq 0$, $i \geq r$, where $J_1 = (1,1\ldots1)$, $J_2 = (-1,1\ldots1)$, $J_3 = (-1-1,\ldots1)$, etc., then $\mathcal{Y} = J_r(I-S)/(I+S)$, $S = (I-J_r\mathcal{Y})/(I+J_r\mathcal{Y})$. One can then show [7] the factorization of $d\mathcal{X}$, when $\mathcal{X}$ is $p \times p$, as

$$d\mathcal{X} = 2^{-p}2^{p(p-1)/2}|\mathcal{X}'\mathcal{X}|^{-1/2}|I+S|^{-(p-1)}d\mathcal{X}'\mathcal{X}\, dS \, , \tag{22}$$

hence the probability distribution of $\mathcal{Y}$, represented by S , as

$$f(S)dS = \frac{2^{p(p-1)/2}}{\prod_{i=1}^{p} \Sigma_{p-i+1}} \cdot \frac{dS}{|I+S|^{p-1}} \, . \tag{23}$$

One can as well obtain this distribution by invariance arguments. Starting with $\mathcal{Y}$, one operates the "rotation" $\mathcal{Y} \to \mathcal{Y}' = 0$, 0 orthogonal, $\mathcal{Y}, \mathcal{Y}'$, and 0 with representations S , S' , and S_0 . One finds $\frac{D(S')}{D(S)} = \frac{|I+S'|^{p-1}}{|I+S|^{p-1}}$, so that $\frac{dS}{|I+S|^{p-1}}$ is invariant under a "rotation," hence the result. As the $\mathcal{Y}$'s constitute a representation of the orthogonal group, the probability distribution of $\mathcal{Y}$ is a representation of the Haar measure on that group.

2. Thompsonization and Thompson's distributions.

One has seen that the integral geometry on the partition $\{\mathcal{X}'\mathcal{X}\}$ is the integral geometry of the p dimensional bases in E_n .

The $\mathcal{X}'\mathcal{X}$ - Thompsonization of $\mathcal{X}$ is $\mathcal{Y} = \mathcal{X}L^{-1}$ and the conditional, as well as marginal, distribution of $\mathcal{X}$, given $\mathcal{X}'\mathcal{X}$, is provided by the distribution of $\mathcal{Y}$ previously studied. $\mathcal{Y}_\ell$ is a coordinate system on $\mathcal{X}'\mathcal{X} = \ell'\ell$. (It is well known that, under the factorization of $\mathcal{X}'\mathcal{X}$ as $\mathcal{X}'\mathcal{X} = \beta'\beta$, β not triangular, the results described so far would not hold true for the reduced form $\mathcal{Y}_\beta = \mathcal{X}\beta^{-1}$, as $\{\beta\}$ is not $\sim \{\mathcal{X}'\mathcal{X}\}$, modulo $\{\mathcal{X}'\mathcal{X}\}$. It can be checked by direct computation that $\frac{d\sigma_{\mathcal{X}'\mathcal{X}}}{K(\mathcal{X}'\mathcal{X})} = \frac{d\mathcal{Y}}{\prod_{i=1}^{p}\Sigma_{n-i+1}} = \frac{dS_{\mathcal{X}'\mathcal{X}}}{S_{\mathcal{X}'\mathcal{X}}}$, as $|\text{grad}\, \mathcal{X}'\mathcal{X}|$ depends only on $\mathcal{X}'\mathcal{X}$.

One may wish to obtain a representation of this integral geometry distribution in terms of $(X_1,\ldots,X_{n-1})'$ ($\mathcal{X}'\mathcal{X}$ - Thompson distribution of order $n-1$) and, more generally, the conditional distribution given $\mathcal{X}'\mathcal{X}$, of a submatrix $(X_1,\ldots,X_k)'$ ($\mathcal{X}'\mathcal{X}$ - Thompson distribution of order k). The latter provides the minimum variance unbiased estimator, MVUE, of the joint distribution of $\xi = (X_1,\ldots,X_k)'$, whatever the hyperspherical $\mathcal{X}$ considered; it provides in particular, the MVUE of the distribution of a sample of k observations of a normally distributed p-dimensional vector. By use of the methods described previously (use of the separated family or use of the $d\mathcal{X}$ factorization) one readily obtains, when $n-k \geq p$,

$$f(\xi|\mathcal{X}'\mathcal{X})d\xi \sim \frac{|\mathcal{X}'\mathcal{X} - \xi'\xi|^{(n-k-p-1)/2}}{|\mathcal{X}'\mathcal{X}|^{(n-p-1)/2}} d\xi \qquad (24i)$$

$$\sim |I - (\mathcal{X}'\mathcal{X})^{-1}\xi'\xi|^{(n-k-p-1)/2}|\mathcal{X}'\mathcal{X}|^{-k/2}d\xi \qquad (24ii)$$

$$\sim |I - \xi(\mathcal{X}'\mathcal{X})^{-1}\xi'|^{(n-k-p-1)/2}|\mathcal{X}'\mathcal{X}|^{-k/2}d\xi\ , \qquad (24iii)$$

all defined on the domain where $\mathcal{X}'\mathcal{X} - \xi'\xi$ is positive definite (the sign $\sim$ indicates proportionality). One may also Thompsonize ξ as $\eta = \xi L^{-1}$, with the corresponding pdf in reduced form:

$$f(\eta)d\eta \sim |I - \eta\eta'|^{(n-k-p-1)/2}d\eta \quad , \qquad (25)$$

valid if $k \geq p$. The pdf of $\xi'\xi$ or $\eta'\eta$ (if $k \geq p$) is, of course, a matric multivariate Beta distribution:

$$f(\eta'\eta) \sim |I - \eta'\eta|^{(n-k-p-1)/2}|\eta'\eta|^{(k-p-1)/2} \ . \qquad (26)$$

One sees that, given $\mathcal{X}'\mathcal{X}/n = t$, when n tends to infinity, the pdf of ξ tends to that of a sample of k vectors normally distributed with covariance matrix t; therefore, the multivariate normal distribution is asymptotically induced by the integral geometry of the partition $\mathcal{X}'\mathcal{X}$.

3. Matric fractional integration and Hankel transforms.

In view of the factorization of $d\mathcal{X}$ with respect to $V = D\mathcal{X}$ and $\mathcal{X}'\mathcal{X}$, where D is $k \times n$ or rank k, the joint pdf of V and

$$f(\mathcal{X}'\mathcal{X},V)d\mathcal{X}\mathcal{X}dV \sim g(\mathcal{X}'\mathcal{X})|DD'|^{-p/2}|\mathcal{X}'\mathcal{X} - V'(DD')^{-1}V|^{(n-k-p-1)/2}d\mathcal{X}'\mathcal{X}dV, \qquad (27)$$

on the domain where the matrix of the determinant is positive definite. This result is the Kabe Lemma [5], 1961, which generalizes the Sverdrup lemma to the multivariate case and whose great importance in multivariate analysis has been shown by Kabe. With, or without ([7], 1959) the use of that lemma one obtains the joint pdf of $\mathcal{X}'\mathcal{X}$ and ξ , $n-k \geq p$, as

$$f(\mathcal{X}'\mathcal{X},\xi)\,d\mathcal{X}'\mathcal{X}\,d\xi \sim |\mathcal{X}'\mathcal{X} - \xi'\xi|^{(n-p-k-1)/2} d\mathcal{X}'\mathcal{X}\,d\xi , \tag{28}$$

therefore, the pdf of ξ as

$$f_k(\xi) = g_k(\xi'\xi) \sim \underset{|\mathcal{X}'\mathcal{X} - \xi'\xi| \geq 0}{\int\cdots\int} |\mathcal{X}'\mathcal{X} - \xi'\xi|^{(n-p-k-1)/2} g_n(\mathcal{X}'\mathcal{X})\,d\mathcal{X}'\mathcal{X} . \tag{29}$$

One sees, therefore, that f_k , which is the projection $f_k = \mathbb{P}_{n,k} f_n$ of $f_n(\mathcal{X}'\mathcal{X})$, is obtained by an integral transform that is a matric generalization of the Weyl transform. This suggests the creation of a matric fractional calculus (note, however, that the case $n - k < p$ requires a separate treatment). In particular, solving in g_n the integral equation above for a given g_k , when such a solution exists, can be interpreted as taking a "derivative" $D_{p,(n-k)/2}$ of order $(n-k)/2$ and dimension p . Such a derivative is an operator of anti-projection and makes it possible, when this is consistent, to generalize hyperspherical distributions of matrices of dimension $k \times p$ to that of dimension $n \times p$; for example, vector models to matrix models. One may define projections by a non integral number of dimensions by considering the operator $D_{p,-\lambda}$, λ not an integer. We have started investigating such a calculus.

If one considers the transform calculus which is used for spherical distributions, one realizes that, in the multivariate case, properties of invariance under projection of some ad hoc "characteristic function" of $\mathcal{X}$ should be used to build a similar calculus. This leads to the -- somewhat classical -- definition of the "characteristic function" of $\mathcal{X}$ as $\phi_{\mathcal{X}}(V) = E[\exp(i \text{ trace } \mathcal{X}V)]$, where V is $p \times n$. The properties of invariance of $\phi_{\mathcal{X}}(U)$ under the transforms $0_1\mathcal{X}$, $\mathcal{X}0_2$, 0_1, 0_2, orthogonal, show that $\phi_{\mathcal{X}}(V) = \Phi_n(VV') = \Phi_n(V'V)$, a characterization of hypersphericity. The characteristic function of $\xi = (X_1,\ldots,X_k)'$ is obtained by replacing by zeros the proper elements of V in $\phi_{\mathcal{X}}(V)$ and is, therefore, invariant undre projection, when considered as a function of the generic argument VV' (or $V'V$). One can write $\phi(V) = I_k\Phi(VV')$, where I_k is an operator to indicate the dimensionality of VV'. One has

$$\phi_{\mathcal{X}}(V) = E_{\mathcal{X}'\mathcal{X}}[\phi_{Dir.\mathcal{X}'\mathcal{X}}(V)] = \int\cdots\int \phi_{Dir.\mathcal{X}'\mathcal{X}}(V)h(\mathcal{X}'\mathcal{X})d\mathcal{X}'\mathcal{X}, \qquad (30)$$

where $h(\mathcal{X}'\mathcal{X})$ is the pdf of $\mathcal{X}'\mathcal{X}$, for $n \geq p$, namely,

$$h(\mathcal{X}'\mathcal{X}) = \prod_{i=1}^{p} \Sigma_{n-i+1}\, g(\mathcal{X}'\mathcal{X})\,|\mathcal{X}'\mathcal{X}|^{(n-p-1)/2},$$

and

$$\begin{aligned}\phi_{Dir.\mathcal{X}'\mathcal{X}}(V) &= (1/\prod_{i=1}^{p}\Sigma_{n-i+1})\int\cdots\int \exp(i\ \text{trace}\ LV)d\mathcal{X} \\ &= (1/\prod_{i=1}^{p}\Sigma_{n-i+1})W(LV), \text{ say.}\end{aligned} \qquad (31)$$

Now, $W(LV)$ is clearly invariant under pre and post multiplication of the argument by orthogonal matrices, and is a function $W^*(V'L'LV) = W^*(V'\mathcal{X}'\mathcal{X}V)$ of V and $\mathcal{X}'\mathcal{X}$. It should be considered a generalized Bessel function as it is the generalization of the average of $\exp[i(t,X)]$ on the unit sphere (the characteristic function of the spherical Dirac distribution), which appears in the study of spherical distributions. It is a Bessel function of argument LV, which can be expressed as a function of $V'\mathcal{X}'\mathcal{X}V$, is a symmetric function of the eigenvalues of the latter, and is an analytic function of the symmetric functions of the eigenvalues of $V'\mathcal{X}'\mathcal{X}V$. The coefficients of the corresponding expansion can be obtained, by different methods, by way of recursive relationships. Methods that use coordinate representations destroy the symmetry of the problem and a procedure of symmetrization has then to take place in order to obtain symmetric formulae. Symmetric methods, directly or indirectly, use the fact that, as $\mathcal{Y}'$ is at random in the E_n space, $\Delta\phi_{Dir.\mathcal{X}'\mathcal{X}}(V) + \phi_{Dir.\mathcal{X}'\mathcal{X}}(V) = 0$, where Δ, the Laplace operator, is taken with respect to Y^1. At any rate, $\phi_{\mathcal{X}}(V)$, at least for $n \geq p$, appears as a matric Hankel transform of $g_n(\mathcal{X}'\mathcal{X})$, which suggests the study of such generalized transform and its relationships with the previously suggested matric fractional differentiation and integration calculus. A unified calculus has not, as yet, been obtained (note that the case $n < p$ requires separate treatment).

4. Applications. Multivariate linear model. Regression residual analysis.

Among the most immediate and direct applications of the matric Thompson distributions, described in previous sections, are the study of the random motion of a p dimensional body in an n dimensional euclidean space; the derivation of confidence regions, prediction regions, and tolerance regions in multivariate analysis; the obtention of MVU estimators, such as those of the probability masses of subsets of the p dimensional space in case the probability density is multivariate normal. Applications to bombing and firing theory were given in [9] and other papers. Other immediate applications are the construction of uniformly most powerful nuisance parameter free tests of hypotheses when such tests exist, the testing of whether a matrix is hyperspherically distributed and, in case the independence of the observations is granted, the testing of multivariate normality. Characterizaton conjectures are also suggested: "if $\mathcal{X}$ is random of dimension $n \times p$, with independent rows and if $\mathcal{Y} = \mathcal{X} L^{-1}$ is a random basis, then the rows of $\mathcal{X}$ are multivariate normal."

As the "errors" $\mathcal{X} - \mathcal{B}\mathcal{Z}$ (where $\mathcal{X}$ is $p \times n$, $\mathcal{B}$ is $p \times q$, $\mathcal{Z}$ is a $q \times n$ matrix with sure columns, $n \geq p + q$, rank $\mathcal{Z} = q$) of a multivariate normal linear model with covariance matrix Σ are hyperspherically distributed, the preceding results have an impact on the study of the multivariate linear model. The maximum likelihood estimators $\hat{\beta}$ and $\hat{\Sigma}$ are a set of AS statistics for the distribution. The conditional pdf of a subsample ξ of size $k \leq n\text{-}p\text{-}q$, given the AS statistics $\hat{\beta}$, $\hat{\Sigma}$, is

$$f(\xi|\hat{\beta},\hat{\Sigma}) \sim |I-(n\hat{\Sigma})^{-1}(\xi-\hat{\beta}Z_{\xi})A(\xi-\hat{\beta}Z_{\xi})'|^{(n-k-p-q-1)/2} \tag{32}$$

and, with $n\hat{\Sigma} = LL'$, L triangular with positive diagonal, the Thompsonized ξ is $\eta = L^{-1}(\xi - \hat{\beta}Z_{\xi})$, with conditional and marginal pdf

$$f(\eta) \sim |I - \eta A \eta'|^{(n-k-p-q-1)/2}, \tag{33}$$

where $A = [I + Z_{\xi}'(ZZ' - Z_{\xi}Z_{\xi}')^{-1}Z_{\xi}]$, and Z_{ξ} is the submatrix of $\mathcal{Z}$ corresponding to ξ. In case $k \geq q$, and with obvious notations

$$f(\xi|\hat{\beta},\hat{\not{Z}}) \sim |I - (k/n)\hat{\not{Z}}^{-1}\hat{\not{Z}}_{\xi} - (1/n)\hat{\not{Z}}^{-1}(\hat{\beta}_{\xi} - \hat{\beta}) E(\hat{\beta}'_{\xi} - \hat{\beta}')|^{(n-k-p-q-1)/2} \qquad (34)$$

where $E = Z_{\xi}Z'_{\xi}(ZZ' - Z_{\xi}Z'_{\xi})^{-1}ZZ'$. Several equivalent forms exist for these distributions. In case β is known and $\hat{\not{Z}}$ denotes the maximum likelihood estimator of $\not{Z}$, these formulae hold true, with $\hat{\beta}$ replaced by β and $q = 0$. These results were given in [8], 1960, where detailed derivations can be found, and reproduced in [9]. They give the probability distribution of the sample residuals of the multivariate linear model, should the errors be hyperspherical, and, as a special case, the distribution of the sample residuals of the classical regression on sure variables model. They provide, of course, in a rather straightforward way, prediction regions for "future" observations, characterization of predictors, and the like, and make it possible to proceed to the analysis of regression residuals, a fact that seems to have been overlooked in the literature of the past fifteen years.

References

[1] Anderson, T. W. (1958). Introduction to Multivariate Statistical Analysis, Wiley, New York.

[2] Barnard, G. A. (1971). Foundations of Statistical Inference, Godambe, V. P. and Sprott, D. A., edits., Holt, Rinehart and Winston, Toronto, p. 415.

[3] Deemer, W. L. and Olkin, I. (1951). The Jacobians of certain matrix transformations useful in multivariate analysis, Biometrika, 28, 345-367.

[4] James, A. T. (1954). Normal multivariate analysis and the orthogonal group, Ann. Math. Statist., 23, 40-75.

[5] Kabe, D. G. (1961). On the generalization of Sverdrup's lemma and its applications to multivariate distribution problems in statistics, W.S.U. Memorandum, 1-36, Wayne State University, Detroit.

[6] Kabe, D. G. (1965). Generalization of Sverdrup's lemma and its application to multivariate distribution theory, Ann. Math. Statist., 36, 671-676.

[7] Laurent, A. G. (1959). Minimum variance unbiased estimates, Generalization of Thompson's distributions, Random orthogonal bases. W.S.U. Technical Report No. 5, NONR-2575(00), 1-34, Wayne State University, Detroit.

[8] Laurent, A. G. (1960). Generalization of Thompson's distributions III, Inst. Math. Statist., April 7-9, 1960, Meeting Handout, W.S.U. Memorandum, 1-4, Wayne State University, Detroit.

[9] Laurent, A. G. (1962). Bombing problems, a statistical approach II, Operations Research, 10, 380-387.

[10] Laurent, A. G. (1973). Remarks of multivariate models based on the notion of structure, in Multivariate Statistical Inference, Kabe, D. G. and Gupta, R. P., edits., North Holland, Amsterdam, 145-168.

[11] Laurent, A. G. (1974). "Applications of fractional calculus to spherical (radial) probability models and generalizations," New Haven Conference on Fractional Calculus, Proceedings.

[12] Mahalanobis, P. C., Bose, R. C. and Roy, S. N. (1937). "Normalization of statistical variates and the use of rectangular coordinates in the theory of sampling distributions," Sankhyā, 3, 1-40.

[13] Olkin, I. (1951). "On distribution problems in multivariate analysis," Institute of Statistics, Mimeo. Series No. 43.

[14] Piaget, J. (1968). Le Structuralisme. P.U.F., Paris.

[15] Thomas, D. J. (1970). Some contributions to radial probability distributions, statistics, and the operational calculi. Ph.D. dissertation. Wayne State University, Detroit.

A. G. Laurent
Department of Mathematics
Wayne State University
Detroit, Michigan 48202
U.S.A.

Applied Statistics, R.P. Gupta, (editor)

NOTE ON ESTIMATION OF PLUME AND PATCH DISPERSION SCALES

by R.H. Loucks, D.J. Lawrence and D.V. Ingraham

Introduction

The purpose of this note is to describe a borrowed technique for estimating plume dispersion from ocean current meter time series data. After the procedure itself is explained, subsequent sections are devoted to discussion of the difficulties with the presence of waves and to the interpretation of 'Range Analysis' in terms of patch diffusion.

Procedure

Initially the procedure was to compile distributions of 'distances made good' and interpret the standard deviation of the distribution as the standard deviation of plume envelope width. This, we find, is equivalent to operating on the velocity signal with the 'variance time function' or VTF (Mandelbrot, 1972). For a velocity component time series $V(t)$, let $V^*(t) = \sum_{s=1}^{t} V(s)$. Then

$$VTF(T) = E[V^*(t+T) - V^*(t)]^2 = \text{var } V^*(T)$$

Mandelbrot has pointed out that the VTF has the form of Taylor's formula for diffusion scale -

$$VTF(T) = T^*C(0) + 2 \sum_{\sigma=1}^{T-1} \sum_{s=1}^{\sigma} C(S)$$

where $C(s)$ is the autocovariance function of $X(t)$ with lag, s, and $VTF(T) = \text{var } V^*(T)$ is the mean square scale of (plume) diffusion after elapsed time, T.

Pasquill (1962) has described another version of Taylor's formula

$$\text{var}_\tau V^*(T) = \left(\text{var}_\tau \overline{V(t)}^T\right) \cdot T^2$$

where τ is sampling time and T is averaging time and age of the plume.

A modification inserts the factor, β , to compensate for the Eulerian nature of the measurements being used to estimate a Lagrangian effect. Thus

$$\mathrm{var}_{\tau}\ V^{*}(T) = \left[\mathrm{var}_{\tau}\ \overline{V(t)}^{T/\beta}\right] \cdot T^{2}$$

where β = ratio of Lagrangian to Eulerian time scales.

In essence then, our procedure is to compile velocity distributions of various averaging times, extract the variance and multiply by T^2 to yield the squared scale of the plume envelope. We typically sample blocks of 2500 data points of data interval 5, 10 or 20 minutes and perform averaging of up to 288 data point strings to produce estimates of the mean square scale of plume diffusion.

Results

Sample results are shown in Figure 1. Data from a series of dye releases in St. Margaret's Bay, Nova Scotia (from Sharaf El Din *et al.* 1970) are included on a graph containing Okubo's (1971) average oceanic diffusion curve as well as plume and patch characteristics. The points depict scale diameters of the dye patches observed on three of the eight days over which the current meter data were sampled and these observations are of the order of magnitude of Okubo's curve. (It was assumed that Okubo's $\sigma_{rc} = \frac{1}{3}{*}\ r_e$, where r_e is the equivalent radius. Thus $\sigma_{rc} = \frac{1}{3}\sqrt{\frac{\text{area}}{\pi}}$.) These patches were released and observed in the neighbourhood of the moored current meter.

The 50 percentile plume characteristic (E) was calculated from the VTF operating on residual data (from which the tidal harmonics have been removed by subtraction). For comparison the 50 percentile plume characteristic (T) was calculated using unaltered current meter data and thus includes the tide.

The patch characteristic (A) is discussed in a later section.

Interpretation

The VTF analyses are interpreted to portray the envelope of an ensemble of plumes due more to the wafting than to diffusion, i.e. due more to the fluctuations of the plume than to the spreading of its elements. Thus the results show only qualitative corroboration of the technique: the 50 percentile plume characteristics are at larger scales than the majority of the dye patch observations and their slopes are approximately the same as for the dye patch observations.

To maintain this rational relationship, the VTF analysis should only be used with a sampling time much larger than (say 10 times) the averaging time. Otherwise the wafting energy may not dominate the diffusion contribution. This is especially important where wave motion is present because waves can legitimately contribute to wafting but give misleading results when included in the diffusion contribution. The results show the enhanced plume envelope scales when tidal wafting is included. The residual curve serves as a lower bound on plume envelope scale.

The 'diffusion' coefficient calculated as $K = \frac{\sigma_{rc}^2}{4T}$ yields the value 2×10^3 cm^2/sec after one hour which is comparable to experimental results of Bowden and Lewis (1973) for plumes at sea.

An interpretation for range analysis in terms of diffusion.

Mandelbrot (1969, 1972) has introduced range R as

$$R(t,T) = (\text{max-min})_{1 \le u \le T} \left(V^*(t+u) - V^*(t) - \frac{u}{T}[V^*(t+T) - V^*(t)] \right)$$

Whereas VTF analysis can be interpreted as plume envelope scale due to wafting, range analysis can be interpreted as patch ensemble scale due to diffusion (provided wave motions are discriminated against). That is, we distiguish between the wafting of the end of the T-point trend line (average distance-made-good) and the range of deviations of distances-made-good from each trend line.

More Results.

Examining Figure 1 again the patch characteristic (A) is the range analysis of the 'appropriate' St. Margaret's Bay current meter data. This characteristic yields scales below those observed. A reason for this may be that the instrument was moored 8 to 10 meters below the surface whereas dye observations were made at the surface. In a water body where wind stress is the dominant driver the attenuating effect of depth would be anticipated to influence the small scale currents of the patch characteristic more than the larger scale contributors of the plume characteristic.

Figure 2 shows data from Head Harbour Passage a tidally dominated channel (the Passamaquoddy region of the Bay of Fundy (Department of the Environment, 1973)). In contrast to a wind-driven body of water, in the tidally-dominated channel, attenuation of turbulence with increasing depth would not be expected. Thus we see the patch and the 50 percentile plume characteristics at more nearly the same scales. (Dye data are not available to accompany these current-meter data.)

The range characteristic (P) for the tide signal grows only slowly for durations exceeding the tidal period. We note this same typical tidal character (the break to a lower slope at the tidal period) still present in the residual patch characteristics. (This residual record is contaminated with tides because the subtracted tidal constants were obtained by averaging over three month's records and so do not 'fit best' at any time.)

The plume characteristic shows a dip at the tidal period. This is also typical of the tidal signal.

Figure 3 shows the major component data taken from Station 4, Head Harbour Passage, at a depth of 5 to 13 metres. The patch characteristic and the 50 percentile plume characteristic do approach the same level. Note that influence due to the tides is discernible in both.

Figure 4 presents the minor component of the same data. In the minor direction, we have a 'near ideal' example of plume and patch characteristics showing the same asymptotic slope.

Figures 5 and 6 show a comparison of patch scales from dye observations and range analysis using data from Lake Ontario kindly made available by Dr. C.R. Murthy of the Canada Centre for Inland Waters.

The dye used (Murthy *et al*, 1974) was for two day patches (ED1 and ED6) at depths of 18 to 26 m and 25 to 50 m respectively. Current meters were moored within a few kilometers at depths of 10, 15, and 50 m and 15, 25, 35 m respectively. Since range is computed on the Cartesian components independently, the appropriate scales from the dye observations are based on the Cartesian component variances, S_x and S_y (using Murthy *et al's* notation, and values from their Table 6) rather than the two-dimensional Cartesian variance S_{xy}, or the radially symmetric polar variance, S_{rc}, where $S_{xy}^2 = S_x^2 + S_y^2$ and $S_{rc}^2 = 2S_xS_y$. From the definition of range, it follows that this is an upper bound patch scale. Therefore we have chosen to estimate the scale of the observed patch as six times S_x or S_y, which in the case of Gaussian distribution axially would encompass 99.7% of the dye.

This comparison of Figures 5 and 6 is superior to that of Figure 1 because the dye and current observations are at similar depths rather than being separated by a sheared layer. The discrepancies between dye observations and range estimates calculated from current meter data are sufficiently small to suggest that the range technique does indeed have applicability as an engineering tool for estimation of patch dispersion. The attempt to verify VTF analysis for plumes dispersion is in progress.

TABLE OF DIFFUSION THEORIES

(Generalized coefficient of diffusion is $ar^{m}f(t)$.)

Investigator	K
Fickian Diffusion	a
Joseph and Sendner (1958)	ar
Ozmidov (1958)	$ar^{4/3}$
Okubo and Pritchard (1960)	ar
Okubo (1962)	$ar^{5/3}$
Obukhov (1959)	$ar^{4/3}$
Fractional Gaussian noise	$ar^{\frac{2H-1}{H}}$ $(0.5 \leq H \leq 1)$

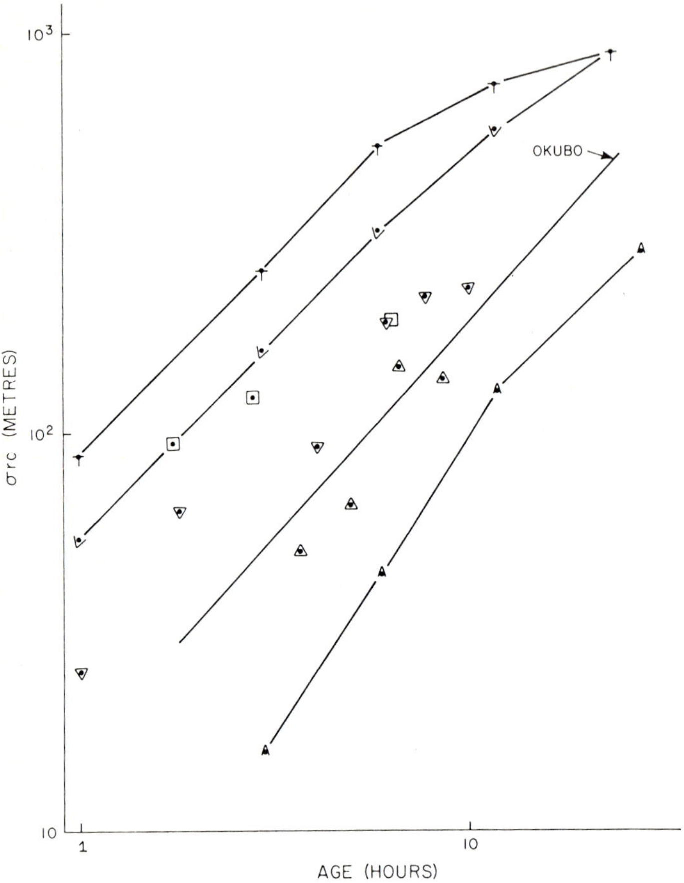

Figure 1. Dye patch observations and plume and patch (current meter) characteristics from Station 17, 8 m depth, St. Margaret's Bay, June 20-28, 1967.

L marks 50%-ile plume characteristic with tides removed.
T marks 50%-ile plume characteristic with tides included
A marks patch characteristic with tides removed.

⊡ marks dye release of June 26.
▲ marks dye release of June 27.
▼ marks dye release of June 28.

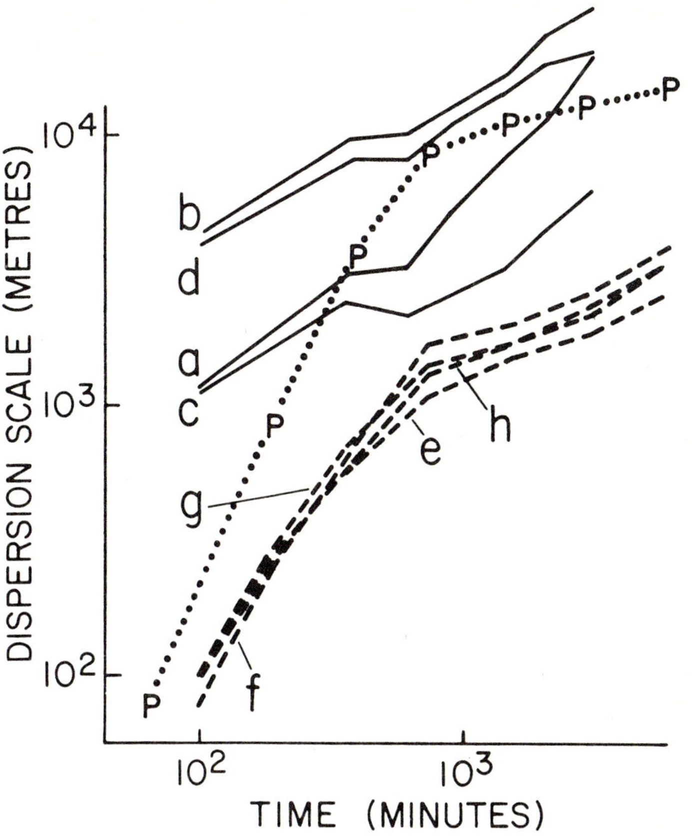

Figure 2. Plume and patch characteristics from current meter station 3, Head Harbour Passage, Bay of Fundy, depth 5 m below low water, longitudinal component, starting date Aug. 23,1973, data interval 10 minutes.

:a marks 50%-ile plume characteristic; data points #1 to #2500.
:b marks 96%-ile plume characteristic; data points #1 to #2500.
:c marks 50%-ile plume characteristic; data points #2500 to #4838.
:d marks 96%-ile plume characteristic; data points #2500 to #4838.

:e marks patch characteristic; data points #1 to #1177.
:f marks patch characteristic; data points #1178 to #2365.
:g marks patch characteristic; data points #2366 to #4838.

:p marks patch characteristic derived from the pure tidal signal as a test.

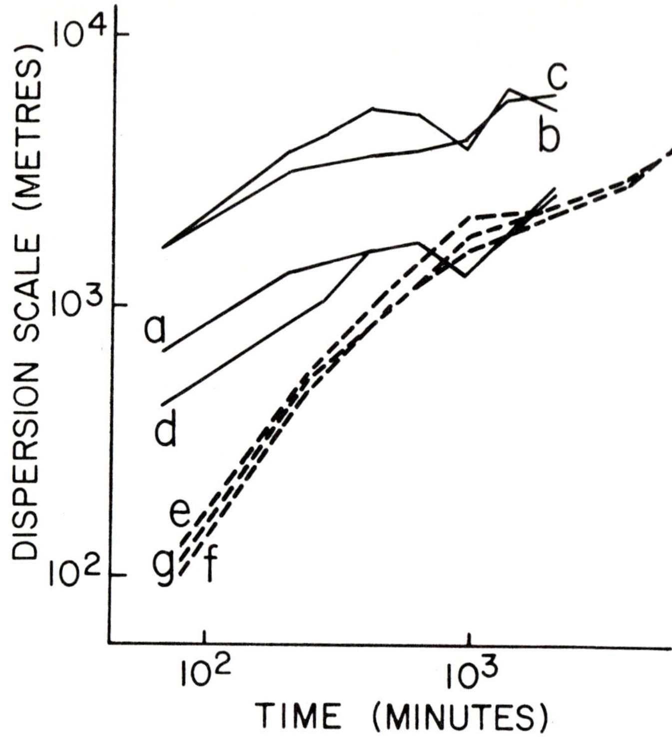

Figure 3. Plume and patch characteristics from current meter station 4, Head Harbour Passage, Bay of Fundy, depth 5 m below low water, longitudinal component, starting date Aug. 23, 1973, data interval 5 minutes.

:a marks 50%-ile plume characteristic; data points #1 to #250
:b " 96%-ile " " " " " " "
:c " 50%-ile " " " " #2501 to #
:d " 96%-ile " " " " " "

:e " patch characteristic; data points #1 to #2378.
:f " " " " " #2379 to #4765.
:g " " " " " #4766 to #7153.

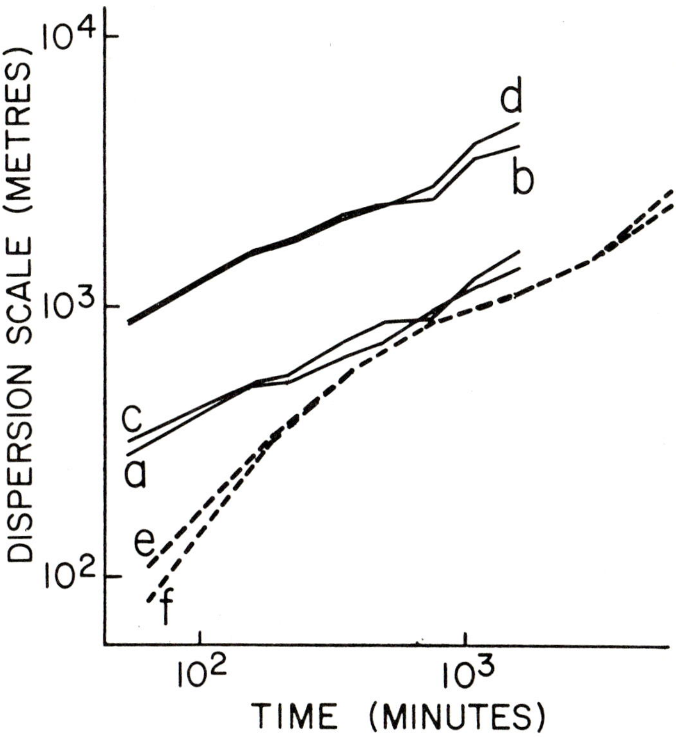

Figure 4. Plume and patch characteristics from current meter station 4 Head Harbour Passage, Bay of Fundy, depth 5 m below low water, transverse component, starting date Aug. 23, 1973, data interval 5 minutes.

:a marks 50%-ile plume characteristic; data points #1 to #2500.
:b " 96%-ile " " " " " " "
:c " 50%-ile " " " " #2500 to #4999.
:c " 96%-ile " " " " " " "

:e " patch characteristic; data points #1 to 2377.
:f " " " " " #2378 to #4765.

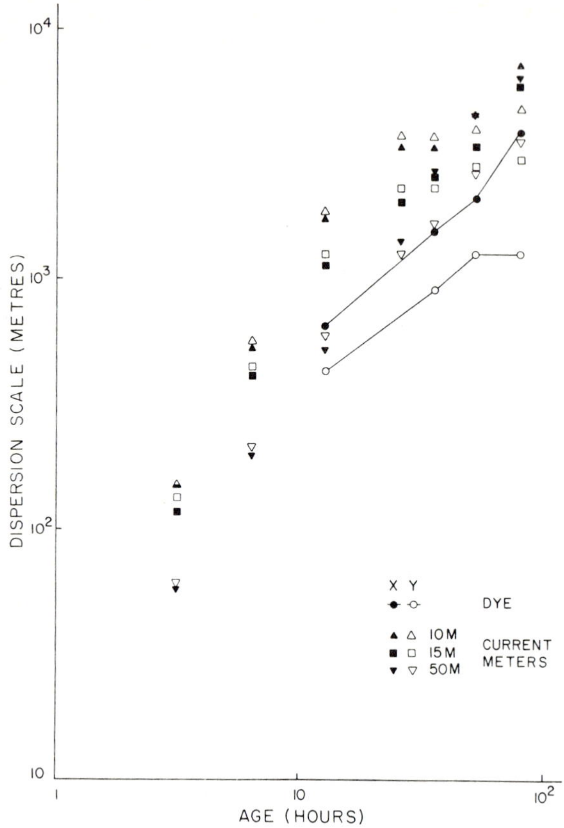

Figure 5. Comparison of dye patch and current meter dispersion scales from Lake Ontario. Dye data from Murthy *et al*, 1974, experiment ED-1, 8 kg dye released 15 August,72, tracked for 79 hours at depths of 18 - 26 m, drift direction (=X axis) 250°. Dispersion scale taken to be 6 x std. deviation of observed concentration distributation or 2 x length scale (Murthy 1974, Table 6). Current meter data (unpublished), from CCIW mooring no. 2-OC-6B, depths 10, 15, 50 m. Points no. 5400 - 6750, covering 01 hrs 12 Aug. - 11 hrs 21 Aug. 72 were subject to range analysis at BIO, with X axis direction = 270°.

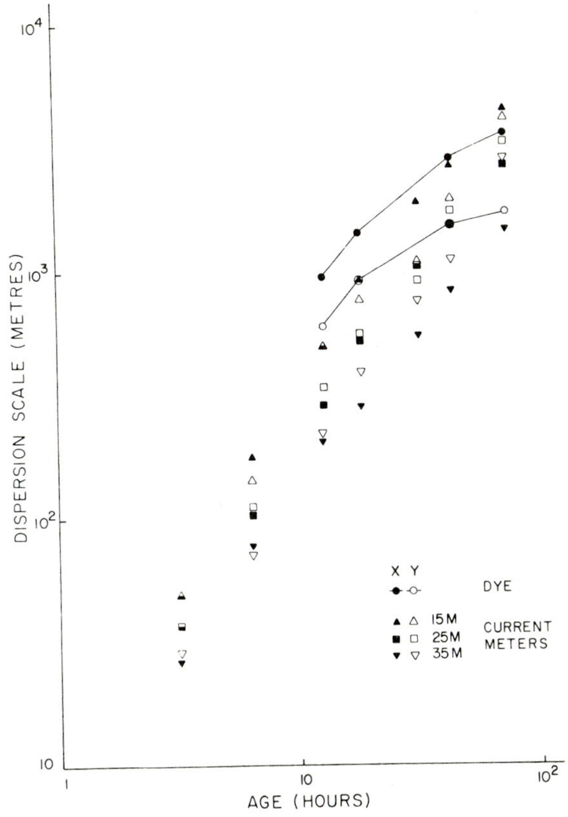

Figure 6. Comparison of dye patch and current meter dispersion scales from Lake Ontario. Dye data from Murthy *et al*, 1974, experiment ED-6, 12 kg dye released 17 Oct. 72, tracked for 72 hours at depths of 25 - 50 m, drift direction (= X axis) 230° - 360°. Dispersion scale taken to be 6 x std. deviation of observed concentration distribution or 2 x length scale (Murthy 1974, Table 6). Current meter data (unpublished), from CCIW mooring no. 2-OC-71A, depths 15, 25, 35 m. Points no. 1 - 1260, covering 20 hrs 16 Oct. - 14 hrs 25 Oct. 72 were subject to range analysis at BIO, with X axis direction = 270°.

References

[1] Bowden, K.F., and R.E. Lewis. (1973), Dispersion in flow from a continuous source at sea, Water Research, 7: 1705-1722.

[2] Department of the Environment. (1973), Summary of physical, biological, socioeconimic and other factors relevant to potential oil spills in the Passamaquoddy region of the Bay of Fundy - preliminary report. (Slated for subsequent release as Technical Report N0. 428 of the Fisheries Research Board of Canada.)

[3] Loucks, R.H., D.J. Lawrence, D. Ingramam, and B. Flemming. (1973) A technique for estimating extreme ocean current vectors, Bedford Institute of Oceanography Report Series No. BI-R-73-5, Dartmouth, Canada.

[4] Mandelbrot, B.B. (1972). Statistical methodology for nonperiodic cycles: from the covariance to R/S analysis, Annals of Economic and Social Measurement, 1: 259-290.

[5] Mandelbrot, B.B. (1971). A fast fractional noise generator, Water Resources Research, 7: 543-553.

[6] Mandelbrot, B.B., and J.R. Wallis, (1969), Some long-run properties of geophysical records, Water Resources Research, 5 (2): 321-340.

[7] Murthy, C.R., G. Kullenberg, H. Westerberg, and K.C. Miners, (1974), Large scale diffusion studies, (IFYGL Project 89 WM), Canadian Centre for Inland Waters, Paper No. 14, Burlington, Ontario, Canada.

[8] Neumann, G., and W.J. Pierson, JR. (1966). Principles of Physical Oceanography. Prentice-Hall, Inc., Englewood Cliffs, New Jersey, 545 pp.

[9] Okubo, A. (1971). Oceanic diffusion diagrams, Deep-Sea Research, 18: 789-802.

[10] Pasquill, F. (1962). Atmospheric Diffusion. Van Nostrand, London, 297 pp.

[11] Sharaf El Din, S.H., E.M. Hassan, and R.W. Trites. (1970). The physical oceanography of St. Margaret's Bay, Fish. Res. Bd. Canada, Tech. Rep. #219, 242 pp.

R.H. Loucks, D.J. Lawrence and D.V. Ingraham
Coastal Oceanography Division
Bedford Institute
Dartmouth, N.S.
Canada

Applied Statistics, R.P. Gupta, (editor)

ESTIMATION OF FINITE DISTRIBUTION MIXTURES

P.D.M. Macdonald

ABSTRACT

Over the past 80 years a bewildering variety of methods have been proposed for statistically estimating the parameters of distribution mixtures. Even in the best-behaved case, when only the mixing proportions are unknown, different authors have derived estimators in many different ways. This paper shows that a number of these estimators are just special cases, or limiting cases, of weighted least-squares. I have found that a weighted least-squares estimator works well in practice; for example, in the analysis of some field data to determine seasonal growth of minnows.

1. INTRODUCTION

A distribution mixture arises when a population, assumed to be infinite, is made up of component populations mixed together in fixed proportions, and the random variable of interest follows a different distribution in each of the component populations. The problem is to "decompose" the mixture by estimating any unknown parameters of the component distributions and mixing proportions, usually given only a sample of observations from the mixed population.

A finite distribution mixture is one with a finite number of components. Let k be the number of components, let F_i denote the distribution function of the random variable in the ith component, and let $\pi_i \geq 0$ be the proportion of the ith component in the mixture $(i=1,\ldots,k)$. The mixed distribution function can then be written $G = \pi_1 F_1 + \ldots + \pi_k F_k$, where $\pi_1 + \ldots + \pi_k = 1$.

A bibliography which M. Brisebois and I are preparing now lists over 200 references directly concerned with finite mixtures, plus an even greater number on infinite mixtures, continuous mixtures, and other more general topics. Fig. 1 shows how the papers on finite mixtures have proliferated over the past 80 or 90 years. A possible explanation for this proliferation is that mixtures problems are encountered by applied statisticians in many different contexts and the problems often appear, at first glance, sufficiently straightforward to be tackled without a thorough search of the literature. For example, the first paper in a fisheries journal (Buchannan-Wollaston & Hodgson, 1929) proposed a graphical analysis using probability paper and, for many years, there was a tendency to consider only graphical methods in fisheries research (for instance, Tanaka, 1962; Bhattacharya, 1967).

Despite the extensive literature, there is still no reliable, general-purpose estimation procedure accessible to users who are not specialists in statistics. Many of the more mathematical papers need to be supplemented by worked examples that would encourage users to try the methods. However, a recent tendency, to illustrate estimation methods with computer-generated data, often gives potential users a false impression of the reliability of the methods, as most estimators are very sensitive to departures from the distributional assumptions. In addition, iterative computer algorithms and graphical methods are notorious for working for no data other than the original author's example.

The difficulty of the estimation problem depends on the number of components, the extent to which they overlap, and the particular combination of parameters to be estimated. When the component distributions are completely specified and only the mixing proportions are to be estimated, there are many estimators that will work, but maximum likelihood will generally give the best results (Macdonald, 1971). Some of these estimators are discussed in §§2 and 3 of this paper. Maximum likelihood will also work well for estimating the location parameters and mixing proportions of a mixture of distributions with known dispersions, especially when the component distributions are separated well enough to show distinct modes. No estimator can be expected to decompose a mixture for which location and dispersion parameters are unknown, unless the component distributions show almost no overlap. Even in the simple case of two univariate normal components, with the mixing proportions and one variance known, and a common mean and the remaining variance to be estimated, Kiefer & Wolfowitz (1956) noted that every observation in the sample generates a singularity in the likelihood surface. Moment estimators may be preferable in such cases (Cohen, 1967), but only if the number of unknown parameters is small. Gridgeman (1970) and Tan & Chang (1972) have examined the efficiency of moment estimators. Maximum likelihood will often give good estimates, however, if sufficient constraints are imposed on the variances of the component distributions; for example, if the variances are assumed to be equal (Macdonald, 1969). Hosmer (1973) and Dick & Bowden (1973) used maximum likelihood to decompose a mixture of two normal distributions, given independent samples from one (or both) component distributions separately. Day (1969) found maximum likelihood to work well for mixtures of multivariate normal distributions.

Kabir (1968) and Saleh (unpublished) devised two-step procedures whereby parameters of the component distributions are estimated first, independently of the mixing proportions, then the mixing proportions are estimated conditionally. Kabir's method applies to mixtures of distributions from the exponential family, with just one unknown parameter per component, and for these cases maximum likelihood would generally be preferable. Saleh's method is derived from minimum chi-square using grouped data, and gives BAN estimates (Neyman, 1949). Rao (1948), Chapman (1952), Hasselblad (1966) and Rennie (1968) have also advocated BAN estimates computed from grouped data.

This paper describes a class of weighted least-squares estimators for the mixing proportions, using either grouped or ungrouped data, and derives, as special cases, a number of estimators proposed by other authors. I believe that a reliable general method for decomposing mixtures can be developed by using one of the estimation criteria in this class and allowing the user to specify as many constraints among the parameters of the component distributions as may be necessary to ensure convergence of the non-linear part of the optimization process.

2. A WEIGHTED LEAST-SQUARES ESTIMATOR

2.1 Grouped data

Suppose that the sample space has been partitioned into m regions $A_1,\dots,A_m$ and let p_j denote the probability that a given observation falls in A_j. For a sample of n observations, let n_j denote the number of observations falling in A_j. It is then well-known that $\hat{p}_j = n_j/n$ is the maximum-likelihood estimate of p_j, and that the distribution of $\hat{\underset{\sim}{p}}$ is asymptotically normal with

$$\mathcal{E}(\hat{\underset{\sim}{p}}) = \underset{\sim}{p} \tag{1}$$

and

$$\mathrm{Var}(\hat{\underset{\sim}{p}}) = n^{-1}\{\mathrm{diag}(\underset{\sim}{p})-\underset{\sim}{p}\underset{\sim}{p}'\}, \tag{2}$$

where $\mathrm{diag}(\underset{\sim}{p})$ denotes the $m\times m$ matrix with elements $p_1,\dots,p_m$ down the diagonal and zeros elsewhere.

Defining the $m\times k$ matrix $\underset{\sim}{\Gamma}$ by

$$\gamma_{ji} = \int_{A_j} dF_i \ , \tag{3}$$

we can write $p_j = \sum_{i=1}^{k} \pi_i \gamma_{ji}$, or

$$\underset{\sim}{p} = \underset{\sim}{\Gamma}\underset{\sim}{\pi} \ . \tag{4}$$

Note that each column of $\underset{\sim}{\Gamma}$ sums to unity; that is, $\underset{\sim}{1}'_m \underset{\sim}{\Gamma} = \underset{\sim}{1}'_k$, where $\underset{\sim}{1}_m$, for example, denotes an m-vector of 1's .

As each p_j is a linear function of $\underset{\sim}{\pi}$, it seems natural to consider estimators of $\underset{\sim}{\pi}$ minimizing the weighted sum of squares

$$\sum_{j=1}^{m} \frac{(\hat{p}_j - p_j)^2}{w_j} \ . \tag{5}$$

Letting $\underset{\sim}{C} = \text{diag}(1/w_1,\ldots,1/w_m)$, (5) can be written as a quadratic form:

$$(\hat{\underset{\sim}{p}} - \underset{\sim}{\Gamma}\underset{\sim}{\pi})' \ \underset{\sim}{C} \ (\underset{\sim}{p} - \underset{\sim}{\Gamma}\underset{\sim}{\pi}) \ . \tag{6}$$

The parameters $\underset{\sim}{\pi}$ are related by one linear constraint

$$\underset{\sim}{1}'_k \underset{\sim}{\pi} = 1 \ . \tag{7}$$

Incorporating this constraint will reduce the dimension of the parameter space, increasing the efficiency of the estimator. It is straightforward to show that the constrained minimum of (6) is given by

$$\tilde{\underset{\sim}{\pi}} = (\underset{\sim}{\Gamma}'\underset{\sim}{C}\underset{\sim}{\Gamma})^{-1}\{\underset{\sim}{\Gamma}'\underset{\sim}{C}\hat{\underset{\sim}{p}} + \frac{1 - \underset{\sim}{1}'_k(\underset{\sim}{\Gamma}'\underset{\sim}{C}\underset{\sim}{\Gamma})^{-1}\underset{\sim}{\Gamma}'\underset{\sim}{C}\hat{\underset{\sim}{p}}}{\underset{\sim}{1}'_k(\underset{\sim}{\Gamma}'\underset{\sim}{C}\underset{\sim}{\Gamma})^{-1}\underset{\sim}{1}_k} \underset{\sim}{1}_k\} \ . \tag{8}$$

A less cumbersome formula can be derived by reparameterizing the model. Let

$$\underset{\sim}{\theta} = \begin{pmatrix} \pi_1 \\ \cdot \\ \cdot \\ \cdot \\ \pi_{k-1} \end{pmatrix} \tag{9}$$

and define an m-vector $\underset{\sim}{y}$ and $m \times (k-1)$ matrix $\underset{\sim}{B}$ by

$$y_j = \hat{p}_j - \gamma_{jk} \qquad (j=1,\ldots,m) \tag{10}$$

and

$$B_{ji} = \gamma_{ji} - \gamma_{jk} \qquad (j=1,\ldots,m;\ i=1,\ldots,k-1)\ , \tag{11}$$

respectively. Note that each column of $\underset{\sim}{B}$ sums to zero; that is, $\underset{\sim}{1}'_m \underset{\sim}{B} = \underset{\sim}{0}_k$, where $\underset{\sim}{0}_k$ denotes a k-vector of 0's . The quadratic form (6) can then be written

$$(\underset{\sim}{y}-\underset{\sim}{B}\underset{\sim}{\theta})'\ \underset{\sim}{C}\ (\underset{\sim}{y}-\underset{\sim}{B}\underset{\sim}{\theta}) \tag{12}$$

and the minimum is given by

$$\tilde{\underset{\sim}{\theta}} = (\underset{\sim}{B}'\underset{\sim}{C}\underset{\sim}{B})^{-1}\ \underset{\sim}{B}'\underset{\sim}{C}\ \underset{\sim}{y} \quad . \tag{13}$$

Letting $\tilde{\pi}_k = 1-\underset{\sim}{1}'_{k-1}\tilde{\underset{\sim}{\theta}}$, it can be verified that (8) and (13) are equivalent.

Since $\mathcal{E}(\underset{\sim}{y}) = \underset{\sim}{B}\underset{\sim}{\theta}$ and $\mathrm{Var}(\underset{\sim}{y}) = \mathrm{Var}(\hat{\underset{\sim}{p}})$, it follows that the distribution of $\tilde{\underset{\sim}{\theta}}$ is asymptotically normal with

$$\mathcal{E}(\tilde{\underset{\sim}{\theta}}) = \underset{\sim}{\theta} \tag{14}$$

and

$$\mathrm{Var}(\tilde{\underset{\sim}{\theta}}) = (\underset{\sim}{B}'\underset{\sim}{C}\underset{\sim}{B})^{-1}\ \underset{\sim}{B}'\underset{\sim}{C}\ \mathrm{Var}(\hat{\underset{\sim}{p}})\underset{\sim}{C}\underset{\sim}{B}(\underset{\sim}{B}'\underset{\sim}{C}\underset{\sim}{B})^{-1} \ . \tag{15}$$

A convenient unbiased estimate of $\mathrm{Var}(\tilde{\underset{\sim}{\theta}})$ is

$$\widehat{\mathrm{Var}}(\tilde{\underset{\sim}{\theta}}) = (n-1)^{-1}(\underset{\sim}{B}'\underset{\sim}{C}\underset{\sim}{B})^{-1}\underset{\sim}{B}'\underset{\sim}{C}\{\mathrm{diag}(\hat{\underset{\sim}{p}})-\hat{\underset{\sim}{p}}\hat{\underset{\sim}{p}}'\}\underset{\sim}{C}\underset{\sim}{B}(\underset{\sim}{B}'\underset{\sim}{C}\underset{\sim}{B})^{-1} \ . \tag{16}$$

Results (14), (15) and (16) are valid only if $\underset{\sim}{C}$ is fixed; the choice of weights must not, for instance, depend on the sample.

One method of choosing weights would be to select an arbitrary $\underset{\sim}{\theta}^\circ$ and set each $w_j = p_j(\underset{\sim}{\theta}^\circ)$. For a given partition $A_1,\ldots,A_m$, the estimator will be fully efficient when $\underset{\sim}{\theta} = \underset{\sim}{\theta}^\circ$, since (15) then reduces to

$$\mathrm{Var}(\tilde{\underset{\sim}{\theta}}|\underset{\sim}{\theta}^\circ) = \underset{\sim}{J}_A^{-1}(\underset{\sim}{\theta}^\circ) \ , \tag{17}$$

where

$$\underset{\sim}{J}_A(\underset{\sim}{\theta}) = n(\underset{\sim}{B}'\underset{\sim}{C}\underset{\sim}{B}) = n\left(\sum_{j=1}^{m} \frac{(\gamma_{j\alpha}-\gamma_{jk})(\gamma_{j\beta}-\gamma_{jk})}{p_j(\underset{\sim}{\theta})}\right)_{(k-1)\times(k-1)}$$

is easily shown to be Fisher's information matrix for $\underset{\sim}{\theta}$.

2.2 Ungrouped data

The weighted least-squares estimator can be extended to handle ungrouped data from a mixture of continuous distributions. For convenience, assume the distributions to be univariate. Let $G(x) = \pi_1 F_1(x)+\ldots+\pi_k F_k(x)$ denote the distribution function of the mixture and let $G_n(x)$ denote the usual empirical distribution function based on a sample of n observations. For an appropriate sequence of progressively finer partitions $A_1,\ldots,A_m$, as $m \to \infty$, the limit of (5) is seen to be

$$\int \frac{(dG_n - dG)^2}{dW} \tag{18}$$

where $W(x)$ is a nondecreasing weight function. This gives the estimator suggested by Bartlett & Macdonald (1968). If W possesses a derivative w, the integral (18) will not actually exist, the denominator of the integrand being everywhere infinitesimal while the numerator is sometimes positive. However, this does not preclude formal minimization. If ϕ is an unknown parameter of G, the least-squares estimate of ϕ will satisfy the equation

$$\frac{1}{n}\sum_{j=1}^{n} \frac{1}{w(x_j)} \frac{\partial g(x_j|\phi)}{\partial \phi} = \mathcal{E}_\phi\left[\frac{1}{w(X)} \frac{\partial g(X|\phi)}{\partial \phi}\right] . \tag{19}$$

Note that the choice $w(x) = g(x|\phi)$ gives the familiar maximum-likelihood equation:

$$\sum_{j=1}^{n} \frac{\partial \log g(x_j|\phi)}{\partial \phi} = 0 . \tag{20}$$

For estimating the proportions of a mixture, the equations (19) are linear. Define the $(k-1)$-dimensional vectors $\underset{\sim}{T}$ and $\underset{\sim}{R}$ and the $(k-1)\times(k-1)$ matrix $\underset{\sim}{J}$ by

$$T_\alpha = \frac{1}{n}\sum_{j=1}^{n} \frac{f_\alpha(x_j) - f_k(x_j)}{w(x_j)} , \tag{21}$$

$$R_\alpha = \int \frac{f_\alpha(x) - f_k(x)}{w(x)} f_k(x)\,dx , \tag{22}$$

$$J_{\alpha\beta} = \int \frac{\{f_\alpha(x) - f_k(x)\}\{f_\beta(x) - f_k(x)\}}{w(x)}\,dx , \tag{23}$$

respectively, and let $\underset{\sim}{\theta} = \begin{pmatrix} \pi_1 \\ \cdot \\ \cdot \\ \cdot \\ \pi_{k-1} \end{pmatrix}$, as before. The estimating equations (19) can be written

$$\underset{\sim}{T} = \underset{\sim}{J}\, \underset{\sim}{\tilde{\theta}} + \underset{\sim}{R} \tag{24}$$

and so the least-squares estimate of $\underset{\sim}{\theta}$ is

$$\underset{\sim}{\tilde{\theta}} = -\underset{\sim}{J}^{-1} \underset{\sim}{R} + \underset{\sim}{J}^{-1} \underset{\sim}{T} . \tag{25}$$

The estimate is asymptotically normal. Since $\mathcal{E}(\underset{\sim}{T}|\underset{\sim}{\theta}) = \underset{\sim}{J}\underset{\sim}{\theta}+\underset{\sim}{R}$ and $\text{Var}(\underset{\sim}{T}|\underset{\sim}{\theta}) = n^{-1}\{\underset{\sim}{H}-(\underset{\sim}{J}\underset{\sim}{\theta}+\underset{\sim}{R})(\underset{\sim}{J}\underset{\sim}{\theta}+\underset{\sim}{R})'\}$, where $\underset{\sim}{H}$ is a $(k-1)\times(k-1)$ matrix defined by

$$H_{\alpha\beta} = \mathcal{E}_{\theta}\left[\frac{\{f_{\alpha}(X)-f_{k}(X)\}\{f_{\beta}(X)-f_{k}(X)\}}{w^{2}(X)}\right] ,$$

it follows that $\underset{\sim}{\tilde{\theta}}$ is unbiased, with exact variance

$$\text{Var}(\underset{\sim}{\tilde{\theta}}) = n^{-1}\{\underset{\sim}{J}^{-1}\underset{\sim}{H}\underset{\sim}{J}^{-1}-(\underset{\sim}{\theta}+\underset{\sim}{J}^{-1}\underset{\sim}{R})(\underset{\sim}{\theta}+\underset{\sim}{J}^{-1}\underset{\sim}{R})'\} . \tag{26}$$

$\underset{\sim}{R}$ and $\underset{\sim}{J}$ may be computed in terms of the $\frac{1}{2}k(k+1)$ integrals $\int\{f_i(x)f_j(x)/w(x)\}dx$ $(i=1,\ldots,1;\ j=i,\ldots,k)$. The grouped-data estimator (13) involves $k(m-1)$ integrals (3) to compute $\underset{\sim}{\Gamma}$; however, efficient numerical methods are available to compute (3) for most standard distributions.

3. RELATION TO OTHER ESTIMATORS

3.1 Grouped data

3.1.1. Minimum chi-square

The weighted least-squares estimator gives a convenient iterative method of computing the minimum chi-square estimate of $\underset{\sim}{\theta}$. Starting with an arbitrary $\underset{\sim}{\theta}^{\circ}$, compute the weights $w_j = p_j(\underset{\sim}{\theta}^{\circ})$ and estimate $\underset{\sim}{\theta}$ by (13). Re-compute the weights with the new estimate of $\underset{\sim}{\theta}$ and continue until the process converges. The resulting estimate, $\underset{\sim}{\hat{\theta}}$, is the minimum chi-square estimate. While (14) and (15) will no longer hold for small samples, it is well-known that $\underset{\sim}{\hat{\theta}}$ is asymptotically normal, unbiased, and fully efficient, with $\text{Var}(\underset{\sim}{\hat{\theta}}) \simeq \underset{\sim}{J}_A^{-1}(\underset{\sim}{\hat{\theta}})$, where $\underset{\sim}{J}_A$ is given by (17).

Hasselblad (1966) describes an iterative scheme for computing the maximum likelihood estimator, which is asymptotically equivalent to the minimum chi-square estimator. His assumption that the sample space has been partitioned into very narrow intervals permits a simple approximation to the integral (3) defining γ_{ij} .

3.1.2 Modified minimum chi-square

The weights $w_j = \hat{p}_j$ give, without iteration, the modified minimum chi-square estimator advocated for mixtures by Chapman (1952) and Maritz (1966). This is asymptotically equivalent to the minimum chi-square estimator but less well-behaved in very small samples because of the possibility of getting a $\hat{p}_j = 0$.

When parameters of the distributions $F_1,\ldots,F_k$ are to be estimated simultaneously with $\underset{\sim}{\theta}$, Saleh (unpublished) has shown that the modified minimum chi-square criterion gives an efficient two-step procedure.

3.1.3 EDA estimator

Macdonald (1969) considered the case $k = m$, where the sample space has been partitioned into as many regions as there are components in the mixture. A discriminant analysis gives such a partition; hence the abbreviation "EDA".

The EDA estimator can be derived from (8) by taking $\underset{\sim}{\Gamma}$ to be a non-singular $k \times k$ matrix. The result is

$$\hat{\underset{\sim}{\pi}} = \underset{\sim}{\Gamma}^{-1}\hat{\underset{\sim}{p}} \tag{27}$$

which does not involve $\underset{\sim}{C}$. For a given partition $A_1,\ldots,A_k$, $\hat{\underset{\sim}{\pi}}$ is the maximum-likelihood estimate of $\underset{\sim}{\pi}$. It is unbiased and fully efficient for any sample size. The variance matrix is

$$\mathrm{Var}(\hat{\underset{\sim}{\pi}}) = n^{-1}\{\underset{\sim}{\Gamma}^{-1}\mathrm{diag}(\underset{\sim}{p})(\underset{\sim}{\Gamma}^{-1})'-\underset{\sim}{\pi}\underset{\sim}{\pi}'\} , \tag{28}$$

which can be estimated by

$$\widehat{\mathrm{Var}(\hat{\underset{\sim}{\pi}})} = (n-1)^{-1}\{\underset{\sim}{\Gamma}^{-1}\mathrm{diag}(\underset{\sim}{p})(\underset{\sim}{\Gamma}^{-1})'-\underset{\sim}{\pi}\underset{\sim}{\pi}'\} . \tag{29}$$

When $m = k = 2$, (27) and (28) give

$$\hat{\pi}_1 = (\hat{p}_1-\gamma_{12})/(\gamma_{11}-\gamma_{12}) \tag{30}$$

$$\mathrm{Var}(\hat{\pi}_1) = \frac{1}{n}\left[\frac{\pi_1\gamma_{11}(1-\gamma_{11})+\pi_2\gamma_{12}(1-\gamma_{12})}{(\gamma_{11}-\gamma_{12})^2} + \pi_1\pi_2\right] ; \tag{31}$$

the estimate (30) seems to have been first suggested by Worlund & Fredin (1962). A similar result was given independently by Boes (1966).

3.2 Ungrouped data

3.2.1 Boes' $\underset{\sim}{\theta}^{\circ}$-efficient unbiased estimator

One way to choose a weight function for (18) would be to select an arbitrary $\underset{\sim}{\theta}^{\circ}$ and set $w(x) = g(x|\underset{\sim}{\theta}^{\circ})$. Let $\underset{\sim}{T}(\underset{\sim}{\theta}^{\circ})$, $\underset{\sim}{R}(\underset{\sim}{\theta}^{\circ})$ and $\underset{\sim}{J}(\underset{\sim}{\theta}^{\circ})$ be the resulting vectors and matrix defined by (21), (22) and (23), respectively. It can be shown that $\underset{\sim}{R}(\underset{\sim}{\theta}^{\circ}) = -\underset{\sim}{J}(\underset{\sim}{\theta}^{\circ})\underset{\sim}{\theta}^{\circ}$. Hence the estimator (25) becomes

$$\underset{\sim}{\tilde{\theta}} = \underset{\sim}{\theta}^{\circ} + \underset{\sim}{J}^{-1}(\underset{\sim}{\theta}^{\circ})\underset{\sim}{T}(\underset{\sim}{\theta}^{\circ}) \tag{32}$$

and the exact variance (26) can be written

$$\mathrm{Var}(\underset{\sim}{\tilde{\theta}}|\underset{\sim}{\theta}) = n^{-1}\{\underset{\sim}{J}^{-1}(\underset{\sim}{\theta}^{\circ})\underset{\sim}{H}(\underset{\sim}{\theta}^{\circ}|\underset{\sim}{\theta})\underset{\sim}{J}^{-1}(\underset{\sim}{\theta}^{\circ}) - (\underset{\sim}{\theta}-\underset{\sim}{\theta}^{\circ})(\underset{\sim}{\theta}-\underset{\sim}{\theta}^{\circ})'\} . \tag{33}$$

The estimator is unbiased and asymptotically normal. If $\underset{\sim}{\theta} = \underset{\sim}{\theta}^{\circ}$, we see that $\underset{\sim}{H}(\underset{\sim}{\theta}^{\circ}|\underset{\sim}{\theta}^{\circ}) = \underset{\sim}{J}(\underset{\sim}{\theta}^{\circ})$ and so

$$\mathrm{Var}(\underset{\sim}{\tilde{\theta}}|\underset{\sim}{\theta}^{\circ}) = n^{-1}\underset{\sim}{J}^{-1}(\underset{\sim}{\theta}^{\circ}) , \tag{34}$$

where $\underset{\sim}{J}(\underset{\sim}{\theta})$ is easily shown to be Fisher's information matrix for $\underset{\sim}{\theta}$. Hence this estimator is fully efficient when $\underset{\sim}{\theta} = \underset{\sim}{\theta}^{\circ}$.

Boes (1966) gave (32) and (34) and derived a number of theoretical results for this estimator. When $k = 2$, he showed that the estimator obtained by partitioning the sample space into two regions $A_1 = (-\infty, x]$ and $A_2 = (x, \infty)$ and taking a suitable weighted average of the estimator (30) over all x-values, is just the $\underset{\sim}{\theta}^{\circ}$-efficient unbiased estimator (32).

3.2.2 Boes' minimax unbiased estimator

Boes (1967) derived a minimax unbiased estimator from the $\underset{\sim}{\theta}^{\circ}$-efficient estimator by choosing $\underset{\sim}{\theta}^{\circ}$ to minimize the maximum of a risk function based on the diagonal elements of $\underset{\sim}{J}^{-1}(\underset{\sim}{\theta}^{\circ})$.

3.2.2 Maximum likelihood

Applying (32) iteratively until $\underset{\sim}{\theta}^{\circ}$ and $\underset{\sim}{\tilde{\theta}}$ converge is just the well-known "scoring" method of iteratively calculating the maximum likelihood estimate $\hat{\underset{\sim}{\theta}}$. It is equivalent to solving (19) iteratively until $w(x) = g(x|\hat{\underset{\sim}{\theta}})$. As the weight function then depends on the sample, the estimate is no longer unbiased in small samples and the variance expression (26) is invalid. However, the estimate is known to be asymptotically normal, unbiased, and fully efficient, with

$$\mathrm{Var}(\hat{\underset{\sim}{\theta}}) \simeq n^{-1}\underset{\sim}{J}^{-1}(\hat{\underset{\sim}{\theta}}) \ . \tag{35}$$

In practice it is usually more convenient to estimate $\underset{\sim}{J}(\underset{\sim}{\theta})$ by its sample analogue

$$\hat{J}_{\alpha\beta}(\underset{\sim}{\theta}) = \frac{1}{n}\sum_{j=1}^{n} \frac{\{f_\alpha(x_j)-f_k(x_j)\}\{f_\beta(x_j)-f_k(x_j)\}}{\{g(x_j|\underset{\sim}{\theta})\}^2} \tag{36}$$

Either way, convergence of the iterations is virtually guaranteed (Macdonald, 1971).

Hasselblad's (1966) maximum-likelihood estimator for finely-grouped data should give results very close to those of the maximum-likelihood estimator for ungrouped data.

4. EXAMPLES

Table 1 compares several grouped-data least-squares estimates of mixing proportions with the maximum likelihood estimates from ungrouped data and with the estimates based on the actual number of observations from each component. The component distributions were assumed to be normal. Note that very little precision was lost by grouping the data. These data were previously used for examples by Macdonald (1969, 1971).

Grouping the data is often advisable in practical work, as a means of smoothing the data and making the estimates less sensitive to the distributional assumptions. For example, if the mixture includes a component in a relatively small proportion, one could choose a region A_j to cover most of that component and fix some or all of its parameters at "reasonable" values while estimating the remaining parameters of the mixture.

This was the method used to separate the components of a minnow population studied by T. J. Pitcher. The minnows were difficult to measure accurately and grouping was necessary to smooth the data. Minimum chi-square estimates were obtained by alternating least-squares estimation of the mixing proportions with direct-search minimization for the parameters of the component distributions,

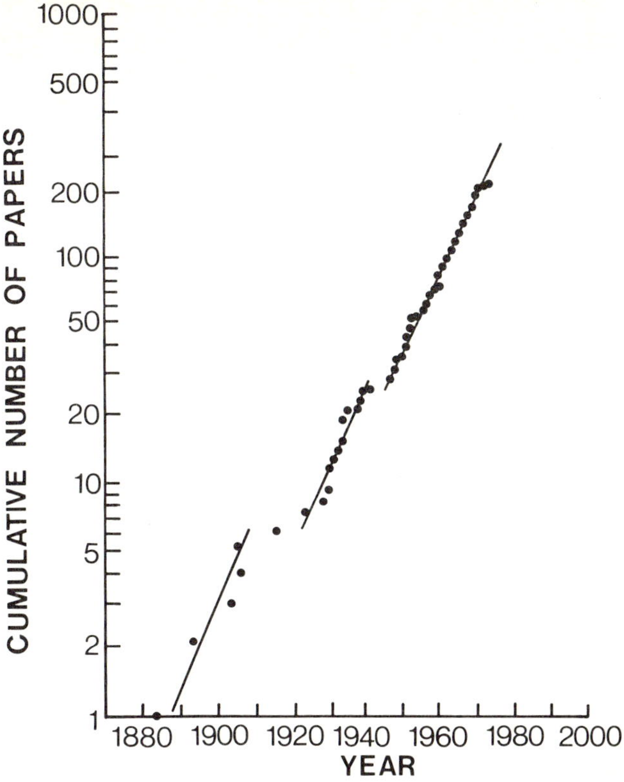

FIG.1 THE CUMULATIVE NUMBER OF PAPERS ON THE ESTIMATION OF FINITE DISTRIBUTION MIXTURES IS PLOTTED ON A LOGARITHMIC SCALE AGAINST TIME.

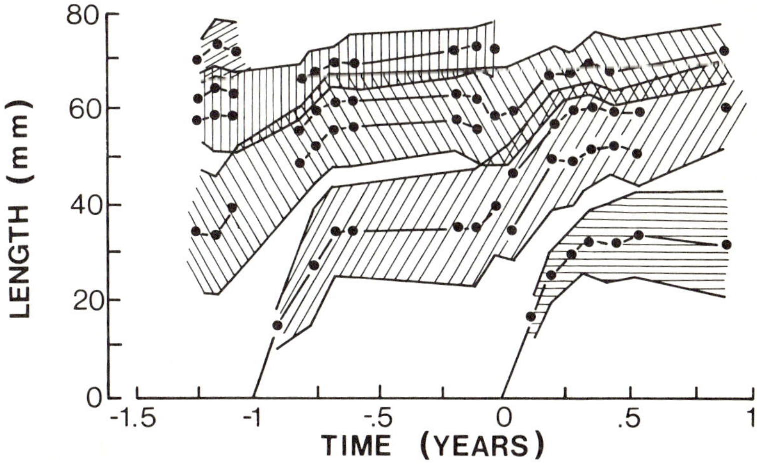

FIG.2 GROWTH CURVES FOR THE COHORTS OF A MINNOW POPULATION. FOR EXPLANATION, SEE TEXT.

α (age)	μ_α (length in cm.)	σ_α	Estimates of π_α					
			Age determination	Maximum likelihood	Least-squares (Grouped data)			
					m=23 minimum chi-square	m=23 $\pi^\circ_\alpha=1/k$	m=11 $\pi^\circ_\alpha=1/k$	m=5 EDA
1	23.3	2.44	.105±.013	.104±.014	.098±.014	.098±.014	.101±.015	.099±.015
2	33.1	3.00	.465±.022	.465±.027	.464±.027	.459±.027	.450±.028	.468±.029
3	41.3	4.27	.298±.020	.299±.029	.300±.029	.311±.030	.321±.032	.299±.033
4	51.2	5.08	.090±.013	.090±.020	.096±.021	.087±.023	.078±.024	.097±.025
5	61.3	7.07	.042±.009	.042±.013	.042±.012	.045±.013	.049±.014	.037±.014

Table 1. For a sample of $n = 523$ pike from a mixture of $k = 5$ age groups, least-squares estimates of $\underset{\sim}{\pi}$ from the mixed length distribution are compared with the maximum-likelihood estimate and the estimate based on the actual number of fish from each age group in the sample. The standard error of each estimate is given.

which were assumed to be normal. The results are shown in Fig. 2. Time is measured in years from June 1st, 1968. "Cohorts" are spawned in June of each year and grow during the summers. In its second year, each cohort splits into two components at the start of the breeding season: these have been identified as early- and late-maturing fish. The components merge again at the start of their third year. The dots indicate mean lengths of the component groups at successive sampling occasions. The bands are intended to include 95% of each cohort. The growth curves from all cohorts were superimposed to give the seasonal growth curve discussed in Pitcher & Macdonald (1973).

ACKNOWLEDGMENTS

Dr. T. J. Pitcher encouraged this work on mixtures and provided the data for Fig. 2. Dr. J. R. H. Dempster suggested the EDA estimator to me as a generalization of Worlund & Fredin's method, in 1966. The National Research Council of Canada and the McMaster University Division of Science and Engineering have supported parts of this research.

Fig. 1. The cumulative number of papers on the estimation of finite distribution mixtures is plotted on a logarithmic scale against time.

Fig. 2. Growth curves for the cohorts of a minnow population. For explanation, see text.

REFERENCES

[1] Bartlett, M. S. and Macdonald, P. D. M. (1968). "Least-squares" estimation of distribution mixtures. Nature, Lond. 217, 195-196.

[2] Bhattacharya, C. G. (1967). A simple method of resolution of a distribution into Gaussian components. Biometrics 23, 115-135.

[3] Boes, D. C. (1966). On the estimation of mixing distributions. Ann. Math. Statist. 37, 177-188.

[4] Buchanan-Wollaston, H. J. and Hodgson, W. C. (1929). A new method of treating frequency curves in fishery statistics, with some results. J. Cons. Int. Explor. Mer. 4, 207-225.

[5] Chapman, D. G. (1952). The analysis of samples from mixed populations (preliminary report)(abstract). Ann. Math. Statist. 23, 479.

[6] Cohen, A. C., Jr. (1967). Estimation in mixtures of two normal distributions. Technometrics 9, 15-28.

[7] Day, N. E. (1969). Estimating the components of a mixture of normal distributions. Biometrika, 56, 463-474.

[8] Dick, N. P. and Bowden, D. C. (1973). Maximum likelihood estimation for mixtures of two normal distributions. Biometrics 29, 781-790.

[9] Gridgeman, N. T. (1970). A comparison of two methods of analysis of mixtures of normal distributions. Technometrics 12, 823-833

[10] Hasselblad, V. (1966). Estimation of parameters for a mixture of normal distributions. Technometrics 8, 431-444.

[11] Hosmer, D. W., Jr. (1973). A comparison of iterative maximum likelihood estimates of the parameters of a mixture of two normal distributions under three different types of sample. Biometrics 29, 761-770.

[12] Kabir, A.B.M.L. (1968). Estimation of parameters of a finite mixture of distributions. J. R. Statist. Soc. B, 30, 472-482.

[13] Kiefer, J. and Wolfowitz, J. (1956). Consistency of the maximum likelihood estimator in the presence of infinitely many incidental parameters. Ann. Math. Statist. 27, 887-906.

[14] Macdonald, P.D.M. (1969). FORTRAN programs for statistical estimation of distribution mixtures: some techniques for statistical analysis of length-frequency data. Fish. Res. Bd. Canada, Tech. Rept. No. 129.

[15] Macdonald, P.D.M. (1971). Comment on "An estimation procedure for mixtures of distributions" by Choi and Bulgren. J. R. Statist. Soc. B., 33, 326-329.

[16] Maritz, J. S. (1966). Smooth empirical Bayes estimation for one-parameter discrete distributions. Biometrika 53, 417-429.

[17] Neyman, J. (1949). Contribution to the theory of the χ^2 test. Proceedings of the 1st Berkley Symposium on Mathematical Statistics and Probability, 239-273.

[18] Pitcher, T. J. and Macdonald, P.D.M. (1973). Two models for seasonal growth in fishes. J. Appl. Ecol. 10, 599-606.

[19] Rao, C. R. (1948). The utilization of multiple measurements in problems of biological classification. J. R. Statist. Soc. B., 10, 159-203.

[20] Rennie, R. R. (1968). Finite mixtures of distributions with common central moments. Ph.D. Thesis, Univ. of British Columbia.

[21] Saleh, A.K.Md.E. (UNPUBLISHED). Decomposition of finite mixture of distributions. Mimeo. report, Carleton Univ.

[22] Tan, W. Y. and Chang, W. C. (1972). Some comparisons of the method of moments and the method of maximum likelihood in estimating parameters of a mixture of normal densities. J. Am. Statist. Ass. 67, 702-708.

[23] Tanaka, S. (1962). A method of analyzing a polymodal frequency distributions and its application to the length distribution of the porgy, Taius tumifrons (T. and S.). J. Fish. Res. Bd. Canada 19, 1143-1159.

[24] Worlund, D. D. and Fredin, R. A. (1962). Differentiation of stocks. In Symposium on Pink Salmon (ed. N. J. Wilimovsky), pp. 143-153. Vancouver: Inst. of Fisheries, Univ. of British Columbia.

[25] Boes, D. C. (1967). Minimax unbiased estimator of mixing distribution for finite mixtures. Sankhyā A, 29, 417-420.

P.D.M. MacDonald
Dept. of Applied Mathematics
McMaster University
Hamilton, Ontario
Canada L8S 4K1

Applied Statistics, R.P. Gupta, (editor)

ANALYSIS OF FISH COUNT DATA OBTAINED USING A DIGITAL ECHO COUNTING SYSTEM

Ian B. MacNeill and Richard G. Dowd

ABSTRACT

Data gathered by a digital echo counting system developed to estimate the distribution of demersal fish populations are subjected to an analysis to ascertain the nature of such data, to examine the properties of the counter and then to determine how closely to theory is the performance of the echo-counter. It is found that gain settings strongly affect the nature of the data. It is found that the apparatus gives consistent results if operated with fixed parameters. It is found that correlation and spectral analysis are relatively robust against changes in fish density but sensitive to changes in system parameters and to trends and transients in the data. The data support the hypothesis that the echo-counter works reliably and consistently and can be used to obtain valid estimates of relative fish density.

INTRODUCTION

This paper presents the analysis of certain data acquired by a digital echo-counting system. The data records produced by the echo-counter consist mainly of sequences of fish counts from overlapping regions located along a transect and at a fixed distance above the sea bed. For each pulse of the transducer, echoes are counted from each of four layers of water about four meters thick stacked one above the other. A detailed account of the digital echo counting system is given in Dowd (1967).

One would like to know the precise number and distribution of fish in the sonified volumes. Then one could determine the output of the echo-counter and tell whether or not it is performing as it is supposed to. Since this precise information is never available we might consider counting the catch from tows over a sonified region and then comparing catch-count with echo-count. If they should coincide we could regard this as evidence that the echo-counter is performing up to expectation. Other evidence to support or discount the hypothesis that the echo-counter is accurately counting the fish in the sound cone can be gained from measures of internal consistency. The main purpose of this paper is to examine a number of these measures using data collected during cruise #39 of the E.E. Prince. The cruise took place on the Cape Breton side of the Cabot straits during January 1969. Although this data was not collected with this use in mind, we are able to make use of some of it in comparing the system's empirical performance with the expected theoretical performance. This data also provides an opportunity to assess the value of various techniques devised for analysing fish count data.

We focus our attention on two sequences of runs over the same fishing grounds. These sequences contain multiple runs back and forth over the same region with certain parameters fixed thus making it possible to examine the system for consistency. Also, multiple runs with different fixed parameter values enable us to explore the effects of changing the parameters of the system. The parameters that were modified during the sequences are 1) gain settings, 2) replication rate and 3) distance of the transducer from the bottom. The A.T. Cameron accompanied the E.E. Prince and made several net tows over the sonified region for the purpose of comparing catch and count.

Our approach to the reduction of the data was first to perform a rough preliminary analysis which consisted of segmenting the data into blocks equivalent to either a full or a half minute's running time. For each of these blocks we computed the number of echoes recorded, the mean and variance of these counts and the accumulated counts, means and variances. This "minute" analysis was performed on each of the 4 four-meter thick layers located one above the other in the sound cone, the bottom one being next to the sea bed. If it appeared, on the basis of the minute analysis, that there was substantial variation over the run in the density of fish, then the run was divided into data sets having reasonably homogeneous fish distribution. Otherwise, the run was analysed as a whole. It was found that there were so few targets in the layers other than the bottom layer that, with two exception, further analysis was confined to the bottom layer. In this way, the eleven runs that were considere provide us with twenty data sets each of which was subjected to a more detailed analysis. This analysis is an application of the methods that have been derived for analysing the data provided by digital echo-counters which, in large part, have been previously reported by MacNeill (1971a, 1971b, 1972).

COLLECTION OF THE DATA

We now present some of the basic parameters of the eleven runs that are analysed in the sequel. Runs 1 and 2 were repetitions over a certain region and runs 3 to 11 were repetitions over another region. On the basis of the minute analysis, run 1 was segmented into two data sets and run 2 into the next three. Run 7 was segmented into two data sets and the two sets (numbered 10 and 11) and the whole (numbered 13) were subjected to an analysis. Run 11 was segmented into the last three data sets. In addition, data sets 12 and 15 represent the second layers for runs 7 and 8. The correspondence between data sets and runs is given as follows:

Data Set	1	2	3	4	5	6	7	8	9	10	11	12	13	14	15	16	17	18	19	20
Run	1	1	2	2	2	3	4	5	6	7	7	7	7	8	8	9	10	11	11	11

We now present some details of the runs as taken from the log of the cruise. Runs 1 and 2 were runs back and forth over the same ground with replication rate of 96 pulses per minute. The gain was adjusted to compensate for absorption and spreading losses. Runs 3 - 11 were taken back and forth along a transect with replication rate of 48 p.p.m. The odd numbered runs were along course 310° and the even numbered runs were along course 130°. Runs 3, 4, 5, 6 were performed with gain settings again adjusted to compensate for absorption and spreading losses. Runs 7 and 8 were performed with gain settings increased by 5 db. above those for runs 3, 4, 5, 6. Each of the runs 3 - 10 was carried out with the transducer at a height of approximately 125 m. above the seabed. Run 11 differs from the previous 10 runs in that the transducer was lowered by about 30 metres and the gains were set to compensate for absorption and spreading loss.

ESTIMATION OF DENSITY

The data required for a discussion of density are to be found in Table 1. Among other things, this table contains the values of the various estimators discussed by MacNeill (1972) for each data set. Both the estimates of density per sonified volume and per $1000m^3$ are computed. Moore's estimator (L = 9) and the mean coincide for densities in the range that we consider, and because the mean is the easier of the two to calculate we discard Moore's estimator in its unmodified form. The saucer censored estimator differs only

Table 1. Density estimates for each data set

Data Set	Av. Depth	Layer Vol.	Mean and Moore	Density (per 10000 m^3)	Saucer Censored	Density From Saucer Censored Estimator	Zero Class Estimator	Density From Zero Class Estimator
1	103.8m	971.4m^3	.322	.332	.328	.338	.289	.298
2	105	994.5	.607	.611	.629	.632	.565	.568
3	105.8	1010.0	.582	.576	.602	.596	.510	.505
4	105.5	1004.1	.251	.250	.255	.254	.251	.250
5	102.5	946.8	.389	.411	.398	.420	.338	.356
6	126.2	1445.9	.564	.390	.583	.403	.551	.406
7	122.4	1358.8	.557	.410	.575	.426	.551	.406
8	118.4	1270.0	.432	.340	.443	.349	.402	.317
9	125.7	1434.3	.517	.360	.532	.371	.490	.342
10	123.3	1379.2	.767	.556	.801	.581	.702	.509
11	122.6	1363.3	1.075	.789	1.145	.840	.896	.657
12	118.9	1280.9	.490	.382	.503	.393	.426	.333
13	122.9	1370.1	.921	.672	.971	.709	.794	.579
14	122.7	1365.6	1.236	.905	1.329	.973	1.047	.767
15	118.7	1276.6	.563	.441	.581	.455	.520	.407
16	121.3	1334.1	.519	.389	.534	.401	.397	.297
17	121.7	1343.0	.518	.386	.533	.397	.364	.271
18	91.8	755.9	.071	.094	.071	.094	.067	.089
19	93.6	786.5	.352	.448	.359	.457	.288	.366
20	97.3	851.4	.153	.179	.154	.181	.126	.148

marginally from the mean for the lower densities but more substantially for the larger densities. The zero class estimator yields considerably smaller estimates of density than the other estimators. In addition, although it usually ranks the run estimates the same as the mean it reverses the order of some pairs. Due to its nature, the zero class estimator has a large variance which might account for some of the changes in rank. However, the substantial differences in its size compared to the mean requires another explanation. A possible cause could be the inadequacy of the Poisson model for fish distribution in that a contagious distribution would yield more zero counts than a Poisson distribution of equal mean. Two of the advantages of the zero class estimator are i) ease of computation and ii) its relative independence of certain distortions to the counts caused by the counter. This independence property is discussed by MacNeill (1972).

Several comments on the data of Figure 1 appear below.

i) Data sets 6, 7, 8, 9 were collected by four successive runs of approximately 15 minutes over regions that coincide as much as could be arranged. The respective densities per 1000m^3 are 0.390, 0.410, 0.340 and 0.360 . Considering the fact that the four runs took about $1\frac{3}{4}$ hours to execute the amount of variation seems quite small. Data sets 16 and 17 which are successive runs back and forth over the same decca line but with lower gain settings than for sets 6, 7, 8, 9 yield almost identical estimates of fish density. However data sets 13 and 14, which are successive runs with higher gain settings than for runs 6, 7, 8, 9 yield substantially different estimates to one another and each substantially larger than those for data sets 6, 7, 8, 9, 16, 17 . One expects that higher gain settings should result in higher estimates but the large increase in variability is disturbing. Some of this increase in variability and echo-count size may be due to the presence of a substantial number of small redfish that would be detected only with high gain setting. The three net tows of the A.T. Cameron contained mainly small redfish and larger cod. The upper layers yielded large counts only for the high gain settings.

ii) Increasing gain settings should result in increasing estimates of density. The average of the mean estimates of density for low gain is 0.388, for standard gain is

0.375 and for high gain is 0.789 . It is not possible to sonify exactly the same region and a substantial amount of time elapses between successive passes so the anomaly for low and medium gain settings could be due to real changes in the estimates. However the great increase in density estimates for high gain suggest that it is important to decide upon an appropriate setting and maintain that setting if one wishes to compare densities over space and time.

TIME SERIES ANALYSIS

For each of the 20 data sets we have computed the observed and theoretical correlation function and the corresponding empirical and theoretical spectral density (spectral mass) and distribution functions. The formulae used are to be found in MacNeill (1971a). The sample spectral density was computed using the Parzen window with a truncation number of 50 and a spectral computation number of 100. The theoretical spectral density was calculated using overlap proportions appropriate to the data set and using the mean of the data set as an estimate of the Poisson parameter.

We first consider the correlation function. The first few lags of both the theoretical and the sample correlation function for the data sets are found in Table 2. It should be noted that the observations used in the computation of the correlation function are counts made at equispaced points. The distance travelled between observations is obtained by dividing the velocity (V) by the pulse rate (P). For data sets 1-5 this distance is $V/P \simeq 200$ m/min $\div$ 96 P/min $\simeq 2.08$ m/P and for the remaining data sets $V/P \simeq$ 200/48 m/P = 4.17 m/P . Thus, for example the correlation of lag 1 (in data sets 1-5) is that for consecutive observations which are taken each time the vessel has moved 2.08 m. We now discuss some of the salient points of these data.

i) Examination of the empirical correlations at all lags for the data sets 6, 7, 8, 9 reveals a very close correspondence Data sets 16, 17 are successive runs over the same region (with lower gain) and the correspondence is even closer than that for 6, 7, 8, 9 . Data sets 13, 14 are successive runs (with high gain) and the correspondence here is somewhat poorer. Data set 11 is a portion of set 13. However, they more closely resemble themselves than any other data set. Data sets 12 and 15 are from the second layers of the

Table 2. Observed and theoretical correlation functions for data sets 1-20 (Significant lags)

Lag	1		2		3		4		5	
0	1.	1.0	1.0	1.0	1.0	1.0	1.0	1.0	1.0	1.0
1	0.71	0.87	0.54	0.88	0.69	0.88	0.60	0.88	0.70	0.87
2	0.60	0.76	0.37	0.76	0.60	0.76	0.50	0.76	0.66	0.75
3	0.52	0.64	0.32	0.64	0.48	0.65	0.35	0.64	0.55	0.63
4	0.37	0.53	0.26	0.53	0.40	0.54	0.25	0.55	0.55	0.52
5	0.30	0.42	0.16	0.42	0.28	0.43	0.15	0.43	0.42	0.42
6	0.24	0.32	0.07	0.32	0.24	0.33	0.06	0.33	0.31	0.31
7	0.18	0.22	0.06	0.23	0.14	0.23	-0.03	0.23	0.24	0.21
8	0.12	0.14	0.01	0.14	0.08	0.15	-0.03	0.15	0.17	0.13
9	0.07	0.07	0.01	0.08	0.05	0.09	0.00	0.08	0.13	0.06
10	0.04	0.01	-0.03	0.02	0.00	0.02	0.00	0.02	0.10	0.01

Lag	6		7		8		9		10		11		12	
0	1.	1.	1.	1.	1.	1.	1.	1.	1.	1.	1.	1.	1.	1.
1	0.62	0.80	0.62	0.80	0.61	0.79	0.57	.80	0.28	.80	0.48	.80	0.06	0.81
2	0.47	0.61	0.44	0.61	0.46	0.59	0.42	.62	0.19	.61	0.31	.61	0.009	0.62
3	0.27	0.44	0.27	0.42	0.27	0.40	0.27	.44	0.15	.43	0.24	.42	0.02	0.44
4	0.14	0.27	0.17	0.25	0.14	0.23	0.15	.27	0.09	.26	0.19	.26	0.06	0.28
5	0.06	0.14	0.06	0.11	0.05	0.09	0.04	.13	0.03	.12	0.15	.12	0.01	0.14
6	0.04	0.03	0.01	0.02	0.02	0.004	0.03	.03	-0.04	.02	0.06	.02	-0.06	0.03

Lag	13		14		15		16		17		18		19		20	
0	1.	1.	1.	1.	1.	1.	1.	1.	1.	1.	1.	1.	1.	1.	1.	1.
1	0.41	0.81	0.52	0.80	0.08	.81	0.76	0.80	0.78	0.80	0.37	0.72	0.62	0.73	0.62	0.74
2	0.28	0.63	0.41	0.61	0.03	.62	0.63	0.60	0.62	0.60	0.09	0.46	0.42	0.47	0.45	0.49
3	0.22	0.46	0.26	0.42	0.02	.44	0.48	0.42	0.47	0.42	0.09	0.23	0.25	0.24	0.21	0.27
4	0.16	0.30	0.18	0.26	0.04	.28	0.36	0.25	0.34	0.25	0.09	0.05	0.18	0.06	0.12	0.09
5	0.12	0.16	0.09	0.12	-0.06	.14	0.25	0.11	0.25	0.11						
6	0.04	0.05	0.004	0.02	0.03	.03	0.18	0.01	0.18	0.01						

successive runs with high gain. Here there is close agreement between the two and very substantial differences with all other data sets even though the mean density is comparable with that of data sets 6, 7, 8, 9.

All these results seem to point to the facts that the correlation function consistently gives the same results if the system parameters remain the same and that it is quite sensitive to changes in certain parameters.

ii) The theoretical correlations remain the same if the overlap proportions remain the same so the close agreement between runs is to be expected.

iii) With the exception of data sets 16 and 17, the agreement between theory and practice leaves much to be desired. Generally speaking the shapes of corresponding theoretical and empirical correlation functions are similar. The empirical correlations are large for small lag and, although they are not shown in Table 2, they are small for large lag. This corresponds to theory. However, except for data sets 16 and 17, there is a substantial difference in size between the theoretical and the observed correlation. A number of possible explanations for this discrepancy could be advanced. a) The supposition that the strength of signal from all parts of the cone is equal probably erroneously enhances the theoretical correlations. b) The depth of the water is such the true overlap is probably substantially less than calculated. One can see MacNeill (1971b) for a discussion of this assumption. c) Noise likely appears in the data thus reducing the empirical correlations below predicted levels. d) Trends and transients in the data can distort the observed correlations. e) The inadequacy of the Poisson model may distort the correlation structure. f) The lowest layer is the one with all the fish in it. If the fish are in the bottom of the lower layer some will be missed by the outer edges in the sound cone. This would make the effective overlap smaller than presumed and would reduce the theoretical correlations.

iv) Unexpected results are effected by changing the gain. Since the order of overlap is determined to a certain extent by gain settings, one expects that low (high) gain and low (high) correlation go together. However, such is not the case. The low gains associated with data sets 16 and 17 result in correlations so high they more or less match the theoretical correlations. The medium gains yield somewhat lower correlation. The high gains for data sets 13 and 14 are associated with low correlation and the upper levels for this run have correlations that are very small. Several points can be made to explain this phenomenon. The first point pertains to the fairly sharp dichotomy of fish size between the redfish and cod caught by the A.T. Cameron. At sufficiently low gain the system will pick up only the large echoes that come from the cod, each fish yielding roughly the same number of echoes. At high enough gain the system also begins to pick up echoes from the small redfish. The number of consecutive transmissions for which an echo is counted for these small redfish is substantially smaller than for the cod. The effect of this is an increase in the variance and the covariances of small lag but no change in the covariances of larger lag. Consequently the correlations for large lag become smaller as the gain settings increase. Secondly, if a large amount of uncorrelated data were introduced into the system by high gain then the result would be a large amount of variability in the data and a similar effect on the correlation structure.

We now turn our attention to the spectral functions. The empirical cumulative spectral distribution function for data sets 6, 7, 8, 9, 12, 15, 16, 17 appear in Figure 1a, those for data sets 1, 2, 3, 4, 5 appear in Figure 1b and those for data sets 10, 11, 13, 14, 18, 19, 20 appear in Figure 1c. The theoretical cumulative spectral distribution functions are in Figure 1d. The empirical and theoretical spectral density functions for data set 6 are sketched in Figure 1e.

The spectral densities, by and large, have a similar shape. They have high power at low frequencies and low power at high frequencies. The tendency is for the empirical density to have less power at low frequency and more at the upper end of the spectrum than the theoretical density. This is illustrated by Figure 1e.

Comparing spectra is much more easily done using the distribution functions than the densities since the inherently erratice behaviour of densities is smoothed out when they are integrated. We use the spectra as plotted in Figures 1a - 1e to confirm certain notions that have been previously noted.

i) Data sets 6, 7, 8, 9 (Figure 1a) are very similar

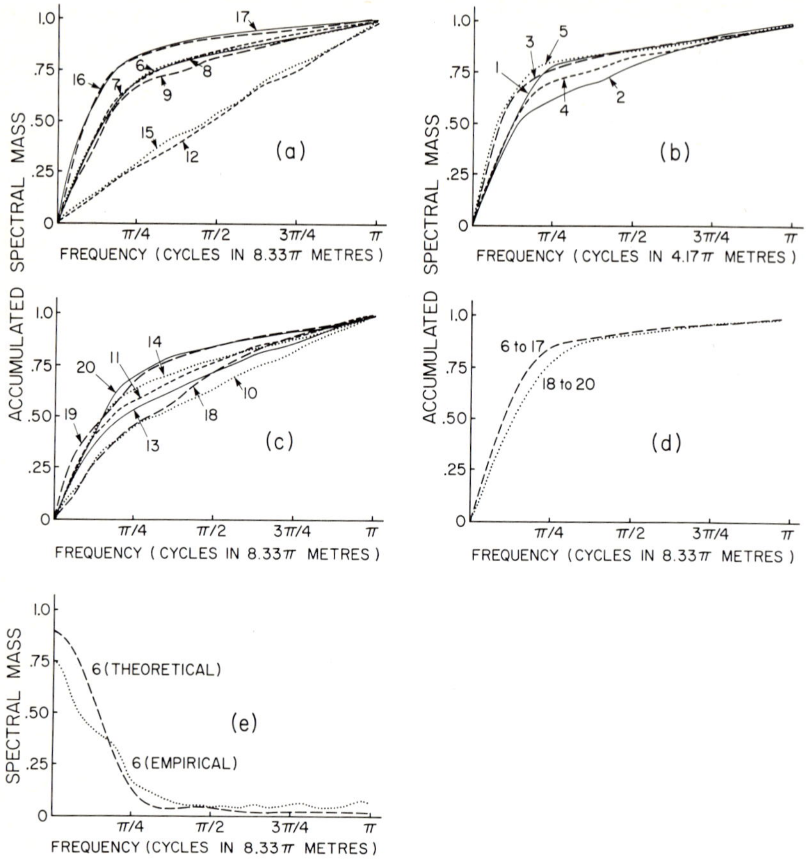

Figure 1 - Spectral Functions

(a) empirical spectral distribution functions for data sets 6, 7, 8, 9, 12, 15, 16, 17.

(b) empirical spectral distribution functions for data sets 1, 2, 3, 4, 5.

(c) empirical spectral distribution functions for data sets 10, 11, 12, 14, 18, 19, 20.

(d) theoretical spectral distribution functions.

(e) typical spectral density functions.

ii) Data sets 16 and 17 (Figure 1a) are almost carbon copies of each other.

iii) Data sets 12 and 15 (Figure 1a) closely resemble each other and both resemble white noise. Very low quality data.

iv) Spectra for data sets 13 and 14 (Figure 1c) have a substantial amount of noise in them but not nearly as much as do 12 and 15. Sets 13 and 14 have spectra that resemble one another but not nearly as well as do those sets in i) and ii).

v) An increase in replication rate results in more low frequency power as one would expect. Compare Figures 1a and 1b.

vi) A lowering of the transducer should result in less overlap and consequently less low frequency power. This is not particularly apparent from a comparison of data sets 6, 7, 8, 9 with data sets 18, 19, 20 .

vii) The theoretical spectra of Figure 1d differ from the empirical spectra by amounts large enough that an explanation is required. The factors that cause these discrepancies also cause the differences between observed and theoretical correlation and we have already discussed this.

OTHER ANALYSES AND COMMENTS

i) Runs 3, 5, 7, 9, 11 were along course 310° and took 21, 19, 20, 21, 27 minutes running time respectively. Runs 4, 6, 8, 10 were in the opposite direction and took 15, 15, 13, 14 minutes respectively. The actual length of the seabed covered was approximately equal for each run (about 10% fluctuation). That there were only 1/2 - 3/4 as many transmissions in the 130° direction as in the 310° direction would suggest that the amount of overlap and the amount of redundancy in the counts should be much less for the 130° direction. As might be expected this does not show up in the spectral analyses even though we expect substantial changes since correlations are functions of overlap proportions which in turn are functions of ship velocity relative to the water. This paradox can be explained by the outgoing tides during runs 3-11 and the currents flowing predominately in the 130° direction in the Cabot Strait. Thus, although the number of transmissions per run in the 130° direction is smaller than in the 310° direction, the ship velocity relative to the water is more or less the same in both directions and so the overlap volumes are not different enough to cause any significant change in the spectral analysis. The effect of the smaller number of transmissions in the 130° direction does show up in the raw echo-count data. If one examines Table 8, where the accumulated recorded echo-count for each of the data sets is to be found, one finds that the 310° direction count is consistently higher than the following 130° direction count. The counts for runs 3-10 are follows.

310°		130°	
Run	Count	Run	Count
3	569	4	401
5	394	6	372
7	884	8	771
9	523	10	348

Using the differences in the pairs to compute a t-statistic yields a value of 3.38 which is significant at the 5% level. The parameters of the system were held fixed for runs 3, 4, 5, 6, for runs 7, 8 and for runs 9, 10.

ii) The effect of transients on the statistical analysis is illustrated by considering layer 2 of runs 9 and 10. The same small aggregation of fish seems to have been encountered during both runs. For run 9, the echo count for the second layer was 26 and 25 coming in 8 successive counts. Thus, out of 1008 transmissions, 999 had no echoes and, of the remaining 9, 8 were consecutive. The 8 consecutive counts were 1, 1, 4, 3, 3, 5, 6, 2. This transient gives a variance estimate that is far larger than one could expect if the 1008 counts had been taken from the Poisson distribution. This phenomenon is even more exaggerated for run 10. When one computes spectra from such data, the transients show up as large power spread evenly over the spectrum.

SUMMARY OF CONCLUSIONS

A major obstacle to coming to firm conclusions on the basis of the data that has been analysed is the unintended and unavoidable mixing of effects. In carrying out experiments at sea it is virtually impossible to keep fixed everything except the parameter(s) of interest

The fish move; the water moves; the same path cannot be precisely repeated; depths are different, velocities are different. There is much variability that cannot be avoided. Large numbers of observations can be used to overcome the effects of some of this variability and we do indeed have large numbers of observation per run. However, we do not have large numbers of runs repeated under similar circumstances. Consequently, on the basis of a detailed examination of the necessarily small number of runs we put forward conclusions some of which are tentative in nature and might better be considered as hypotheses to be verified, modified or discarded on the basis of other findings, past, present and future. This cautiousness is further motivated by the need to be careful in data-sifting that one does not make too much of certain patterns and phenomena in that something is bound to be peculiar about every set of data.

Keeping in mind the caveat implied by the preceeding remarks we make the following statements.

i) The choice of gain settings is an important consideration in the use of the device. Excessively high gain settings inject noise into the data and result in low quality observations.

ii) If gain settings are not excessively high, the echo-counting apparatus gives consistent results if operated with fixed parameters. Even if the calibration of the apparatus is not accurate, one should still be able to do a good job of making relative comparisons of fish density for various places and times.

iii) The correlation and spectral analyses are relatively robust against changes in fish density but sensitive to changes in system parameters and to trends and transients in the data.

REFERENCES

[1] Dowd, R. G., 1967. An echo counting system for demersal fishes. FAO conference on fish behaviour in relation to fishing techniques and tactics. FB/67/7: 1-6.

[2] MacNeill, I. B., 1971. a) On the estimation of fish population distributions using acoustic methods, p. 553-582. In G. P. Patil [ed.] Proceedings of the international symposium on statistical ecology. Pennsylvania State Press, College Park, Penn.

[3] MacNeill, I. B., 1971. b) Quick statistical methods for analysing the sequences of fish counts provided by digital echo counter. J. Fish. Res. Bd. Canada 28: 1035-1042.

[4] MacNeill, I. B., 1972. Censored overlapping Poisson counts. Biometrika 59: 427-434.

I. B. MacNeill
Dept. of Applied Mathematics
University of Western Ontario
London, Ontario
Canada

R. G. Dowd
Bedford Institute of Oceanography
Dartmouth, Nova Scotia
Canada

Applied Statistics, R.P. Gupta, (editor)

ECOLOGICAL MODELS ON AN ENVIRONMENTAL GRADIENT[1]

E.C. Pielou

The term "statistical ecology" has come to be used to describe the labours of two quite different groups of people. On the one hand are mathematical statisticians who use ecological data as illustrative material; they do not care whether the ecological problems they solve in the process are humdrum or not. On the other hand are ecologists who use statistics as a necessary part of their research; their objectives are to find solutions to problems of genuine ecological interest and if only run-of-the-mill statistics are required to solve a problem, they feel (or should feel) no need to apologize; new statistics was not their target.

In short, the motives and loyalties of the two groups differ. And each should realize that it is asking too much to expect any one piece of research to yield both good, new ecology and good, new statistics. Occasionally this happens, but it cannot be counted upon. Therefore each group should do what it does best, but should avoid two complementary pitfalls. One is for statisticians to attack laborious but uninteresting mathematical problems that they believe (erroneously) are of interest to ecologists; a statistician should ensure that his chosen problem really is of interest to ecologists before becoming embroiled in it unless, of course, the statistics is worth doing for its own sake. The converse pitfall awaits unwary ecologists: they run the risk (and sometimes succumb to it) of "solving", by expensive computer simulations, problems that would yield (and perhaps already have yielded) to direct mathematical analysis.

The great virtue of meetings at which the two groups of workers are brought together is that they provide chances for statisticians to discover what is currently of interest to ecologists; and for ecologists to discover what new statistical insights and methods are available for them to exploit. In this paper I shall describe one of the central theoretical problems of academic ecology and discuss ways of tackling it. The purpose is to obtain by means of statistical arguments clear, unambiguous answers to simple questions. The questions, and the methods for answering them, have to be so planned that gathering the needed field data is practicable; indeed, the easier it is, the better.

[1] Supported by a grant from the National Research Council of Canada.

The ecological problem is as follows. It is well known that particular types of natural communities (forests, for instance, or tide pool communities, or freshwater zooplankton, or grassland vegetation --- examples are endless) are rich in species in some localities and poor in species in others. As a general rule, the lower the lattitude, the more numerous the species in a given kind of community. It is also known that there is a limit to the number of different species (say of trees, or of littoral invertebrates, or of zooplankters, or of grass species) that can coexist at one place. If too many species compete for the same or similar resources, natural selection must lead to the exclusion of the unsuccessful competitors. The problem is: why can a greater number of competing species coexist at some places than at others? What determines the number in a given case? And, where numerous species with apparently similar resource requirements do coexist in equilibrium, it is because their respective tolerance ranges are narrow and non-overlapping; or because they can (for unknown reasons) persist in spite of considerable overlap in their tolerances and requriements?

It occurred to me that plant communities growing in places with strong, unidirectional environmental gradients would provide good observational material. In such places the vegetation usually shows clear zonation; the zoned vegetation of mountainsides, of salt-marshes, and of intertidal seaweed beds are examples. The environmental factor that varies along the gradient (length of frost-free season; soil salinity and frequency of inundation; hours per tide cycle submerged, in the foregoing examples respectively) determines the position and width of each species' zone. Any one species can grow only in a belt where the environmental conditions are within its tolerance range; and it may be restricted to a narrower belt than this because of the presence of competitors with overlapping tolerance ranges or environmental requirements.

Consider, therefore, the zone patterns of a few chosen species that are found sometimes in species-poor communities with only a few, if any companion species, and at other times in species-rich communities with many companion species. For concreteness, consider just three species, A, B and C. The question now arises: do the zones of species A, B and C overlap to the same extent in both communities? Or, alternatively, does one find that in the species-rich community A, B and C tend to be confined to narrower zones because of the demands of the more numerous competitors? Thus suppose in the species-poor community the three species' zones can be portrayed as on the left below. The question is, is the relative pattern of these three species the same in the species-rich community, or is it as on the right, where the zones are narrower?

(Each bar represents the width of a species' zone. The zones extend left and right across the page and the gradient is up and down the page).

A way of answering the question is as follows. Suppose one lays a transect across the zones and lists the order of occurrence along the transect of the following events: first occurrence of species A, last occurrence of species A, and likewise for species B and C. Denote by A1 and A2 respectively the first and last occurrences of species A, and use corresponding symbols for species B and C. Then clearly the sequence of events as one travels down the gradient in the example on the left is

A1 B1 C1 A2 B2 C2 ;

whereas on the right it is

A1 A2 B1 B2 C1 C2 .

A test to determine whether the rankings in two different communities differ significantly from each other is therefore required.

I shall now describe a specific example. The saltmarshes occurring here and there along the southeastern shore of Nova Scotia provide excellent specimens of zoned vegetation. The zonation in these marshes is clear and obvious and all but the lowest zone (of pure *Spartina alterniflora*) support several species of plants. Two such marshes (hereafter called marsh X and marsh Y) were to be compared to judge whether what may be called the pattern of zone overlaps of eleven species* that occurred in both marshes was the same.

*The species were: *Atriplex patula*, *Spartina alterniflora*, *S. patens*, *Triglochin maritima*, *Solidago sempervirens*, *Agropyron repens*, *Juncus Gerardi*, *J. balticus*, *Poa pratensis*, *Lathyrus palustris* and *Festuca* spp.

In obtaining data on the zonation patterns of all the plants present, both marshes had been sampled by laying several transects accross them (19 across marsh X and 21 across marsh Y). Each transect consisted of a row of contiguous quadrats, each 1 m long and 0.5 m wide, and lists were made of all the species in each quadrat. It was found that the 11 species of interest had been found in 6 of the transects in each marsh. Thus each marsh yielded a sample of 6 rankings of 22 events, the starts and finishes of the 11 species. Within each marsh the correspondence among the rankings was strong, as one would expect. Measuring the degree of correspondence by Kendall's coefficient of concordance, W, it was found that $W_X = 0.946$ and $W_Y = 0.890$; (the subscripts refer to the two marshes). This coefficient is defined as

$$W = \frac{\Sigma(R_j - \bar{R})^2}{k^2 N(N^2-1)/12}$$

where R_j is the sum over all transects of the ranks of the jth event; k is the number of transects; and N is the number of observed events (which is twice the number of species). That is, W is the ratio of the observed variance of the summed ranks to the variance they would have if all k rankings were identical. We do not, of course, wish to test whether W_X and W_Y differ from zero. Obviously they must since the events ranked cannot be independent: the start of a species must always precede its finish. What is to be tested is whether the concordance between marshes was significantly less than that within marshes, as it would be if their patterns of zone overlaps differed.

One way of judging would be to determine W_{tot}, say, the concordance among all twelve transects from both marshes, but even if it were found that $W_{tot} < \text{Min}(W_X, W_Y)$ there would be no way of testing the significance of the result.

I therefore devised a procedure I have called "splicing" (because it bears a distant resemblance to jack-knifing). This consists in selecting, at random, half the transects from marsh X and half the transects from marsh Y, combining (or splicing) them, and calculating their concordance, say W_s. Since each marsh had yielded $k = 6$ transects, there were 20 ways of choosing a trio of transects from each marsh and hence 20 possible splicings in which each trio was used only once. Thus any one run of the test yielded 20 different values of W_s; and different runs could be performed which spliced these trios in different combinations. The test was run five times. Denoting by n the number of W_s values out of each set of 20 for which $W_s < \mathrm{Min}(W_x, W_y)$, the five values of n obtained were 5, 10, 9, 8, 9. Only if these were consistently much less than 10 would one reject the null hypothesis that the patterns of zone overlaps differed in the two marshes. The conclusion is therefore clear. In spite of the marked difference between these marshes in much of their species complements, the patterns of zone overlaps of the 11 shared species considered were closely similar at the two localities.

The test just described is not the only one to which the transect data from these marshes have been subjected. My object has been to wrest from the data, which are comparatively easy to collect, all that they can be made to yield in the way of clear answers to simple questions. Three other interesting conclusions have been reached, using very elementary statistical arguments.

One is an answer to this question: do the zones of the various species in a marsh tend to occur in batches (as on the left, overleaf); or do they overlap regularly like tiles on a roof (as on the right) ? There is a third possibility, of course; the arrangement of the zones may be entirely haphazard. In any case, no clear answer can be arrived at merely by looking at a marsh; the various plant species vary greatly in abundance, size and conspicuousness, and we now wish to take all species, however inconspicuous, into consideration.

(Each bar represents the width and location of one species' zone. The zones extend left and right across the page and the gradient is up and down the page.)

The following is one way of answering the question. Recall that each transect consisted of a row of meter-long quadrats. Let the number of quadrats in a given transect be T ; (T varies among the transects because of the varying widths of the marshes). And let the number of different plant species encountered in this transect be r . Obviously as one travels across the marsh from dry land towards the sea, inspecting each quadrat in turn, each one will either contain, or lack, one or more "new" species, new, that is, to the transect. (To avoid end effects, a quadrat at the start of each transect was disregarded; it was not treated as one of the T examined and its species were not included among the r whose first occurrences were observed.)

Denote by U the number of quadrats in which new species do <u>not</u> occur. Then

$$Pr(U=u) = \frac{\binom{T}{u}\binom{r-1}{T-u-1}}{\binom{T+r-1}{r}}$$

$$= \frac{T!\ (r-1)!\,r!\ (T-1)!}{u!\ (T-u)!\ (T-u-1)!\ (r+u-T)!\ (T+r-1)!}$$

By evaluating and summing these probabilities, the expected median value, $M(u)$ say, for every observed combination of T and r was found and compared with the corresponding observed u . The results were as follows:

Marsh X: $u > M(u)$ in 11 of 19 transects;

Marsh Y: $u > M(u)$ in 7 of 21 transects.

Obviously there is no reason to suppose that the species-zones started in batches or that they were regularly overlapped; they appear to have been at random.

The foregoing conclusion applies only to the starts of the zones as one goes from land to sea, that is, to the upslope zone boundaries. Now consider their downslope boundaries and write V for the number of quadrats in which new species do not occur as one travels from the seaward to the landward end of a transect. In this case the results were:

Marsh X: $v > M(v)$ in 6 of 19 transects;

Marsh Y: $v > M(v)$ in 4 of 21 transects.

Combining the results gives $v > M(v)$ in 10 of 40 transects with probability (two-tailed) of 0.002. It thus appears that the downslope boundaries of the zones tended to be evenly spaced. At their seaward ends (though not at their landward ends) the zones overlapped regularly as on the right in the text figure above.

Another interesting result was obtained by compiling, for each of the 40 transects, the following 2×2 table:

		Quadrat contains a species "start"	
		Yes	No
Quadrat contains a species "finish"	Yes	a	b
	No	c	d

We can evaluate the hypergeometric probability of each possible set of cell frequencies consistent with the given marginals and hence obtain $M(a)$, the expected median value of a . Clearly, if $a > M(a)$ improbably often one can infer that, as one crosses a marsh, there is a tendency for the beginnings of new zones to coincide with the ends of old zones. The results were:

Marsh X: $a > M(a)$ in 16 of 19 transects;

Marsh Y: $a > M(a)$ in 14 of 21 transects.

The combined result, that $a > M(a)$ in 30 of 40 transects, has probability 0.001 ; (a one-tailed test is appropriate since there is only one reasonable alternative to the null hypothesis of independence). There thus appears to be a very definite tendency for one species-zone to begin where another ends. The implication is that there are many cases in which two species occupy contiguous non-overlapping zones separated by an abrupt boundary along which, presumably, the species complete with each other.

The fourth conclusion that has so far been derived from these data has resulted from using them to test the so-called "overlapping niche" hypothesis proposed by MacArthur (1957). (The hypothesis was not suggested as a likely description of actual species interrelationships but as a bound to what could conceivably happen). In terms of organisms confined to zones on an environmental gradient, the hypothesis is that the positions of the upper and lower limits of each zone could be simulated by placing two points at random on a line down the gradient. In the context of the saltmarsh plants , the hypothesis could not be entertained in regard to all species, of course, as it had already been shown that they were not mutually independent. But it did seem worth while to test whether it applied to the "conspicuous" species, that is, the species whose presence made their zones clearly distinct on visual inspection. There were five such species in marsh X and seven in marsh Y*.

According to the hypothesis the widths of the species' zones are taken to be i.r.v.'s from the distribution of the range of two variates having identical rectangular distributions. Since the observed transects consisted of rows of metre-long quadrats, the observed zone widths were discrete variates. The width, Z, of any one species' zone, as judged from its occurrences in a single transect, was taken to be the number of quadrats between and including those containing its uppermost and lowermost occurrences. Then, for given T, the number of quadrats in the transect,

$$Pr(Z=z) = (T+1-z)\Big/\binom{T+1}{2}$$

whence $E(Z) = (T+2)/3$, $\text{Var}(Z) = (T-1)(T+2)/18$ and $\text{Var}(Z)/E(Z) = (T-1)/6$, a linear function of transect length. The results are shown in Figure 1. The straight line shows the hypothesized relationship between $\text{Var}(Z)/E(Z)$ and T. The scattered points are observed values of $\text{var}(z)/\bar{z}$; the solid and hollow dots represent transects from marsh X and marsh Y respectively.

Fig. 1.

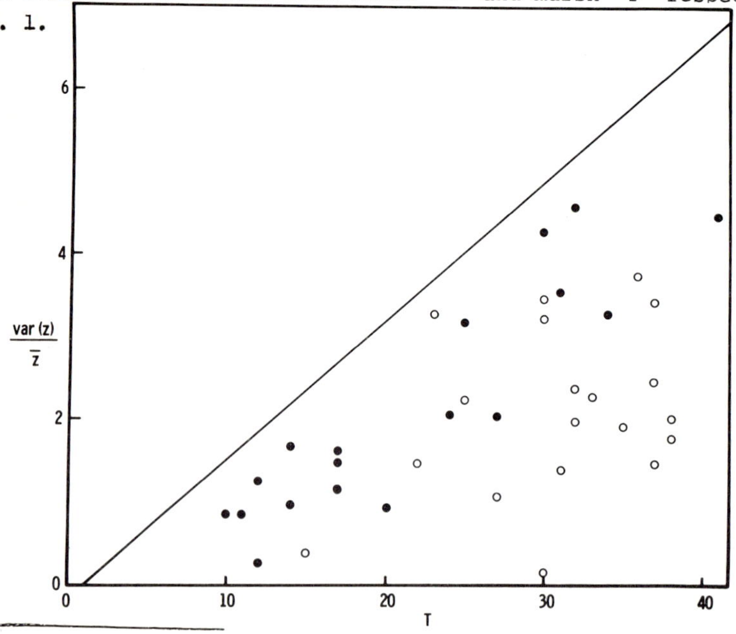

*In marsh X: *Ligusticum scothicum*, *Solidago sempervirens*, *Juncus Gerardi*, *Carex palaeacea* and *Spartina patens*.
In marsh Y: *Solidago sempervirens*, *Potentilla anserina*, *Juncus Gerardi* *Atriplex patula*, *Glaux maritima*, *Plantago maritima* and *Spartina patens*.

The conclusion is in no doubt. So far as these sets of species are concerned (recall that they were selected on the basis of their visibility as "zone-marking" species) the coefficient of variation of their zone widths is obviously less than the hypothesis predicts.

A tantalizing question remains. Does a species occupy the whole of the zone to which it is adapted, or is its "realized" zone narrower than its "evolutionary" zone because of competition with the occupants of neighbouring zones ?

Acknowledgements I thank Richard Routledge for useful discussions and mush help with the field work; and Professor John Tukey for a good suggestion.

Reference

1. MacArthur, R.H. (1957). On the relative abundance of bird species. *Proc. Nat. Acad. Sci. Wash.* 43:293-295.

E. C. Pielou
Biology Department
Dalhousie University
Halifax
Nova Scotia
Canada.

Applied Statistics, R.P. Gupta, (editor)

DESIGN OF EXPERIMENTS AND DATA ANALYSIS: A SCIENTIFIC APPROACH

J. R. Rutherford

Often statistical methods of experimental design and data analysis are recommended by enthusiasts as substitutes for ordinary scientific methods. The purpose of this paper is to acquaint statisticians with some of those scientific methods which are useful in design and analysis. The justification for doing this is that we consider that a statistician who regularly helps experimenters plan experiments and analyse data is often expected to become an expert in the scientific aspects as well as the statistical aspects of design and analysis.

As a reflex reaction some statisticians object to experiments involving one-at-a-time variation of factors. Science has made fantastic advances using one-at-a-time experiments and major advances are still being made. Cuthbert Daniel (3) has a recent paper in which he tries to get statisticians to give some respectability to one-at-a-time experiments. One purpose of the present paper is to get statisticians to think more like scientists in this respect.

A second reflex action of some statisticians is to advocate the use of a large number of factors in an experiment because they have what they consider to be appropriate tools to handle multifactor experiments. Good scientists always to try to reduce the factors to as few as possible to make interpretation easy; their ideal seems to be a reduction to only one factor. A second purpose of this paper is to show how an experienced statistician can guide an engineer or scientist through a thinking process which often results in a major reduction in the number of factors to be considered.

This paper is in two parts: the first presents an approach to the planning of experiments which is quite informal and the second presents a more formal approach based on dimensional analysis. Both approaches require substantial insight into the phenomenon being examined but we consider that both can be mastered by experienced statisticians who will extract the requisite physical insights from the scientist by skillful, directed questioning.

The third purpose of this paper is to show that statisticians and scientists do not mean the same thing when they use the phrases "planning of experiments" and "data analysis." This paper will help statisticians to develop a better understanding of the scientific meaning of these phrases. The use of the techniques and approaches discussed will help a consulting statistician to discharge the broader responsibilities often thrust upon statisticians by scientists, broader responsibilities which have been described by Alvan Feinstein in an earlier paper in this volume.

PART I - EXAMPLES OF AN INFORMAL SCIENTIFIC APPROACH

ANTHRACENE CRYSTALS: POSTULATION OF THE MECHANISM LEADS TO A NEW PROCESS

Thin crystals of anthracene were required for a product development program. The crystals were to be about 1/2 mm. in thickness and as large in area as possible. The method used to make them was as follows:

Anthracene and a half atmosphere of helium were sealed in a glass tube. The glass tube was hung in a furnace (Figure 1) at about 250°C. Slowly (one inch per hour) the tube was withdrawn from the furnace and crystals would form in the upper reaches of the tube. When the tube had cooled, the tube was broken and the crystals retrieved.

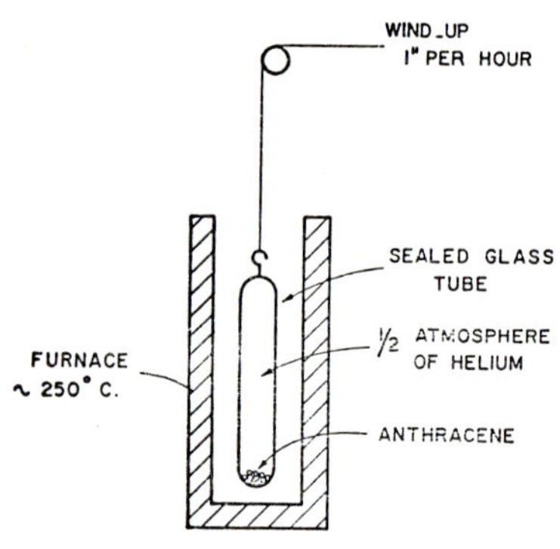

Figure 1: Schematic of the oven and tube for making large thin anthracene crystals.

The variables which were considered to be important were: surface type of tube, amount of anthracene, volume of tube, pressure of helium, temperature of the oven, speed of withdrawal, and the surroundings into which the tube is drawn.

The crystal making process was to be improved so that more, larger, flatter crystals would be made in each batch and so there were many responses of interest and these included: size of crystals and distribution of sizes in a batch, thickness and thickness variation, and number of crystals in a batch.

After some discussion a mechanism of crystal formation was postulated. At some point in the withdrawal of the tube, the top has a temperature which is below the crystallization point of anthracene vapours. Crystals then begin to form at active sites on the walls of the tube and some grow quite large and thin, others grow to be small and thick. Instead of trying to improve the existing process, it was decided to develop a new process which made it easy to maximize the time of crystal growth. A two temperature oven was designed and the concept of moving the tube was abandoned. (See Figure 2 for a representation.)

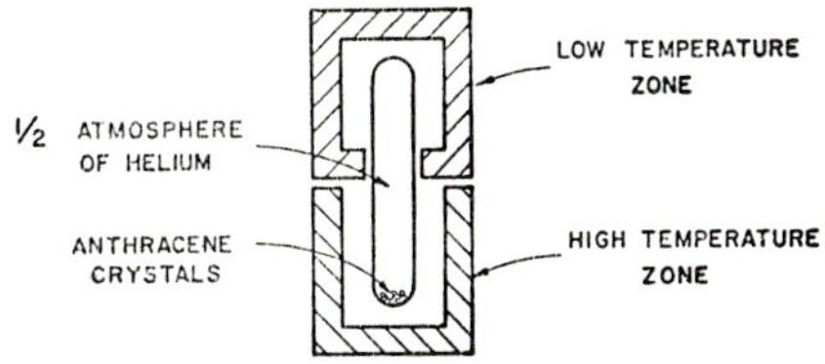

Figure 2: Schematic of the New Anthracene Crystal Making Process.

A new experimental program was planned to optimize the temperatures, helium pressure, anthracene concentration and glass type. However, the program was never completed as the first run, at what the scientist considered to be good conditions, provided enough large, thin crystals to last for a year.

THE ORIENTING OF FILM: FUNDAMENTAL VARIABLES AND MULTIPLE RESPONSE

The objective of the program was to take ordinary polyolefin film, stretch it in one direction to increase its strength making it suitable for a special application.

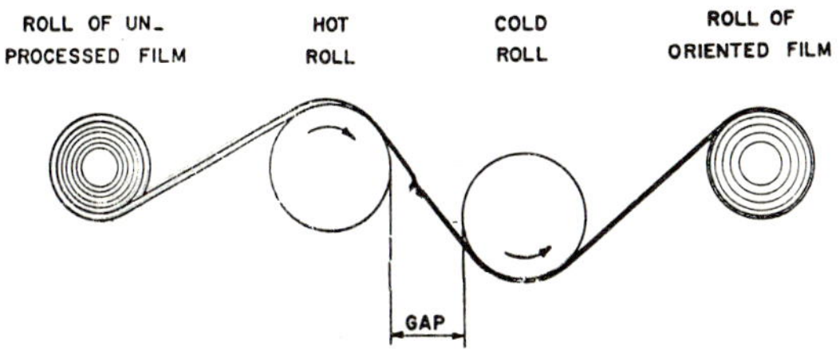

Figure 4: Schematic Diagram of the Orientation Process

In Figure 4 is a schematic representation of the process. Film is pulled over the hot roll where it is almost melted, then it is continuously transferred to the cold roll which runs several times faster than the hot roll. In the gap between the rolls the film is stretched, or drawn, several times its original length. This causes the film to be stronger because the molecules in the film have been oriented in a direction parallel to the draw direction (the strength in the transverse direction is reduced coincidently). The draw ratio is defined as ratio of final length to initial length.

The factors which were considered originally to be important included (together with number of levels): density of resin used to make film (4); molecular weight distribution of resin - broad or narrow (2); temperatures of hot roll and cold roll (3×3); length of contact of film on hot and cold rolls (2×2); film thickness - initial and final (3×2); roll speeds, hot and cold, the ratio of the ratio of the speeds is called the draw ratio (3×2); and gap, distance between rolls (2).

A complete factorial experiment, $2^5\ 3^4\ 4$, requiring 2688 runs was completely out of the question as only three months was available for development and each run would have required about eight hours to complete on the pilot plant equipment. The magnitude of the experimental effort was reduced in stages in the following way.

Changing the Objective

Background information existed and it was well known how resin density and MW distribution type would influence strength, so the program objective was changed from that of finding the best resin to that of finding a satisfactory resin. The engineer then chose a resin which he thought would be satisfactory.

Decomposition of the Process

Further discussion led to the view that the process has three basic components or phases - heating the film, drawing the film and cooling the film. It was considered that these phases were relatively independent and that manipulation of process variables, or machine setting, which controlled one phase was unlikely to influence subsequent or preceeding phases in a major way.

Heating Phase: The use of a Fundamental Variable - The objective of the first phase is to heat the film to a temperature near the melting point, thus film temperature is the fundamental variable. The machine settings: length of contact on the hot roll, hot roll speed, hot roll temperature and film thickness are, in a sense, nuisance variables. For example, so long as the contact length, roll speed and film thickness are properly coordinated, the film would be raised uniformly to a temperature near the hot roll temperature. Thus, what had appeared to be four variables could be reduced to one variable, hot roll temperature, that was very closely related to the fundamental variable important in the process.

Drawing Phase: Multiple Criteria and One Variable at a Time Sub-Experiment - The draw ratio and gap were the two machine variables capable of adjustment in this phase of the process. The two are mechanically independent and affect the final product in different ways.

The strength of the film is dependent upon how parallel the molecules are lined up; in general, the more the film is drawn the greater will be the parallel orientation of the molecules and greater will be the film strength. The draw ratio is the ratio of initial to final length and this is controlled by the ratio of the speeds of the hot and cold rolls. That is, for fixed hot roll speeds, high draw ratios are associated with high cold roll speeds.

Hence the ratio of the roll speeds could be related to the fundamental variable, draw ratio.

On the other hand, the importance of gap lay in the yield of flat film. In the middle of the rolls, the film experiences almost perfect unidirectional drawing. However, at the edges of the roll, there is some narrowing down of the width so that the edges of the stretched roll are thicker than at the centre. For a representation of the effect see Figure 5.

Section of Original Film

Section of Drawn Film

Figure 5: Hypothetical Cross Section of Original Undrawn Film and of Drawn Film With No Edge Trim Showing the Bead

It is not possible to wind-up on nice tight rolls film which has a thick edge and so the rolls have to be trimmed, and the edges are wasted. The gap controls completely the amount which has to be trimmed - that is, a large gap will give rise to a lot of trim waste.

At this stage of the experimental planning session, the engineers were asked what yields were considered satisfactory and the answer was 98%. Thus, it was planned that the first experimental runs would involve only the continuous manipulation of the gap until the trim was reduced to the allowable limit. For the rest of the program, the gap would not be adjusted.

Cooling Phase: A Second Fundamental Variable - The elongated, semi-melted film is cooled on the cold roll thus locking in the molecular orientation achieved during the drawing operation. Rapid cooling, which is associated with a low cold roll temperature, was thought to have two effects. A rapid cooling (low cold roll temperature) results in a lower density of the resin in the film and hence a lower strength for a fixed draw ratio. At the same time rapid cooling makes the film less hazy. Thus the requirement of high strength and low haze are competitive.

If the cold roll contact length, film thickness, the roll speed and cold roll temperature were such that the film was solidified before being wound up, the only real variable was the temperature to which the film had cooled by the time the film was taken off the cold roll. Thus it was concluded that the third phase was controlled by one fundamental variable, cold roll temperature, instead of three machine settings and one product variable.

Actual Experiment - The important variables had been reduced to three: hot roll temperature, draw ratio and cold roll temperature. An extra variable was added - hot roll speed - to reassure the engineer that the results to be obtained were capable of being generalized to a full sized unit. For a further discussion of this experiment see Part II in which dimensional analysis is applied.

PART II - EXAMPLES OF A FORMAL SCIENTIFIC APPROACH USING DIMENSIONAL ANALYSIS

The basic premise of dimensional analysis is that the solution of a practical problem is expressable, or representable, by means of a dimensionally homogeneous equation in terms of the characteristic parameters of the problem. The basic theorem (7; p. 18), called Buckingham's π Theorem, states that if an equation is dimensionally homogeneous then it can be reduced to a function of dimensionless groups. The meaning of these statements is best shown by an example.

THE PRESSURE DROP IN A UNIFORM STRAIGHT PIPE CONVEYING AN INCOMPRESSIBLE LIQUID (7; p. 22)

The characteristic parameters are length and diameter of the pipe (L and D), surface roughness (ϵ), average velocity of liquid (V), density and viscosity of the liquid (ρ and μ). The pressure drop is denoted by ΔP. The dimensions of these parameters are given in terms of fundamental quantities mass (M), length (L) and time (T); for example, the dimensions of ΔP are $ML^{-1}T^{-2}$.

For some arbitrary unknown function, ϕ, a dimensionally homogeneous equation is

$$\phi(\Delta P, L, D, \epsilon, V, \rho, \mu) = 0 .$$

The π theorem says that this function can be reduced to

$$\phi'(P, R, L/D, \epsilon/D) = 0$$

where $P = \dfrac{\Delta P}{\rho V^2}$ (pressure coefficient)

$R = \dfrac{DV\rho}{\mu}$ (Reynolds number)

and the four arguments of ϕ' are all dimensionless groups.

We are interested in the pressure drop so we may write

$$\frac{\Delta P}{\rho V^2} = \phi^*(R, L/D, \epsilon/D)$$

for some, still unknown, function ϕ^*. Under further assumptions, which are justified by an insight into the problem, the representation is further modified to give

$$\frac{\Delta P}{\rho V^2} = \frac{L}{D} f\ (R, \epsilon/D)$$

or

$$\Delta P = \rho V^2\ L/D\ f\ (R, \epsilon/D) \qquad (1)$$

where the function $f(R, \epsilon/D)$ is called the pipe friction factor.

ADVANTAGES OF DIMENSIONAL ANALYSIS

1. There is a reduction in the number of variables, in the above example, from six to two. The rule is as follows: If there are n characteristic parameters and k fundamental quantities, there will be n - k dimensionless groups. (k is not always

equal to three as some problems have temperature and/or electric current as fundamental quantities.)*

2. The form of the dimensionless groups often suggest ways in which the variables can be manipulated to give a maximum of information. Varying the velocity and pipe diameter can provide a complete characterization of ΔP since $R = DV\rho/\mu$ and if the same type of pipe is used, e.g. drawn copper, the surface roughness, ϵ , will remain about constant and the ratio ϵ/D will vary with $DV\rho/\mu$. This will provide a complete characterization of the pipe friction factor.
3. The use of dimensionless groups often suggests very economical ways to summarize data. In this case, figure 6 shows how the pipe friction factor, $f(R, \epsilon/D)$, varies

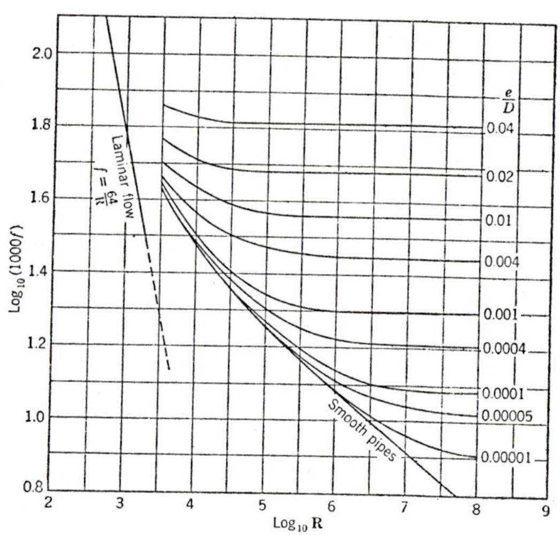

Figure 6: Pipe friction factor as a function of Reynolds number.

with Reynolds number, this chart along with equation (1) completely summarizes pressure drop of incompressible fluids in straight pipe.

---

*The reduction of six to two arose here because of the assumption that $\Delta P \propto L$ which gave f from $\emptyset$ * .

4. Extrapolation is more certain. In the example, once $f(R, \epsilon/D)$ is determined for one liquid (e.g. water) the results will apply for any incompressible fluids in any kind of pipe, with any velocity, of any length, etc.

5. The dimensionless groups sometimes have a physical interpretation which suggests interesting ways to conduct experiments.

6. Sometimes it is possible to deduce the complete form of the solution to provide insight into the underlying phenomenon. Einstein (1; p. 89) in the early days of the study of the specific heat of solids and their connection with quantum phenomena asked the question whether the same forces between the atoms which determine ordinary elastic behaviour of a solid might not also be the forces concerned in the infra-red characteristic optical frequencies. The characteristic parameters were: characteristic frequency (ν), compressibility (k), number of atoms per unit volume (N) and mass of an atom (M). Only one dimensionless group results and it is $k \nu^2 N^{1/3} M$, or $k = \text{constant}\ \nu^{-2} N^{-1/3} M^{-1}$. On substituting the known values for copper of these parameters the constant was evaluated as being .18. Debye considered this a very plausible value which lent support to a positive answer to Einstein's question. Debye then planned an experiment to prove the affirmative result.

For more information on dimensional analysis see the references which are partially annotated.

DIMENSIONAL ANALYSIS OF THE ORIENTATION OF FILM

In figure 7 is a simple representation of the process with symbols for the various characteristic parameters: initial film thickness (t), velocity of rolls (V_H and V_C) temperatures (T_H and T_C), angles of contact (θ_H and θ_C), roll diameters (D) and distance between (δ).

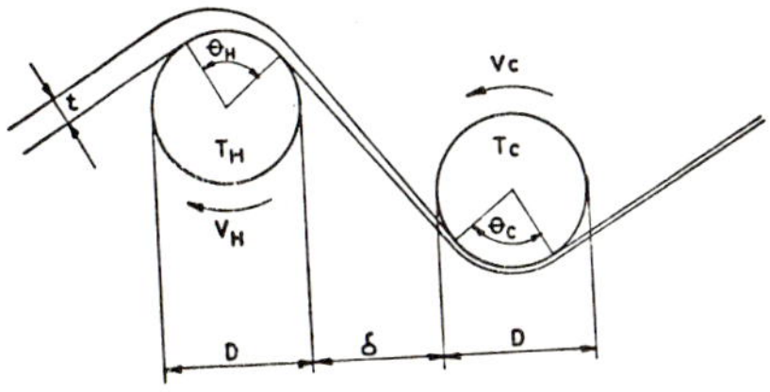

Figure 7: A simple representation of the film drawing process.

By inspection the dimensionless groups were determined as: $\pi_1 = V_H/V_C$; $\pi_2 = \delta/D$; $\pi_3 = t/D$; $\pi_4 = T_H/T_C$; $\pi_5 = \theta_H/\theta_C$ and $\pi_6 = \theta_H$.

The strength of the film is thus determined by the dimensionally homogeneous equation:

$$\sigma = \rho\, V_H^2\, \phi(\pi_1,\ \delta/D,\ \pi_2,\ \pi_3,\ \pi_4,\ \pi_5,\ \pi_6) \qquad (2)$$

where ρ is the film density and σ is film strength.

In the actual experiments on the pilot plant, the gap, δ , was determined by a one-variable-at-a-time experiment and the film velocity was virtually ignored. On scale-up, the experiments had to be repeated because the hot and cold rolls in the factory had a diameter different from those in the pilot plant and the film velocities were substantially different. If equation (2) had been known at the time of the original experiments on the pilot plant, in all likelihood the experiments would have been quite different, with the difference being in a way to have reduced the scale-up difficulties.

An important point can be made here: if there is a physical relationship between two characterizing parameters, then there will exist a dimensionally homogeneous relation. The converse is not true, that is, two parameters can be in the same dimensionless group but this does not imply a physical relation. This means that physical interpretation of dimensionless groups can lead to non-sensical inferences.

AN EXAMPLE OF SCIENTIFIC DATA ANALYSIS

Sand waves can have three forms: dunes, ripples and antidunes. The waves are formed when a fluid flows over a particulate bed. Commonly the fluid is air or water and the particles are of sand or of mud. Yalin (11) has developed an explanation of the mechanisms which lead to the development of each of the three wave forms. The postulated mechanism for ripples was arrived at by an elimination of plausible mechanisms, a procedure which was made possible because dimensional analysis had reduced very greatly the number of variables and relations which had to be considered (11; p. 228).

The characterizing parameters include particle size (D), or sand roughness (K_s) , fluid density and viscosity (ρ and μ), depth of flow (h) and various velocities: fluid bulk velocity (Um) or a shear velocity at the surface (U_*) . To describe the wave length (Λ) a sufficient set of dimensionless groups is

$$X = \frac{U_* D\rho}{\mu} \quad \text{- Reynolds number,}$$

$$Z = \frac{h}{D} \quad \text{- a roughness factor}$$

and $$Fr = \frac{Um}{\sqrt{gh}} \quad \text{- the Froude number}$$

Dunes and ripples appear only in tranquil flow, that is, for $F_r < 1$, and consequently the equation which describes their wave lengths has the form

$$\frac{\Lambda}{D} = \phi\ (X,Z)\ . \tag{3}$$

An analysis of the data of fifteen authors was performed by Yalin, the basis of the analysis being the three dimensionless groups in equation 3. The theoretical model is three dimensional and a flat representation is given in figure 8. In figure 9 is a cross section, for a set of X values, on which is plotted some of the raw data. There is a reasonable fit considering the fact that particle size, an important characteristic parameter, was not well controlled in the experiments that were reported.

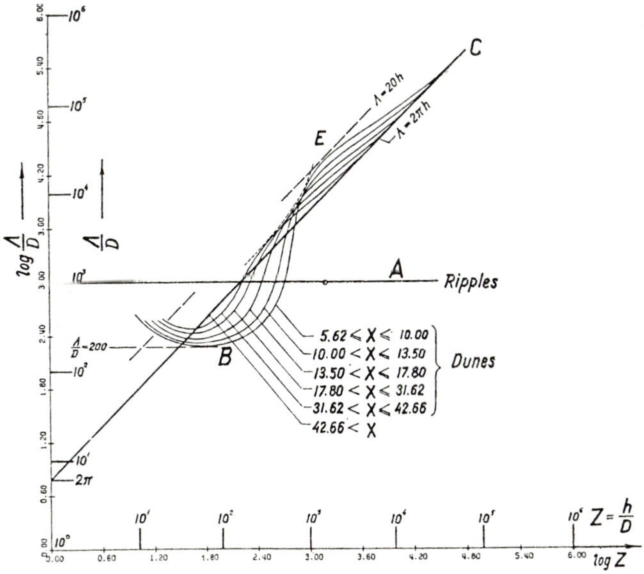

Figure 8: A flat representation of the dimensionless relation (3) which related dune and ripple wave length to a Reynolds number and a roughness factor.

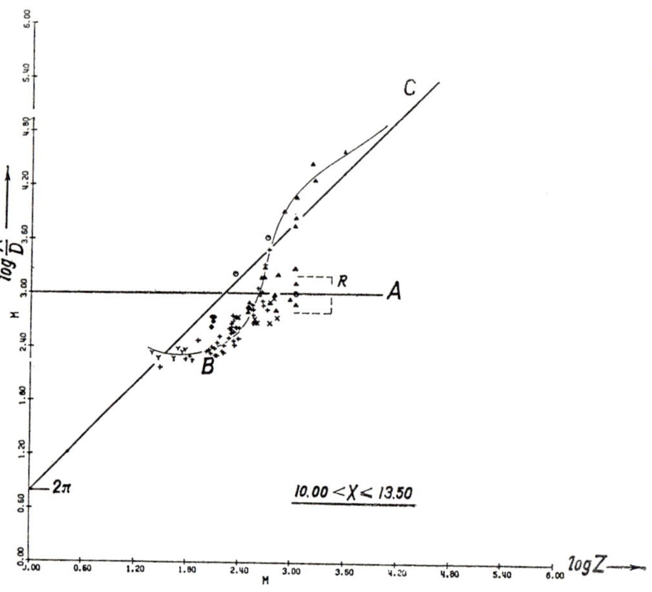

Figure 9: A cross section of the function in Figure 8 showing some of the data from the fifteen sources.

COOKING A PIE IN AN OVEN

An entirely different approach to the problem of selection of factors is given by Johnstone and Thring (4). Their method is based upon the differential (or integral) equations representation of the phenomenon being examined. For example, consider an experiment to investigate the cooking of a pie in an ordinary electric oven. An oven is very stagnant so most of the heat is transferred by radiation from the black walls of the oven.

Radiation of heat follows the integral equation (4; p. 55).

$$\frac{dQ}{dt} = \sigma e(T_1{}^4 - T_2{}^4)L^2$$

Q = net quantity of heat transferred

t = time

T_1, T_2 = absolute temperatures of hot and cold surfaces

e = combined emissivities of the hot and cold surfaces

L^2 = radiating surface area

σ = Stephan-Boltzman constant

The consequent dimensionless equation is

$$\phi\left(\frac{Q}{\sigma e L^2 T^4 t}, \frac{T_1}{T_2}\right) = 0$$

This shows what the fundamental variables might be and how the process could be expected to be modified by changing physical characteristics of the system.

In a data analysis situation, such relations could suggest to statisticians what transformations of the raw data might lead to meaningful reductions in the dimensionality of the problem. That is, the importance of the fifth order interaction $Q/\sigma e L^2 T^4 t$ would be unlikely to be discovered by a statistician doing a routine post-experiment analysis.

NON-UNIQUENESS OF THE DIMENSIONLESS GROUPS

The dimensionless groups which are derivable because of Buckingham's π Theorem are usually not unique.

DRAG ON A SHIP (7; p. 20)

The characteristic parameters are velocity (V), a shape factor (L), the drag force (F), liquid density and viscosity (ρ and μ) as well as the gravitational constant (g). A complete set of dimensionless groups is:

$$P = \frac{F}{\rho V^2 L^2} ; \quad R = \frac{VL\rho}{\mu} ; \text{ and } \frac{V^2}{Lg} .$$

Thus the work (W) in moving a ship a distance D is

$$W = F.D = \rho V^2 L^2 \phi\left(\frac{VL\rho}{\mu}, \frac{V^2}{Lg}\right) .D .$$

In water of known temperature ρ and μ would be constant and the work will depend upon a function of two arguments VL and V^2/L . This function might be approximated by a polynomial in VL and V^2/L .

However, a different set of dimensionless groups can be derived:

$$\pi_1 = V\left(\frac{\rho}{\mu g}\right)^{\frac{1}{3}} ; \quad \pi_2 = L\left(\frac{\rho^2 g}{\mu^2}\right)^{\frac{1}{3}} ; \text{ and } \pi_3 = F\frac{\rho}{\mu^2} .$$

Thus the work can be completely specified by

$$W = F.D = \mu^2/\rho \; \emptyset^* \; (V(\frac{\rho}{\mu g})^{\frac{1}{3}} \; , \; L(\frac{\rho^2 g}{\mu^2})^{\frac{1}{3}}) \; .D \; .$$

This could be approximated with a polynomial in V and L at a fixed water temperature.

This latter representation has the usual statistical form and its simplicity makes it preferable to that based on a polynomial in VL and V^2/L . Further, there is no reason to believe that the more complicated form has greater scientific validity than does the simple form.

DIMENSIONAL ANALYSIS IN SAMPLING

Drobot and Warmus (4) have looked at sampling inspection from the viewpoint of dimensional analysis. In this view, the individual elements of a lot are not considered, only an averall picture of the phenomenon of sampling is required. To explain this, consider the historical development of the laws of thermodynamics. First "certain facts were discovered and the connections between them were described phenomenologically. Thus, for example, Boyle and Marriotte's law of gases was discovered and then the first and second laws of thermodynamics. The phenomenological theory of the phenonoma did not make use at all of the atomic properties of matter and of probability methods." The subsequent application of statistical methods provided insight into more subtle problems but the basic phenomenological laws have remained, at least statistically, true.

The opposite sequence has been followed in the development of sampling theory; the theory of probability has lead to many results but a different approach is possible and it leads to new relations between lot size and value and sample size. The characterizing parameters used by Drobot and Warmus are:

size of the lot	N	
size of the sample	n	
price of the lot	C	
price of the sample	c	
conversion coefficient	q	(converts sample values to lot values)
dispersion in the lot	S	
dispersion in the sample	s	
sampling costs	k	

Two results from Cochrane (2) in the above notation are (p. 83 and 86):

for a loss function $\lambda(\hat{y} - \bar{Y})^2$ $n = \sqrt{\frac{\lambda}{k}}\, s$

for a loss function $\lambda|\hat{y} - \bar{Y}|$ $n = (\frac{\lambda s}{k\sqrt{2\pi}})^{2/3}$

where λ is "like" the conversion coefficient q.

The dimensional result for the case where sample size n depends only on lot size (N), sample cost (k), sample variance (s) and conversion coefficient (q) is

$$n \propto (\frac{Nqs}{k})^{2/3} .$$

If the sampling costs are low or negligible then the result is

$$n \propto \frac{\sqrt{NqS}}{C} .$$

If the only factors which influence the sample size are lot size, lot price and sample costs then

$$n \propto \frac{CN}{k} .$$

These last three results are interesting in that they show that the relation of sample size, n, to lot size, N, varies from $N^{2/3}$, $N^{1/2}$ to N depending upon the risk. The first two are not consistent with current statistical practice: the CGSB 105-GP-1* Inspection by Attributes has a non-constant power of N, that is, the log-log plot of n vs N is curved, but the 2/3 power is a good approximation for $N < 10{,}000$. Hald (5) has derived a relation in $n \propto \ln(N/\sqrt{\ln(N)})$ which gives a plot similar to that for the CGSB 105-GP-1 on both semi-log and log-log paper.

The relation, $n \propto N$, is consistent with practice in at least two areas. An old rule of thumb for sample size determination is called "the ten percent rule" in which the $n = .1\ N$ for all lots. A second is in stratified sampling in cases where one of the strata contains few items that have very high values. Then $n = N$ is commonly used.

CONCLUSIONS

Statisticians who are working regularly in applications can use the methods described in this paper to help scientists in the planning of experiments and in data analysis.

The statistician should learn to ask questions of the scientist, questions which will reveal the nature of the problem. In the selection of the objective or in the definition of the problems

* The Canadian version of MIL STD 105D.

some good questions have the form:

- What are you trying to do? Why?
- What information are you looking for? Why?
- What would you like to be able to do? Why?
- Will you be able to do it when you have completed this experiment? i.e. Why does the logic of the scientific inference problem require these results?
- Can you measure the thing you are interested in?
- Is what you can measure related in any way to what you would really like to measure?

In analysing the actual physical problem some good question types are:

- What is going on? What are the characteristic parameters?
- What are plausible mechanisms?
- Would you have to do the whole experiment if the mechanism were actually such and such?
- How can the process be described very briefly?
- Are the multiple responses all dependent on the same process variables?
- Can the process variables be segregated by the responses they affect? If so, could the whole experiment be broken up into a number of small, essentially independent sub-experiments?
- Do the dimensionless groups all have a reasonable physical interpretation? Does this suggest plausible mechanisms?
- Is there a fundamental variable? Or a set of fundamental variables?
- Can the fundamental variables be varied in a way to prevent interactions amongst the process variables composing the fundamental group?

These questions can be asked in a way to make any scientist cringe and become defensive. This will prevent effective communication and the statistical consultant will become ineffective. The ability to ask these types of questions without giving offence is essential for a statistician to function effectively in the broader area of science and statistics described in this paper.

References

[1] Bridgman, P. W. (1933), Dimensional Analysis, Revised Edition, Yale University Press.

This book is filled with many examples which provide great insight into the method. In this respect it is superior to Langhaar but it is quite weak on the theory and algorithms.

[2] Cochrane, W. G. (1963), Sampling Techniques, 2nd Edition, Wiley

[3] Daniel, Cuthbert (1973), One-at-at-time plans, J. Am. Stat. Assoc., 68, p. 353.

[4] Drobot, S., and M. Warmus (1954), Dimensional Analysis in Sampling Inspection, Rozprawy Matematyczne, V., Warsaw.

[5] Hald, A. (1965), Bayesian Single Sampling Attribute Plans for Discrete Prior Distributions, Mat. Fys. Sur. Dan. Vid. Selsu, 3, No. 2 Copenhagen.

[6] Johnstone, R. E. and M. W. Thring (1957), Pilot Plants and Scale-Up Methods in Chemical Engineering, McGraw-Hill.

[7] Langhaar, H. L. (1951), Dimensional Analysis and Theory of Models, Wiley.

This is an excellent book which provides a thorough discussion of the algorithms of one approach to dimensional analysis. There are also many examples.

[8] Pankhurst, R. C. (1964), Dimensional Analysis and Scale Factors, Rheinhold, New York.

This book is aimed at experimental physicists.

[9] Sedov, L. I. (1959), Similarity and Dimensional Methods in Mechanics, Infosearch Ltd., London.

[10] Yalin, M. S. (1971), Theory of Hydraulic Models, MacMillan.

This book has several good chapters on the theory of dimensional analysis which are the best of all read.

[11] Yalin, M. S. (1972), Mechanics of Sediment Transport, Pergamon Press.

J. R. Rutherford
Dupont of Canada
Kingston, Ontario
Canada

Applied Statistics, R.P. Gupta, (editor)

ON THE EXISTENCE OF AFFINE μ-RESOLVABLE BIBDS

S.M. Shah[1]

μ-resolvable and affine μ-resolvable balanced incomplete block designs were introduced by Shrikhande and Raghavarao [5]. In this paper we give some necessary conditions for the existence of an affine μ-resolvable BIBD and prove some other results.

1. Introduction

If a BIBD with parameters v, $b = mt$, $r = \mu t$, k and λ can be arranged in t sets of m blocks each such that every treatment occurs exactly μ times in each set, then the design is called μ-resolvable [5]. Further, if any pair of blocks belonging to the same set contain q_1 treatments in common, while any pair of blocks belonging to different sets contain q_2 treatments in common, then the design is called affine μ-resolvable. For a μ-resolvable BIBD, the parameters satisfy the relations:

$$(1.1) \qquad \begin{aligned} &bk = vr,\ r(k-1) = \lambda(v-1),\ b \geq v \\ &b = mt,\ r = \mu t\ . \end{aligned}$$

For an affine μ-resolvable BIBD, the parameters satisfy the relations [5],

$$(1.2) \qquad \begin{aligned} &bk = vr,\ r(k-1) = \lambda(v-1),\ b \geq v \\ &b = mt,\ r = \mu t, \\ &q_1 = (\mu-1)k/(m-1) = k + \lambda - r \\ &q_2 = \mu k/m - k^2/v \end{aligned}$$

Shrikhande and Raghararao [5] proved that the necessary and sufficient condition for a μ-resolvable BIBD to be affine μ-resolvable is that $b = v + t - 1$. Kageyama [1] and Raghavarao [4] proved that a necessary condition for the existence of a μ-resolvable BIBD with parameters v , $b = mt$, $r = \mu t$, k and λ is that $b \geq v + t - 1$.

The purpose of this paper is to give some necessary conditions for the existence of an affine μ-resolvable BIBD and to prove some other results. The results are given in section 2.

[1] Research supported by NRC Grant, Canada, No. A-4018.

2. Results

THEOREM 2.1. Each of the following conditions is necessary for the existence of a BIBD with parameters v, $b = mt$, $r = \mu t$, k, λ and $b = v + t - 1$:

(i) $b/(b-v+1) =$ integer

(ii) $r/(b-v+1) =$ integer

(iii) $b \leq v+r-1$

(iv) $\mu(\mu-1)/(m-1) =$ integer

Proof. From $b = v+t-1$, we obtain $t = b-v+1$. Hence $m = b/t$ and $\mu = r/t$ prove conditions (i) and (ii). Condition (iii) follows from condition (ii). We now prove condition (iv). Note that μ can be equal to 1, but $m > 1$.

We write

$$(2.1) \qquad \lambda = \frac{r(k-1)}{v-1} = \frac{\mu t(k-1)}{b-t}$$

$$= \frac{\mu(k-\mu)}{(m-1)} + \frac{\mu(\mu-1)}{(m-1)}$$

Now putting $b = mt$, $r = \mu t$ in $b = v + t - 1$, we obtain

$$(2.2) \qquad k - \mu = (m-1)(r-k)$$

Substituting from (2.2) in (2.1), we get

$$(2.3) \qquad \lambda = \mu(r-k) + \frac{\mu(\mu-1)}{(m-1)}$$

which proves condition (iv).

Corollary 2.1. Each of the conditions mentioned in Theorem 2.1 is necessary for the existence of an affine μ-resolvable BIBD with parameters v, $b = mt$, $r = \mu t$, k and λ.

THEOREM 2.2. The necessary and sufficient condition for the inequality $b \gtreqless v + r - 1$ to hold for any BIBD is that $r \gtreqless \lambda + k$.

Proof. Khatri and Shah [3] have proved that for any BIBD,

$$(2.3) \qquad b = v + r - 1 + k^{-1}(v-1)(r-k-\lambda) .$$

Theorem 2.2 follows from (2.4).

For a BIBD with parameters v, $b = mt$, $r = \mu t$, k, λ and $b = v + t - 1$, we have from Theorem 2.1, $b \leq v + r - 1$. Hence, applying Theorem 2.2, we obtain

Corollary 2.2. A necessary condition for the existence of a BIBD with parameters v , $b = mt$, $r = \mu t$, k , λ and $b = v + t - 1$ is that $r \leq \lambda + k$.

Corollary 2.3. A necessary condition for the existence of an affine μ-resolvable BIBD is that $r \leq \lambda + k$.

THEOREM 2.3. The necessary and sufficient condition for a BIBD with parameters v , $b = mt$, $r = \mu t$, k and λ to satisfy the inequality $b \gtreqless v + r - 1$ is that

$$\mu r - \lambda m \gtreqless \mu \quad .$$

Proof. We write

$$(2.5) \quad b = \frac{vr}{k} = \frac{(\mu r - \lambda m)(v-1)}{\mu} + r$$

$$= v + r - 1 + \mu^{-1} (v-1)(\mu r-\lambda m-\mu) \quad .$$

Theorem 2.3 follows from (2.5).

From condition (iii) and Theorem 2.3, we obtain

Corollary 2.4. A necessary condition for the existence of an affine μ-resolvable BIBD with parameters v , $b = mt$, $r = \mu t$, k and λ is that $\mu r - \lambda m \leqq \mu$.

Now it is known that the necessary and sufficient condition for the existence of an affine 1-resolvable BIBD is that $b = v + r - 1$. Hence applying Theorem 2.3, we obtain, taking $\mu = 1$

Corollary 2.5. The necessary and sufficient condition for the existence of an affine 1-resolvable BIBD is that $r = \lambda m + 1$.

THEOREM 2.4. The necessary and sufficient condition for a BIBD with parameters v , $b = mt$, $r = \mu t$, k , λ to satisfy the inequality $b \gtreqless v + t - 1$ is that $(m-1)(\mu r-\lambda m) \gtreqless \mu(m-\mu)$.

Proof. From (2.5), we have

$$(2.6) \quad b = \frac{(\mu r-\lambda m)(v-1)}{\mu} + r$$

We write (2.6) as

$$(2.7) \quad b = v + t - 1 + \frac{(v-1)(\mu r-\lambda m-\mu)+r(\mu-1)}{\mu}$$

Now $r(k-1) = \lambda(v-1)$ gives

(2.8) $k = \mu(r-\lambda)/(\mu r-\lambda m)$

Hence, we get

(2.9) $v - 1 = \frac{mk}{\mu} - 1 = r(m-\mu)/(\mu r-\lambda m)$

Substituting from (2.9) in (2.7), we get

(2.10) $b = v + t - 1 + \frac{r[(m-1)(\mu r-\lambda m)-(m-\mu)]}{\mu(\mu r-\lambda m)}$

As $\mu r - \lambda m \geq 1$ [2], the theorem follows from (2.10).

It has been proved [1] and [4] that a necessary condition for the existence of a μ-resolvable BIBD with parameters v, $b = mt$, $r = \mu t$, k and λ is that $b \geq v + t - 1$. Hence applying Theorem 2.4, we obtain

Corollary 2.6. A necessary condition for the existence of a μ-resolvable BIBD with parameters v, $b = mt$, $r = \mu t$, k and λ is that

$$(m-1)(\mu r - \lambda m) \geq \mu(m-\mu) .$$

It has been proved [5], that the necessary and sufficient condition for a BIBD with parameters v, $b = mt$, $r = \mu t$, k and λ to be affine μ-resolvable is that $b = v+t-1$. Hence applying Theorem 2.4, we obtain

Corollary 2.7. The necessary and sufficient condition for a μ-resolvable BIBD with parameters v, $b = mt$, $r = \mu t$, k and λ to be affine μ-resolvable is that

$$(m-1)(\mu r-\lambda m) = \mu(m-\mu) .$$

We now consider a BIBD with parameters v, $b = mt$, $r = \mu t$, k, λ and $b > v + t - 1$.

THEOREM 2.5. If in a BIBD with parameters v, $b = mt$, $r = \mu t$, k and λ, $b > v + t - 1$, then $\mu r - \lambda m \geq 2$.

Proof. From Theorem 2.4, we get

(2.11) $(m-1)(\mu r-\lambda m) > \mu(m-\mu)$

From (2.11), we obtain

(2.12) $(m-1)(\mu r-\lambda m-1) > (\mu-1)(m-\mu-1)$

Now $\mu \geq 1$ and $m > \mu$, hence from (2.12) we get

(2.13) $\mu r - \lambda m - 1 \geq 1$,

which paroves the theorem.

We now apply Theorem 2.5 to obtain lower bound for the number of blocks in a non-affine μ-resolvable BIBD. We use the following theorem due to Kageyama [2].

THEOREM 2.6. For a BIBD with parameters v , $b = mt$, $r = \mu t$, k and λ , if $\mu r - \lambda m > p$ for a non-negative integer p , then

$$b \geq [\lambda m^2 + (p+1)m]/\mu^2 \geq (p+1)(v-1)/\mu + r$$

In Theorem 2.5, we have proved that for a BIBD with parameters v , $b = mt$, $r = \mu t$, k , λ and $b > v + t - 1$, $\mu r - \lambda m \geq 2$. Hence applying Theorem 2.6 for $p = 1$, we obtain

Corollary 2.8. For a BIBD with parameters v , $b = mt$, $r = \mu t$, k , λ and $b > v + t - 1$,

$$b \geq (\lambda m^2 + 2m)/\mu^2 \geq 2(v-1)/\mu + r$$

Corollary 2.9. For a non-affine μ-resolvable BIBD with parameters v , $b = mt$, $r = \mu t$, k and λ ,

$$b \geq (\lambda m^2 + 2m)/\mu^2 \geq 2(v-1)/\mu + r \quad .$$

Corollaries 2.8 and 2.9 are respectively corollaries 2.1 and 2.2 proved by Kageyama [2].

References

[1] Kageyama, S. (1973). On μ-resolvable and affine μ-resolvable balanced incomplete block designs. Ann. Statist. 1, 195-203.

[2] Kageyama, S. (1973). On the inequality for BIBDS with special parameters. Ann. Statist. 1 , 204-207.

[3] Khatri, C.G. and Shah, S.M. (1963). An inequality for balanced incomplete block designs. Ann. Inst. Statist. Mathematics, 14, 95-96.

[4] Raghavarao D. (1971). Constructions and Combinatorial Problems in Design of Experiments. Wiley , New York.

[5] Shrikhande, S.S. and Raghavarao D. (1964). Affine μ-resolvable incomplete block designs. Contributions to Statistics. Volume presented to Professor P.C. Mahalanobis on his 70th birthday. Pergamon Press. Oxford and Statistical Pub. Society, Calcutta.

S.M. Shah
Department of Mathematics
St. Mary's University
Halifax, Nova Scotia
Canada

Applied Statistics, R.P. Gupta, (editor)

DERIVATION OF A PAIRED COMPARISON MODEL

by Jagbir Singh and Radhey S. Gupta

1. Introduction.

In a paired comparison experiment each object is compared with every other object. Whenever a direct comparison among three or more objects is not feasible, the method of paired comparisons presents itself naturally. The method is attributed to Fechner [3] who used it in investigations of Weber's law. Paired comparisons have been used as an indirect technique for scaling stimuli.

Thurstone [10] postulated that the frequencies with which "discriminal processes" are associated with any given stimulus form a normal distribution on the psychological scale, and the discriminal process most often associated with a given stimulus is defined as the model discriminal process. Restating Thurstone's postulates in statistical terminology we say that sensations X_i and X_j evoked in an individual by a pair of stimuli S_i and S_j are assumed to be normally distributed with modes μ_i and μ_j respectively as true sensations located on the psychological scale. Let p_{ij} be the probability that S_i is reported as greater than S_j. Thurstone's model is summarized as

$$p_{ij} = \int_{-(\mu_i-\mu_j)/\sigma}^{\infty} \phi(x)\,dx \tag{1.1}$$

where $\phi(x)$ is the standard normal density and σ depends on the variances and covariance of X_i and X_j. Equation (1.1) is known as the "Law of comparative judgement". For this model, Mosteller [7] gave least square estimates of Stimuli positions using paired comparison data. He also proposed a goodness of fit test for the model.

An alternative to the law of comparative judgement is known as Bradley-Terry [1] model. There it is postulated that

$$p_{ij} = \frac{\pi_i}{\pi_i+\pi_j} \; ; \; \pi_i \geq 0 \quad ,$$

where π_i denotes the "True rating" of stimulus S_i. Luce [5], using an "axiom of choice", derived the above model.

Thompson and Singh [9] using some probability limit theorems show that the two basic mechanisms of discrimination, "addition" and "substitution" suggested by Stevens [8] lead to respectively Thurstone's and Bradley-Terry-Luce model.

The aim of this paper is to develop a new paired comparison model based on psychophysical ideas of sensory perception. Bradley-Terry-Luce model is a special case of ours.

2. Two Hypotheses.

Our approach is based on psychophysical ideas of sensory perception. The beginning of sensory perception is the activation of many sense receptors by a stimulus. Each activated receptor sends a signal to the brain which synthesizes them into an over-all measure of sensation. Due to momentary fluctuations, a stimulus does not always excite the same member of receptors. This is our first hypothesis.

Hypothesis 1. The number of signals $N_i(t)$ registered at the brain counter during time t as a result of sense receptors activated by stimulus S_i is a stochastic process.

We understand that the actual process of converting signals into an overall measure for decision making is quite complicated. However, it is appealing to assume that signals corresponding to a weaker stimulus will reach at the brain counter less frequently than those corresponding to a stronger stimulus. The arrival time for successive signals from a process corresponding to a stronger stimulus ought to be less than those signals which come from a process corresponding to a weaker stimulus. What we have just said is, of course, an over-simplification of a complicated phenomenon, but it is a basis of our second hypotheses whose consequences we wish to study. The hypothesis is to be stated in terms of the joint waiting times of the bivariate process $N_{ij}(t) = (N_i(t), N_j(t))$. Beginning at any moment, we define the waiting time T_i to the next signal in $N_i(t)$ together with the time T_j to the next signal in the $N_j(t)$ process.

Hypothesis 2. $p_{ij} = P[T_j > T_i]$.

It should be pointed out here that in a comparison of stimuli S_i and S_j, the two signal processes $N_i(t)$ and $N_j(t)$ may not function independently owing to the "similarity" or "dissimilarity" of the stimuli. In other words the process $N_{ij}(t)$ may involve a correlation parameter. Moreover, if the joint distribution of T_i and T_j is absolutely continuous then $p_{ij} + p_{ji} = 1$, and consequently a comparison of S_i and S_j would result in a tie only with probability zero. In this study, however, the resulting joint distribution of T_i and T_j would be such that $P[T_i=T_j] = q_{ij}$, say, is not necessarily zero.

3. A Bivariate Poisson Process.

Let $p(n_1,n_2,t)$ denote the density of the random vector $N_{ij}(t)$. The following three assumptions pertain to the mechanism generating the process $N_{ij}(t)$.

1. a. $p(1,0,h) = \pi_i h+0(h); \quad \pi_i > 0$

   b. $p(0,1,h) = \pi_j h+0(h); \quad \pi_j > 0$

   c. $p(1,1,h) = \pi_{ij} h+0(h); \quad \pi_{ij} > 0$

2. $\sum_A p(n_1,n_2,h) = 0(h)$, where $A = \{(n_1,n_2);\ n_i \geq 2$ for at least one $i = 1,2\}$

3. Signals arrive independently over non-overlapping time intervals.

If $G(z_1,z_2,t)$ denote the probability generating function of $p(n_1,n_2,t)$ for $|z_i| \leq 1,\ i = 1,2,$ then we can prove that a consequence of the three assumptions is that $G(z_1,z_2,t) = \mathrm{Exp}[\{\pi_i(z_1-1)+\pi_j(z_2-1) + \pi_{ij}(z_1z_2-1)\}t]$. This expression is indeed the probability generating function of a bivariate Poisson process with $p(n_1,n_2,t) =$

$$\mathrm{Exp}[-(\pi_i+\pi_j+\pi_{ij})t]\ \Sigma \frac{(\pi_i t)^{n_1-k}}{(n_1-k)!} \cdot \frac{(\pi_j t)^{n_2-k}}{(n_2-k)!} \cdot \frac{(\pi_{ij} t)^k}{k!}$$ where the sum

is over k from zero to $\min(n_1,n_2)$.

An equivalent representation from the distribution point of view of the bivariate Poisson process $N_{ij}(t)$ is the following. There exists three independent Poisson processes $X_i(t)$, $X_j(t)$, and $X_{ij}(t)$ with parameters $\pi_i t$, $\pi_j t$, and $\pi_{ij} t$ respectively so that the joint density of $X_i(t) + X_{ij}(t)$ and $X_j(t) + X_{ij}(t)$ is $p(n_1,n_2,t)$ (see [4]). We may associate a possible interpretation to this representation. When a subject is comparing stimulus S_i and S_j, his brain is identifying signals as coming from three independent processes. Signals governed by $X_i(t)$ and $X_j(t)$ carry information about stimulus S_i and stimulus S_j respectively, whereas the signals from $X_{ij}(t)$ carry information relating to the similarity or dissimilarity of the stimuli. Regardless, an advantage of the above representation is that it is so easy to compute the following probability.

For $s,t > 0$,

$$\begin{aligned} P[T_i > S, T_j > t] &= P[N_i(s) = 0,\ N_j(t) = 0] \\ &= P[X_i(s) + X_{ij}(s) = 0, X_j(t) + X_{ij}(t) = 0] \\ &= P[X_i(s) = 0, X_j(t) = 0,\ X_{ij}(\max(s,t)) = 0] \\ &= \text{Exp}[-\pi_i s - \pi_j t - \pi_{ij}\max(s,t)]. \end{aligned} \tag{3,1}$$

In a different context, this distribution has been derived by Marshall and Olkins [6]. There they noted that T_i and T_j have the joint distribution (3.1) if and only if

$$\begin{aligned} T_i &= \min(u,w), \\ T_j &= \min(v,w) \end{aligned} \tag{3.2}$$

where u,v and w are independent exponential random variables with parameters π_i, π_j and π_{ij} respectively. That is, the probability density of u is $\pi_i \text{Exp}[-\pi_i u]$ for $u > 0$, and zero otherwise.

4. A New Paired Comparison Model.

Now we are ready to derive a new model. Using representation (3.2), we get

$$\begin{aligned} p_{ij} &= P[T_j > T_i] \\ &= P[\min(v,w) > \min(u,w)] \\ &= P[v > w > u \text{ or } w > v > u] \\ &= P[T_j > u] . \end{aligned}$$

Since T_j and u are independent, it is a matter of simple integration to find

$$p_{ij} = \frac{\pi_i}{\pi_i + \pi_j + \pi_{ij}} \tag{3.3}$$

Also, $p_{ij} + p_{ji} + q_{ij} = 1$ implies

$$q_{ij} = \frac{\pi_{ij}}{\pi_i + \pi_j + \pi_{ij}} \tag{3.4}$$

A physical interpretation for π_i is that it is a measure of the strength of stimulus S_i. On the other hand, π_{ij} a measure of extent to which stimuli S_i and S_j are similar or dissimilar.

Finally, the following two special cases of our model are immediate.

Davidson [2] in extending the Bradley-Terry-Luce model makes an assumption that the probability of tie, q_{ij} , is proportional to the geometric mean of p_{ij} and p_{ji}. He also assumes that the constant of proportionality does not depend upon any stimuli pair. Under these assumptions, regardless of their justification, we find from our model that $\pi_{ij} = \nu\sqrt{\pi_i \pi_j}$, $\nu > 0$, a constant of proportionality. For this representation of π_{ij}, our model is the one studied by Davidson. When $\nu = 0$, we get the Bradley-Terry-Luce model.

5. Summary.

Both Thurstone and Bradley-Terry-Luce models and their extensions to accommodate ties are essentially advanced on intuitive grounds. A new paired comparison model is developed using psychophysical ideas of sensory perception. Two hypotheses are advanced. One of them postulates that the number of signals registered at the brain during time t as a result of sensory receptors activated by a stimulus is governed by a stochastic process. Thus, in a paired comparison task, a bivariate process would govern the number of signals corresponding to a pair of stimuli. The second hypothesis pertains to the probability of reporting one stimulus greater than the other in terms of the joint waiting times associated with the bivariate process. An advantage of this approach is that physical interpretation of the parameters involved in the model presents itself naturally. Bradley-Terry-Luce model is a special case of this model.

Acknowledgement.

This research was partially sponsored by the Air Force Office of Scientific Research, United States Air Force, under Grant No. AF-AFOSR-72-2329. The United States Government is authorized to reproduce and distribute reprints for Governmental purposes notwithstanding any copyright notation hereon.

References

[1] Bradley, R.A. and Terry, M.E. (1952. "Rank analysis of incomplete block designs I: the method of paired comparisons", Biometrika, Vol. 39, pp.324-345.

[2] Davidson, R.R. (1970). "On extending the Bradley-Terry model to accommodate ties in paired comparison experiments", J. Am. Stat. Assoc., Vol. 65, pp. 317-328.

[3] Fechner, G.T. (1960). Elemente der Psychophysik, Breitkopf and Hartel, Leipzig.

[4] Holgate, P. (1964). "Estimation for the bivariate Poisson distribution", Biometrika, Vol. 51, pp. 241-249.

[5] Luce, R.D. (1959). Individual choice behavior, John Wiley and sons, Inc., New York.

[6] Marshall, A.W. and Olkins, I. (1967). "A multivariate exponential distribution", J. Am. Stat. Assoc., Vol. 62, pp. 30-34.

[7] Mosteller, F. (1959a). "Remark on the method of paired comparisons I: the least square solution assuming equal standard deviations and equal correlations", Psychometrika, Vol. 16, pp. 3-9.

[8] Stevens, S.S. (1957). "On the psychophysical laws", Psychophysical Review, Vol. 64, pp. 153-181.

[9] Thompson, Jr. W.A. and Singh, J. (1967), "The use of limiting theorems in paired comparison model building", Psychometrika, Vol. 32, pp. 255-264.

[10] Thurstone, L.L. (1927), "Psychophysical analysis", Am.J. Phych. Vol. 38, pp. 368-389.

J. Singh
Department of Statistics
Ohio State University
Columbus, Ohio 43210
U.S.A.

R. S. Gupta
Department of Statistics
Ohio State University
Columbus, Ohio 43210
U.S.A.

Applied Statistics, R.P. Gupta, (editor)

APPLICATION OF LINKED BLOCK DESIGNS IN SUCCESSIVE SAMPLING

Rajinder Singh[1] and D. Raghavarao[2]

Abstract

The close relationship between experimental designs and sampling techniques is only partly explored in literature (see Chakrabarti (1963), Mohanty (1971), Raghavarao and Rajinder Singh (1972) and Rajinder Singh, Raghavarao and Federer (1973)). Linked Block (LB) designs were not so far used in sampling situations and the purpose of this paper is to study the applications of LB designs to successive sampling.

1. Introduction

A linked block (LB) design is defined as follows:

Definition 1.1. A LB design is an arrangement of v symbols in b sets each of size k such that

1. Every symbol occurs at most once in a set.
2. Every symbol occurs in exactly r sets.
3. Every pair of sets has exactly μ symbols in common.

The parameters of a LB design satisfy

$$vr = bk \quad , \quad \mu(b-1) = k(r-1) \ . \tag{1.1}$$

A symmetrical BIB design with $v = b$ and consequently $r = k$ is a particular case of LB designs. For the classification and further details of LB designs we refer to Raghavarao (1971).

In sampling from finite populations, it is sometimes advantageous to make use of some auxiliary information available from the units. Ratio, Regression and Difference estimators are examples of well known statistics using auxiliary information in addition to the measurement on the study variable. When information on an auxiliary variable is not available, then one may use the observations of the study variable made on a previous occasion as auxiliary information. The technique of successive sampling was developed to meet the situation where sampling was done on two occasions having a portion of the sample common at both occasions. For further details on

---

[1] The work of this author was financially supported by an ICAR fellowship.

[2] On leave from Punjab Agricultural University. The work of this author was financially supported by NSF grants GU-2059 and 19568, while he was at the University of North Carolina, Chapel Hill.

successive sampling, we refer to Sukhatme and Sukhatme (1970).

2. Use of LB Designs in Successive Sampling Where Regression Estimator is Used on the Matched Portion

Let a Linked Block Design with parameters $v = N, b, r, k = n, \mu = m$ exists. Let the N units in the population be identified with the symbols of the design. To draw a successive sample, we first select one set of the LB design out of the total of b sets with equal probability and consider it as the sample at first occasion. Then out of the remaining $(b-1)$ sets we again select one set with equal probability and take it as the sample at second occasion. Then, clearly there are m units common between the two samples of size n each. Let the observations at first occasion give auxiliary information on the variable X and observations at second occasion represent the study variable Y.

We define the following:

$\bar{x}_m$ and $\bar{y}_m$ be the means of m common units on X and Y variables;

$\bar{x}_n$ and $\bar{y}_n$ be the means of X and Y variables on the two occasions;

$\bar{y}_{n-m}$ be the mean of Y variable based on the unmatched sample;

$s_x'^2$, $s_y'^2$ and s_{xy}' be the sample variance on X, sample variance on Y and sample covariance between X and Y based on the m common units;

$\hat{\beta} = s_{xy}'/s_x'^2$;

$\bar{X}$ and $\bar{Y}$ be the population means at first and second occasions respectively;

S_x^2, S_y^2 and S_{xy} denote the population variance at first occasion, population variance at second occasion and population covariance between first and second occasions;

$\beta = S_{xy}/S_x^2$.

Further, let

$$\bar{y}_1 = \bar{y}_m + \hat{\beta}(\bar{x}_n - \bar{x}_m) , \tag{2.1}$$

$$\bar{y}_2 = \bar{y}_{n-m}, \tag{2.2}$$

$$\bar{y}_{LB} = W_1\bar{y}_1 + W_2\bar{y}_2 , \tag{2.3}$$

where $W_i \propto 1/V(\bar{y}_i)$, for $i=1,2$, such that $W_1 + W_2 = 1$.

Let $E_1(.)$ and $E_2(.)$ denote the expectations, $V_1(.)$ and $V_2(.)$ denote the variances, $Cov_1(.,.)$ and $Cov_2(.,.)$ denote the covariances of the estimators in parenthesis for the first or second occasions indicated by the subscripts. Now

$$E(\bar{y}_1) = E(\bar{y}_m) + E[\hat{\beta}(\bar{x}_n - \bar{x}_m)]$$

$$= E_1E_2(\sum_{i=1}^{m} y_i/m) + E_1[E_2\{\hat{\beta}(E_2(\bar{x}_m) - \bar{x}_m)\}]$$

$$= E_1[\frac{r-1}{m(b-1)} \sum_{i=1}^{n} y_i] - E_1 \; Cov_2 \; (\hat{\beta},\bar{x}_m)$$

$$= \bar{Y} - E_1 \; Cov_2 \; (\hat{\beta},\bar{x}_m) \; . \tag{2.4}$$

The expression in (2.4) shows that $\bar{y}_1$ is a biased estimator for $\bar{Y}$ and the amount of bias is $E_1 \; Cov_2 \; (\hat{\beta},\bar{x}_m)$. However, the bias becomes zero if the joint distribution of Y and X is bivariate normal (see Sukhatme and Sukhatme (1970), page 195).

It can easily be shown that

$$E(\bar{y}_2) = \bar{Y} \quad , \tag{2.5}$$

and hence we have the following theorem.

Theorem 2.1.

$$E(\bar{y}_{LB}) = \bar{Y} - W_1E_1 \; Cov_2(\hat{\beta},\bar{x}_m) \; . \tag{2.6}$$

To derive the variance of $\bar{y}_{LB}$, we let

$$s'_{xy} = S_{xy} + \varepsilon_1 \; ,$$

$$s'^2_x = S_x^2 + \varepsilon_2 \; ,$$

$$\bar{x}_n = \bar{X} + \varepsilon_3 \; ,$$

$$\bar{x}_m = \bar{X} + \varepsilon_4 \; ,$$

where $E(\varepsilon_i) = 0$, $i = 1,2,3,4$.

If $V(.)$ and $Cov(.,.)$ denote the variance and covariance of the estimator in parenthesis, respectively then

$$V(\bar{y}_m) = E_1V_2(\bar{y}_m) + V_1E_2(\bar{y}_m) \tag{2.7}$$
$$= E_1V_2(\bar{y}_m) + V_1(\bar{y}_n) .$$

After a little algebra, we have

$$E_1V_2(\bar{y}_m) = \frac{1}{Nm}\sum_{i=1}^{N} y_i^2 + \frac{1}{m^2b(b-1)}\sum_{\substack{i,j=1\\ i\neq j}}^{N} y_iy_j\lambda_{ij}(\lambda_{ij}-1) - V_1(\bar{y}_n)-(\bar{Y})^2 , \tag{2.8}$$

where λ_{ij} denotes the number of sets in which the i-th and j-th symbols occur together in the LB design.

Substituting from (2.7) and (2.8) in (2.6), we get

$$V(\bar{y}_m) = \frac{1}{Nm}\sum_{i=1}^{N} y_i^2 + \frac{1}{m^2b(b-1)}\sum_{\substack{i,j=1\\ i\neq j}}^{N} y_iy_j\lambda_{ij}(\lambda_{ij}-1) - (\bar{Y})^2 \tag{2.9}$$

The following expressions can also be easily obtained:

$$V(\bar{y}_{n-m}) = \frac{1}{N(n-m)}\sum_{i=1}^{N} y_i^2 + \frac{1}{(n-m)^2b(b-1)}\sum_{\substack{i,j=1\\ i\neq j}}^{N} y_iy_j\lambda_{ij}(b-2r+\lambda_{ij})-(\bar{Y})^2, \tag{2.10}$$

$$V(\bar{x}_m) = \frac{1}{Nm}\sum_{i=1}^{N} x_i^2 + \frac{1}{m^2b(b-1)}\sum_{\substack{i,j=1\\ i\neq j}}^{N} x_ix_j\lambda_{ij}(\lambda_{ij}-1)-(\bar{X})^2, \tag{2.11}$$

$$Cov(\bar{x}_n,\bar{x}_m) = V(\bar{x}_n)$$
$$= \frac{1}{Nn}\sum_{i=1}^{N} x_i^2 + \frac{1}{n^2b}\sum_{\substack{i,j=1\\ i\neq j}}^{N} x_ix_j\lambda_{ij} - (\bar{X})^2 , \tag{2.12}$$

$$Cov(\bar{x}_n,\bar{y}_m) = Cov(\bar{x}_n,\bar{y}_n)$$
$$= \frac{1}{Nn}\sum_{i=1}^{N} x_iy_i + \frac{1}{n^2b}\sum_{\substack{i,j=1\\ i\neq j}}^{N} x_iy_j\lambda_{ij} - \bar{X}\bar{Y} , \tag{2.13}$$

$$Cov(\bar{x}_m,\bar{y}_m) = \frac{1}{Nm}\sum_{i=1}^{N} x_iy_i + \frac{1}{m^2b(b-1)}\sum_{\substack{i,j=1\\ i\neq j}}^{N} x_iy_j\lambda_{ij}(\lambda_{ij}-1)-\bar{X}\bar{Y} \tag{2.14}$$

The $V(\bar{y}_{LB})$ is given by the following theorem:

Theorem 2.2.

$$V(\bar{y}_{LB}) \cong W_1^2[V(\bar{y}_m) + \beta^2\{V(\bar{x}_m) - V(\bar{x}_n)\}$$

$$- 2\beta\{\mathrm{Cov}(\bar{x}_m,\bar{y}_m) - \mathrm{Cov}(\bar{x}_n,\bar{y}_n)\}] + W_2^2 V(\bar{y}_{n-m}), \qquad (2.15)$$

where the variances and covariances of the right hand side are given in equations (2.9) to (2.14).

Proof: Clearly

$$V(\bar{y}_{LB}) = V(W_1\bar{y}_1 + W_2\bar{y}_2)$$

$$= W_1^2 V(\bar{y}_1) + W_2^2 V(\bar{y}_2) + 2W_1W_2\,\mathrm{Cov}(\bar{y}_1,\bar{y}_2)$$

$$= W_1^2 V(\bar{y}_1) + W_2^2 V(\bar{y}_2)\ . \qquad (2.16)$$

Now

$$V(\bar{y}_1) = V[\bar{y}_m + \hat{\beta}(\bar{x}_n - \bar{x}_m)]$$

$$= V[\bar{y}_m + \frac{s'_{xy}}{s'^2_x}(\bar{x}_n - \bar{x}_m)]$$

$$= V[\bar{y}_m + \frac{S_{xy}+\varepsilon_1}{S_x^2+\varepsilon_2}(\varepsilon_3 - \varepsilon_4)]$$

$$= V[\bar{y}_m + \beta(1 + \frac{\varepsilon_1}{S_{xy}})(1 + \frac{\varepsilon_2}{S_x^2})^{-1}(\varepsilon_3 - \varepsilon_4)]\ ,$$

from which by neglecting higher order terms and retaining the first term, we get

$$V(\bar{y}_1) \cong V[\bar{y}_m + \beta(\varepsilon_3 - \varepsilon_4)]$$

$$= V(\bar{y}_m) + \beta^2[V(\bar{x}_n) + V(\bar{x}_m) - 2\,\mathrm{Cov}(\bar{x}_n,\bar{x}_m)]$$

$$+ 2\beta[\mathrm{Cov}(\bar{x}_n,\bar{y}_m) - \mathrm{Cov}(\bar{x}_m,\bar{y}_m)]$$

$$= V(\bar{y}_m) + \beta^2[V(\bar{x}_m) - V(\bar{x}_n)]$$

$$- 2\beta[\mathrm{Cov}(\bar{x}_m,\bar{y}_m) - \mathrm{Cov}(\bar{x}_n,\bar{y}_n)]\ . \qquad (2.17)$$

Again

$$V(\bar{y}_2) = V(\bar{y}_{n-m})\ . \qquad (2.18)$$

Substituting (2.17) and (2.18) in (2.16), we get the required result.

By using a symmetrical BIB design as the LB design to draw the samples, we get more neater expression for the variance of the extimator $\bar{y}_{LB}$.

3. Estimate of $V(\bar{y}_{LB})$

If $\hat{V}(.)$ denotes an unbiased estimator of $V(.)$, the following results can be easily established:

$$\hat{V}(\bar{y}_m) = \frac{N-m}{Nm^2}\sum_{i=1}^{m} y_i^2 - \frac{b(b-1)}{N^2}\sum_{\substack{i,j=1\\ i\neq j}}^{m} \frac{y_i y_j}{\lambda_{ij}(\lambda_{ij}-1)} + \frac{1}{m^2}\sum_{\substack{i,j=1\\ i\neq j}}^{m} y_i y_j, \quad (3.1)$$

$$\hat{V}(\bar{x}_m) = \frac{N-m}{Nm^2}\sum_{i=1}^{m} x_i^2 - \frac{b(b-1)}{N^2}\sum_{\substack{i,j=1\\ i\neq j}}^{m} \frac{y_i y_j}{\lambda_{ij}(\lambda_{ij}-1)} + \frac{1}{m^2}\sum_{\substack{i,j=1\\ i\neq j}}^{m} y_i y_j, \quad (3.2)$$

$$\hat{V}(\bar{y}_{n-m}) = \frac{N-n+m}{N(n-m)^2}\sum_{i=1}^{n-m} y_i^2 - \frac{b(b-1)}{N^2}\sum_{\substack{i,j=1\\ i\neq j}}^{n-m} \frac{y_i y_j}{\lambda_{ij}(b-2r+\lambda_{ij})}$$

$$+ \frac{1}{(n-m)^2}\sum_{\substack{i,j=1\\ i\neq j}}^{n-m} y_i y_j , \quad (3.3)$$

$$\hat{V}(\bar{x}_n) = \frac{N-n}{Nn^2}\sum_{i=1}^{n} y_i^2 - \frac{b}{N^2}\sum_{\substack{i,j=1\\ i\neq j}}^{n} \frac{y_i y_j}{\lambda_{ij}} + \frac{1}{n^2}\sum_{\substack{i,j=1\\ i\neq j}}^{n} y_i y_j , \quad (3.4)$$

$$\hat{Cov}(\bar{x}_m,\bar{y}_m) = \frac{N-m}{Nm^2}\sum_{i=1}^{m} x_i y_i - \frac{b(b-1)}{N^2}\sum_{\substack{i,j=1\\ i\neq j}}^{m} \frac{x_i y_i}{\lambda_{ij}(\lambda_{ij}-1)}$$

$$+ \frac{1}{m^2}\sum_{\substack{i,j=1\\ i\neq j}}^{m} x_i y_j , \quad (3.5)$$

$$\hat{Cov}(\bar{x}_n,\bar{y}_n) = \frac{N-n}{Nn^2}\sum_{i=1}^{n} x_i y_i - \frac{b}{N^2}\sum_{\substack{i,j=1\\ i\neq j}}^{n} \frac{x_i y_j}{\lambda_{ij}} + \frac{1}{n^2}\sum_{\substack{i,j=1\\ i\neq j}}^{n} x_i y_j . \quad (3.6)$$

Taking into consideration the above unbiased estimators of the variances, we get the following theorem:

Theorem 3.1. An unbiased estimator $\hat{V}(y_{LB})$ of $V(\bar{y}_{LB})$ is given by

$$\hat{V}(\bar{y}_{LB}) = W_1^2[\hat{V}(\bar{y}_m) + \beta^2\{\hat{V}(\bar{x}_m) - \hat{V}(\bar{x}_n)\}$$
$$- 2\beta\{\hat{Cov}(\bar{x}_m,\bar{y}_m) - Cov(\bar{x}_n,\bar{y}_n)\}] + W_2^2\hat{V}(\bar{y}_{n-m}), \quad (3.7)$$

where the values of various terms are given from (3.1) to (3.7).

4. Use of LB Designs in Successive Sampling Where Difference Estimator is Used on the Matched Portion.

Let the Linked Block design with parameters $v=N,b,r,k=n,\mu=m$ exist. Let the N symbols of the design be identified with the population units. Then we draw successive samples by the procedure discussed in Section 2. Following the notations of Section 2, we form the following estimator of population mean. If B is a predetermined constant, then

$$\bar{y}_1' = \bar{y}_m + B(\bar{x}_n - \bar{x}_m) , \quad (4.1)$$

$$\bar{y}_2' = \bar{y}_{n-m} , \quad (4.2)$$

$$\bar{y}_{LB}' = W_1'\bar{y}_1' + W_2'\bar{y}_2' \quad (4.3)$$

where $W_i' \propto 1/V(\bar{y}_i')$, for $i=1,2,$ such that $W_1' + W_2' = 1$. Then analogous to the results of Section 2, the following can be established.

Theorem 4.1. $\bar{y}_{LB}'$ is an unbiased estimator of $\bar{Y}$.

Theorem 4.2.

$$V(\bar{y}_{LB}') = W_1'^2[\frac{1}{Nm}\sum_{i=1}^{N} y_i^2 + \frac{1}{m^2b(b-1)}\sum_{\substack{i,j=1\\ i\neq j}}^{N} y_iy_j\lambda_{ij}(\lambda_{ij}-1) - (\bar{Y})^2$$
$$+ B^2\{\frac{1}{Nm}\sum_{i=1}^{N} x_i^2 + \frac{1}{m^2b(b-1)}\sum_{\substack{i,j=1\\ i\neq j}}^{N} x_ix_j\lambda_{ij}(\lambda_{ij}-1)-V(\bar{x}_n)-(\bar{X})^2\}$$
$$- 2B\{\frac{1}{Nm}\sum_{i=1}^{N} x_iy_i + \frac{1}{m^2b(b-1)}\sum_{\substack{i,j=1\\ i\neq j}}^{N} x_iy_j\lambda_{ij}(\lambda_{ij}-1)$$

$$- \bar{X}\bar{Y} - \left(\frac{1}{Nn}\sum_{i=1}^{N} x_i y_i + \frac{1}{n^2 b}\sum_{\substack{i,j=1\\ i\neq j}}^{N} x_i y_j \lambda_{ij} - \bar{X}\bar{Y}\right)\}]$$

$$+ W'^2_2 \left[\frac{1}{N(n-m)} \sum_{i=1}^{N} y_i^2\right.$$

$$\left. + \frac{1}{(n-m)^2 b(b-1)} \sum_{\substack{i,j=1\\ i\neq j}}^{N} y_i y_j \lambda_{ij}(b-2r+\lambda_{ij}) - (\bar{Y})^2\right]. \quad (4.4)$$

The estimated variance of $\bar{y}'_{LB}$ can be easily obtained.

5. Concluding Remarks

In the sampling procedure discussed in this paper, there is controlled selection as the matched portions of the sample and the unmatched portions do not take all possible combinations of the population units. With a suitable choice of the design and suitable numbering of the population units, we can avoid the inclusion of a set of population units of cardinality greater than 2, in the samples.

The ideas can be extended and by taking the duals of tactical configurations $C[K,\ell,\delta,v]$, we can develop sampling procedures for successive sampling in more than two occasions.

References

[1] Chakrabarti, M. C. (1963). On the use of incidence matrices of designs in sampling from finite populations. J. Indian Statist. Assoc., 1, 78-85.

[2] Mohanty, S. (1971). Unpublished Ph.D. thesis, Institute of Agricultural Research Statistics, New Delhi.

[3] Raghavarao, D. (1971). Constructions and Combinatorial Problems in Design of Experiments. John Wiley.

[4] Raghavarao, D. and Rajinder Singh (1972). Applications of PBIB designs in cluster sampling. Paper presented at the symposium on 'Combinatorial Mathematics and its applications,' Dec. 1972, New Delhi.

[5] Rajinder Singh, Raghavarao, D. and Federer, W. T. (1973). Applications of higher associate class PBIB designs in multidimensional cluster sampling. Mimeo No. BU-463-M. of Biometrics unit, Cornell University.

[6] Sukhatme, P. V. and Sukhatme, B. V. (1970). Sampling theory of surveys with applications. Asia Publishing House, Bombay, India.

R. Singh
Department of Mathematics
Punjab Agricultural University
Ludhiana, Punjab
India

D. Raghavarao
Department of Mathematics and Statistics
Guelph University
Guelph, Ontario
Canada

Applied Statistics, R.P. Gupta, (editor)

AXIAL AND BIMODAL DATA ON A SPHERE

M. A. Stephens

1. INTRODUCTION. DIRECTED AND AXIAL DATA.

1.1. In recent years techniques have been developed for dealing with statistical data where the observations are unit vectors. It is convenient to classify these vectors according to the type of data, as follows:

(a) Vectors denoting directions, say of magnetic axes of rocks, or of flights of migratory birds, for which often the parent population is considered to have a single mode. Such vectors will be called directed, and a typical vector can be represented by $\underset{\sim}{OP}$, where P is a point moving on the surface of a sphere or the circumference of a circle, with unit radius and centre O. Sometimes directed data arises with several modes, and a bimodal (or multimodal) distribution should be used to describe the data.

(b) Vectors in which the direction is not important, e.g. normals to cleavage planes of rocks, where either direction of the normal is used only to fix the plane at right angles to it. A typical vector might be recorded by $\underset{\sim}{OP}$ as above, or equivalently, by $\underset{\sim}{OT}$, where POT is a diameter of the sphere or circle. This type of data is called axial data; any treatment of such data must not depend on whether P or T is used.

In three dimensions, the distribution used to describe unimodal directed data is Fisher's (1953) distribution, and in two dimensions, it is the von Mises distribution. Considerable progress has been made in providing statistical techniques for use with these distributions. For axial data, in two dimensions, a commonly proposed technique is to double the recorded angles of the sample vectors; then the points derived from P and T will be identical, and the von Mises distribution is used to discuss the derived points. For three dimensions, this simple technique will not work. Watson (1966) has analysed the analogous distribution on the sphere, and has given useful approximate tests based on an analysis of variance technique he first developed for the Fisher distribution. Nevertheless exact tests seem to be very difficult to develop (see, e.g. Anderson and Stephens (1972) for a test of randomness against the Watson alternative).

The techniques described above for axial data can also be used for bimodal directed data, provided it is recognized that the distribution used has modes of equal weight, at opposite ends of a single diameter, and with equal concentration around each mode.

In the present paper we offer techniques of estimation and testing, for three-dimensional vectors, which are applicable to both axial data and to bimodal directed data; for the latter, the modes need not be of equal weight. It should also be possible to adapt the procedures to modes which are not opposite each other. The procedures for testing hypotheses make use of tables already prepared for the Fisher distribution. These methods are somewhat approximate, and it is hoped that they will be further tested in practical applications; nevertheless, it might be remarked that the analogous procedures have been used with success on two-dimensional data. Further, we give illustrations below which show that, where comparisons can be made, as for axial data, the techniques give good results when compared with those given by the Watson distribution.

1.2. Notation. Directed data in three dimensions will be recorded by a unit vector OP, from the centre O of a sphere of unit radius to a moving point P on its surface. An arrowhead at P gives its sense. The vector OP, or equivalently the point P, can be located by spherical polar coordinates θ, ϕ; $\theta = 0$ will often be assumed to point to the north pole, so that $\theta < \pi/2$ defines the northern hemisphere. In practical work, ϕ is sometimes called the orientation, and OP is located by dip angle δ below the equator; then $\theta = \pi/2 + \delta$. It is convenient to introduce rectangular coordinates; the z-axis along $\theta = 0$, the x-axis along $\theta = \pi/2$, $\phi = 0$, and the y-axis along $\theta = \pi/2$, $\phi = \pi/2$. Then components of OP are

$$x = \sin\theta\cos\phi\ ,\ y = \sin\theta\sin\phi\ ,\ z = \cos\theta\ .$$

For a sample of size N, let the components of OP_i be x_i, y_i, z_i, $i = 1,2,\ldots,N$; let $X = \Sigma x_i$, $Y = \Sigma y_i$, $Z = \Sigma z_i$, and let $\underset{\sim}{R}$ be the vector with components X, Y, Z; this is the vector sum or resultant of the sample, and its length is R given by $R^2 = X^2 + Y^2 + Z^2$. When several samples are present, subscript r denotes the r-th sample, e.g. N_r, $\underset{\sim}{R}_r$, R_r; N then denotes $N_1 + N_2 + \ldots + N_s$, where s is the number of samples.

2. THE FISHER DISTRIBUTION.

2.1. The Fisher density for a directed vector OP is proportional to $\exp(\kappa\cos\theta_1)$, where θ_1 is the angle between OP and a modal vector $\underset{\sim}{A}$. If for convenience we put $\underset{\sim}{A}$ along $\theta = 0$, θ_1 becomes the polar coordinate θ for P as described above, and, since an element of surface area is $\sin\theta \, d\theta \, d\phi$, where θ and ϕ are in radians, the density of θ becomes

$$(1) \qquad f(\theta) = \frac{\kappa \sin\theta}{2 \sinh\kappa} \exp(\kappa \cos\theta) , \qquad 0 \le \theta \le \pi .$$

The angle ϕ is uniform between 0 and 2π, independent of θ. The positive parameter κ measures the concentration around $\underset{\sim}{A}$; when $\kappa = 0$ the distribution reduces to the uniform distribution over the sphere. Clearly an orthogonal transformation of coordinates may be used to place $\underset{\sim}{A}$ along any vector with coordinates (θ_o, ϕ_o), though for practical purposes the more complicated expression for the density of θ and ϕ is not needed.

2.2. Estimation of $\underset{\sim}{A}$ and of κ. The maximum likelihood estimator of $\underset{\sim}{A}$ is given by the direction of $\underset{\sim}{R}$, i.e. by the unit vector $\underset{\sim}{R}/R$, and that of κ is given by the solution of

$$(2) \qquad \coth\kappa - 1/\kappa = R/N ,$$

when $\underset{\sim}{A}$ is not known; when $\underset{\sim}{A}$ is known, C replaces R in equation (2), C being the component of $\underset{\sim}{R}$ on $\underset{\sim}{A}$. A table for solving (2) is in Stephens (1967, Table 3). Note that many of the tables subsequently referred to are collected together in Vol. 2 of Biometrika Tables for Statisticians, (Pearson and Hartley, 1972).

3. BIMODAL EXTENSION OF THE FISHER DISTRIBUTION.

3.1. The natural extension of the Fisher distribution to cover bimodal data is found by superimposing two Fisher densities with opposite modal vectors. The simplest form, obtained when the line along $\theta = 0$ and $\theta = \pi$ is the modal axis (recorded by a vector $\underset{\sim}{A}$ in either sense) is

$$(3) \qquad f(\theta) = \frac{\kappa \sin\theta}{2 \sinh\kappa} (a \exp(\kappa \cos\theta) + (1-a)\exp(-\kappa \cos\theta)), \qquad 0 \le \theta \le \pi ;$$

angle Φ has a uniform distribution between 0 and 2π as before. The relative strength of the two modes is measured by the parameter a, which lies between 0 and 1. When $a = 0.5$, (3) becomes

$$(4) \qquad f(\theta) = \frac{\kappa \sin\theta}{2 \sinh\kappa} (\cosh(\kappa \cos\theta)) , \qquad 0 \le \theta \le \pi .$$

Distribution (3) will be used to describe directed data with unequal modes; distribution (4) will be used in analyzing axial data, if both ends are recorded. In general, the modal axis $\underset{\sim}{A}$ will not be known, and must be estimated. When $\underset{\sim}{A}$ is known, joining say, South and North poles, we could choose to record axial data by one point only, in the northern hemisphere; the density of θ would then be

$$(5) \qquad f(\theta) = \frac{\kappa \sin \theta}{\sinh \kappa} (\cosh(\kappa \cos \theta)) , \qquad 0 \leq \theta \leq \pi/2 .$$

This situation is probably rather unlikely to occur. An alternative distribution for axial data has been proposed by Watson (1966); the density for θ is proportional to $\exp(\lambda \cos^2 \theta)$ (if the modal vector $\underset{\sim}{A}$ is along $\theta = 0$), and $\emptyset$ is uniform as before. An example of the use of this density is given in Section 5.1.

3.2. Estimate of $\underset{\sim}{A}$ for the bimodal distribution. The usual maximum likelihood technique gives very intractable equations for the estimate of $\underset{\sim}{A}$ for distributions (3) and (4), so we give a procedure below, motivated as follows. Suppose a plane M is chosen through 0, the center of the sphere, and fixed by its normal $\underset{\sim}{n}$. If the data is axial, one end only of each recorded diameter is chosen to give a directed vector, such that all vectors lie on the same side of M. For directed data, some vectors are reversed if necessary so that, again, all vectors are on the same side of M. The resultant $\underset{\sim}{R}$ of these directed vectors clearly depends on the position of M. When M is at right angles to the modal vector $\underset{\sim}{A}$, the expected length R of $\underset{\sim}{R}$ is a maximum. This suggests that we find the plane M which gives maximum R for the data, and estimate $\underset{\sim}{A}$ by the normal to M. The steps in this procedure are then as follows:

(a) For axial data start by choosing one end only to be P; then for both axial and directed data the current direction of the vector will be denoted by OP.

(b) If a good estimate $\underset{\sim}{V}$, of length V, exists for the modal vector, find unit vector $\underset{\sim}{v} = \underset{\sim}{V}/V$, and let its components be ℓ, m, n. If no good estimate exists, start with $(\ell,m,n) = (1,0,0)$ and follow steps (c) and (d) below; then repeat with $(\ell,m,n)=(0,1,0)$ and again with $(\ell,m,n) = (0,0,1)$. Of these three, choose as initial vector that which gives largest $\underset{\sim}{R}$ in step (d) below.

(c) Suppose a typical sample vector OP_i has components x_i, y_i, z_i. Calculate $S = \underset{\sim}{v} \cdot OP_i = \ell x_i + m y_i + n z_i$. If S is negative, reverse the signs of x_i, y_i, z_i, thus reversing the direction of OP_i. It now makes an angle less than $\pi/2$ with $\underset{\sim}{v}$.

Follow this procedure for all sample vectors.

(d) Calculate the resultant $\underset{\sim}{R}$, length R , of the final set of vectors given by (c).

(e) Let the unit vector in the direction of $\underset{\sim}{R}$ be the new $\underset{\sim}{v}$, i.e. $\underset{\sim}{v} = \underset{\sim}{R}/R$. Repeat steps (c) and (d) until two successive values of $\underset{\sim}{v}$ are identical. Let this final $\underset{\sim}{v}$ be called $\underset{\sim}{v}_o$ and the final resultant be $\underset{\sim}{R}^o$, length R^o . The line along which $\underset{\sim}{v}_o$ lies is the estimate of $\underset{\sim}{A}$.

3.3. Estimate of κ - axial data. For estimating κ , we must now distinguish between distributions (3) and (4). For illustration, we take distribution (4), applicable to axial data. There are two situations to consider, when the modal axis is known (probably rather unlikely), and when it is not known. When $\underset{\sim}{A}$ is known, the maximum likelihood equation for κ is

$$(6)\qquad \coth\kappa - \frac{1}{\kappa} = \frac{1}{N}\sum_{i=1}^{N} \cos\theta_i \tanh(\kappa\cos\theta_i) ,$$

where θ_i is the angle between OP_i and $\underset{\sim}{A}$. Note that $\underset{\sim}{A}$ may point in either direction; the right hand side remains the same if all $\cos\theta_i$ change sign.

When $\underset{\sim}{A}$ is not known, we shall solve (6) with $\underset{\sim}{A}$ replaced by its estimate as found above. When $\underset{\sim}{A}$ is estimated by $\underset{\sim}{v}_o = (\ell_o, m_o, n_o)$ let the final components of a typical OP_i be x_i, y_i, z_i . Then $\cos\theta_i$ will be positive, with value $\ell_o x_i + m_o y_i + n_o z_i$, and is already available from step (c) in section 3.2. Equation (6) is then solved iteratively by the following steps.

(a) Let $\sum_i \cos\theta_i/N$ be Y_o ; let the right hand side of (6), as a function of κ , be $M(\kappa)$. Solve $\coth\kappa - 1/\kappa = Y_o$ to give an initial estimate κ_1 for κ . This may be done using e.g. Table 3 in Stephens (1967); if $Y_o \geq 8$, k_1 is $1/(1-Y_o)$.

(b) Solve $\coth\kappa - 1/\kappa = M(\kappa_1)$ for κ ; call the solution κ_2.

(c) Solve $\coth\kappa - 1/\kappa = M(\kappa_2)$; call the solution κ_3 , etc., and repeat this procedure; the sequence for k converges, and the limit is $\hat{k}$, the estimate of κ . In practical situations, κ-values are often greater than, say, 4, and the technique then converges very rapidly.

The above procedure is proved convergent as follows. The estimate $\hat{\kappa}$ is the solution of (6). Since $\tanh x < 1$, $M(\kappa_1) < Y_o$; and since $\coth\kappa - 1/\kappa$ is monotonic in κ , $\kappa_2 < \kappa_1$. But $\tanh x$ is monotonic in x , so $M(\kappa_2) < M(\kappa_1)$; so by repetition of the argument, $\kappa_3 < \kappa_2$. Similarly $\kappa_4 < \kappa_3$, etc. Also $Y_o > M(\hat{\kappa})$, so $\kappa_1 > \hat{\kappa}$, then $M(\kappa_1) > M(\hat{\kappa})$, and so $\kappa_2 > \hat{\kappa}$: similarly $\kappa_i > \hat{\kappa}$ for all i . Thus the sequence of solutions κ_i is decreasing, bounded below by $\hat{\kappa}$, and so converges. But $\coth \kappa_n - 1/\kappa_n = M(\kappa_{n-1})$, and the limiting value of κ , as $n \to \infty$, must be $\hat{\kappa}$.

3.4. Examples. In Table 1 are given measurements of vectors denoting axial data; they are normals to pebbles, considered as planes, deposited in moraines, and have been kindly supplied by Professor C. King of the Department of Geography at the University of Nottingham, England. The observations have been measured to the nearest 5 degrees; the effect of lack of precision in measurement will not be considered in this paper. The table gives also the steps in the estimation procedure. It is divided into several parts:

(a) The 25 values of θ , Φ are listed first, in degrees. The degree symbol will be omitted.

(b) Estimation of $\underset{\sim}{A}$. Unit vectors along the x, y and z axes are taken as starting values of $\underset{\sim}{v}$, assuming no initial approximation is known for the modal axis. The resultant length is given, together with its coordinates θ , Φ , and in the last column are listed those vectors which must be reversed from the original given direction to lie at an angle less than 90 with the current $\underset{\sim}{v}$.

(c) The largest $\underset{\sim}{R}$ obtained from (b) is 20.20, with vector No. 1 reversed; this $\underset{\sim}{R}$, reduced to unit length, becomes the next $\underset{\sim}{v}$ with direction cosines $l = -0.329$, $m = 0.921$, $n = -0.208$. Now vector No. 1 must be returned to its original sense (so all the vectors, as finally given, are within 90 degrees of the current $\underset{\sim}{v}$) and the new $R = 20.54$; on using this to make the new $\underset{\sim}{v}$, we get no change in $\underset{\sim}{R}$, so that this is the final resultant, $\underset{\sim}{R}^o$. Then $\underset{\sim}{v}_o$ is $\underset{\sim}{R}^o/\underset{\sim}{R}^o$; here $\underset{\sim}{v}_o$ has components -0.418, 0.889, -0.188, and polar coordinates $\theta = 100.8$, $\emptyset = 115.2$. When necessary, a subscript will denote a particular sample; Table 1 contains sample 1, and its R^o is then R_1^o .

(d) Estimation of κ . The first estimate of κ , derived from $Y_o = \underset{\sim}{R}_1^o/N$, is $\kappa_1 = 5.61$. The successive estimates converge rapidly to $\hat{\kappa} = 5.56$, following the procedure in Section 3.3. Table 2 gives only the final results for $\underset{\sim}{R}^o$ and $\hat{\kappa}$ for three other samples, labelled samples 2,3,4. Also given, for later use, are results for samples 1 and 2 taken together as one sample of 50 observations, and for samples 3 and 4 taken together. X, Y, Z are the components of $\underset{\sim}{R}^o$; θ , Φ are its coordinates in degrees.

Thus $X = R^{o} \sin\theta \cos\Phi$, $Y = R^{o} \sin\theta \sin\Phi$, $Z = R^{o} \cos\theta$. The techniques converge very rapidly for these values of κ . Even for sample 3, only 3 iterations are needed to obtain $\hat{\kappa} = 3.92$.

3.5. Estimates of κ and a, bimodal data. When density (3) is to be used, an obvious estimate $\hat{a}$ of a will be that proportion of vectors which is not reversed in the final sample used to estimate $\underset{\sim}{A}$. This is easily found from a comparison of original and final components of each vector. The equation for estimating κ becomes

$$(7) \quad \coth\kappa - \frac{1}{\kappa} = \frac{1}{N}\sum_{i=1}^{N} \cos\theta_i \frac{a\exp(\kappa\cos\theta_i) - (1-a)\exp(-\kappa\cos\theta_i)}{a\exp(\kappa\cos\theta_i) + (1-a)\exp(-\kappa\cos\theta_i)}.$$

This is solved using exactly the same iteration procedure as in section (3.3), with $M(\kappa)$ now the right-hand side of (7). When a is not known, $\hat{a}$ replaces a .

4. TESTS OF HYPOTHESES.

4.1. We now consider tests for bimodal or axial data. Tests for the unimodal Fisher distribution (1) are now extensively developed, and are given in Stephens, (1962, 1967, 1969); they are based extensively on earlier work of Watson (1956) and Watson and Williams (1956). They are collected together in Section 9 of Pearson and Hartley (1972). The tests are based on resultants $\underset{\sim}{R}$ of samples, or the components C on $\underset{\sim}{A}$ when this is known. The important test that s different samples have the same modal vector, for example, is based on the conditional distribution of $Z = R_1 + R_2 + \ldots + R_s$, given R , the length of the resultant $\underset{\sim}{R}$ obtained when all s samples are pooled together as one. The tests to be proposed below are devised to make use of the tests and tables already prepared for the Fisher distribution.

4.2. The adjusted resultant $\underset{\sim}{S}$. For the bimodal distribution when $\underset{\sim}{A}$ is not known, it would be natural to consider tests based on the distribution of R^{o} . When $\underset{\sim}{A}$ is known, possible test statistics might be C^{o} , the component of $\underset{\sim}{R}^{o}$ on $\underset{\sim}{A}$, or the sum C of the components on $\underset{\sim}{A}$ of all the given vectors, each one pointing so that its component is positive. (Note that these two statistics are not necessarily the same. The value of C^{o} is the sum of the components of the individual vectors, as finally oriented to obtain $\underset{\sim}{R}^{o}$. One of these might be at 89 degrees to $\underset{\sim}{R}^{o}$, and if A were at 10 degrees to $\underset{\sim}{R}^{o}$, this vector would be at more than

90 degrees to $\underset{\sim}{A}$, thus giving a negative component in the $\underset{\sim}{A}$ direction. This vector would have to be reversed to calculate C, which would then be larger than C^o). However, even the distribution of R^o is not known, and we consider an alternative statistic. Suppose, for a given sample of size N, we have found the estimates $\hat{A}$ and $\hat{\kappa}$ by the above methods; we then ask for the vector $\underset{\sim}{S}$ which would have given the same estimates, of modal vector $\underset{\sim}{A}$ and of κ, on the assumption $\underset{\sim}{S}$ is the resultant of a <u>Fisher-distributed sample</u> of the same size. Clearly $\underset{\sim}{S}$ will lie along $\underset{\sim}{R}^o$, and its length S is easily found from the calculations leading to $\hat{\kappa}$ for the bimodal sample. Consider, for example, Sample 1. The final $\hat{\kappa}$ is 5.56, obtained by solving $\coth \kappa - 1/\kappa = 0.8200$; comparison with equation (2) for the Fisher distribution shows that 0.8200 must be the value of S/N, so that $S = 0.8200 \times 25 = 20.50$. In general, then, if the final estimate of κ is $\hat{\kappa}$, $\underset{\sim}{S}$ is placed along the direction of $\underset{\sim}{R}^o$, and its length S is given by

$$S = N(\coth \hat{\kappa} - 1/\hat{\kappa}) .$$

We call $\underset{\sim}{S}$ the <u>adjusted</u> resultant of the bimodal sample; note that it might point along either direction of the estimated modal axis, according to the final direction taken by $\underset{\sim}{R}^o$. In effect we have imagined constructing a Fisher sample from the original data, with the same N, κ, and modal vector (based on our estimates), and $\underset{\sim}{S}$ would then be its resultant. The values of $\underset{\sim}{S}$ are included in Table 2. Since we have tests and tables available for the Fisher distribution, based on the Fisher resultant $\underset{\sim}{R}$, we could now, as an approximate procedure for the bimodal distribution, use the same tests, with the adjusted resultant $\underset{\sim}{S}$ replacing $\underset{\sim}{R}$.

4.3. <u>One sample tests</u>. This procedure will first be illustrated with one-sample tests on Sample 1.

<u>Test for the modal axis $\underset{\sim}{A}$</u>. Suppose the null hypothesis is H_o : $\underset{\sim}{A}$, for Sample 1, is along $\theta = 90$, $\Phi = 90$, i.e. the y-axis. The appropriate Fisher distribution test is based on the conditional distribution of R given the component C on the hypothesized modal vector; if R is too large, H_o is rejected. Here, the value of S is 20.50; it lies along the direction of $\mathbf{R}^o$, and its component C is then 18.22. The test, for $\alpha = 0.05$, uses Figure 2 of Stephens (1962), or approximations which accompany it for the case when the component, called X in the Figure, is beyond the range given. In this case, the approximation in Section 3.3 of Stephens

(1962) applies; the critical value of S, say S_o, will be calculated from $\frac{S_o - C}{N - S_o} = \frac{F_{2,48}(\alpha)}{24}$; $F_{2,48}(\alpha)$ is the usual F-distribution, here with 2 and 48 degrees of freedom, at the level α. For $\alpha = 0.05$, we get $S_o = 19.016$. Since S is greater than S_o, H_o is rejected at the 5% level.

Note that for reasonably large κ, R^o very nearly equals S, as here, and the test could be made using R^o with no change in the result.

4.4. Tests for several samples.

The following procedure is suggested for tests on s samples, to make use of existing tables for Fisher - distributed samples.

(a) Let the S values of the individual samples be S_i, $i = 1,2,\ldots,s$;

(b) Pool the s samples into one, and calculate its $\underset{\sim}{S}$, called $\underset{\sim}{Q}_{12\ldots s}$, with length $Q_{12\ldots s}$.

(c) Treat the $\underset{\sim}{S}_i$ like individual Fisher resultants, and $\underset{\sim}{Q}_{12\ldots s}$ like the overall Fisher resultant when s samples are pooled, and use multi-sample test procedures given by Stephens (1969) and Watson and Williams (1956). (We demonstrate in Section 4.5 that $\underset{\sim}{Q}_{123}$, say, is not the same as $\underset{\sim}{S}_1 + \underset{\sim}{S}_2 + \underset{\sim}{S}_3$; hence the new letter Q). We give one illustration of a multisample test.

Test for a Common Modal Axis. A test for H_o : that s samples with the same unknown κ, have a common modal axis, will be based on the arithmetic sum $S_1 + S_2 + \ldots + S_s$ given $Q_{12\ldots s}$. Suppose we test this hypothesis, at the 5% level, for samples 1 and 2. $S_1 = 20.50$, $S_2 = 20.78$, and $Q_{12} = 40.47$. The test follows the procedure in Stephens (1969). The steps are as follows:

(a) Calculate $W = Q_{12}/N = 0.810$.

(b) Calculate $Z = (S_1+S_2)/N = 0.826$.

(c) Use Section 2 of Stephens (1969) to find the critical value z for $N = 50$, $W = 0.810$, and $\alpha = 0.05$; using the F-approximation there given, we find $z = 0.821$.

(d) Since Z exceeds z, reject H_o at the 5% level. Z is in fact just significant at the 2.5% level (critical value $z = 0.824$).

Following Section 4.3, we might consider using values of R^o instead of values of $\underset{\sim}{S}$. First note that when samples are pooled together, the pooled $\underset{\sim}{R}^o$ is not necessarily the same as $\underset{\sim}{R}_1^o + \underset{\sim}{R}_2^o$, as it would be for Fisher samples, even when the individual resultants are directed to give maximum vector sum. For example, consider samples 3 and 4. To get maximum vector sum, we would need to reverse one resultant, say $\underset{\sim}{R}_4^o$. The vector sum would then have X component 16.74 + 12.25 = 28.99. But the pooled resultant has X = 28.16, so the pooled resultant is not the same as the vector sum; for this reason we call it $\underset{\sim}{P}_{34}^o$, length P_{34}^o . The Y and Z components display similar disparities. The difference occurs because one vector (there could be more) in sample 1 finishes in one sense when $\underset{\sim}{R}_3^o$ is found, and is reversed during the calculation of $\underset{\sim}{P}_{34}^o$. For samples 1 and 2, it so happens that $\underset{\sim}{P}_{12}^o = \underset{\sim}{R}_1^o + \underset{\sim}{R}_2^o$, as can easily be checked from the components in Table 2. We use these samples to illustrate how $\underset{\sim}{Q}_{12}$ is not $\underset{\sim}{S}_1 + \underset{\sim}{S}_2$. $\underset{\sim}{S}_1$ lies along $\underset{\sim}{R}_1^o$, with length 20.50 whereas R_1^o is 20.54. Thus components of S are 20.50/20.54 of those for R^o , namely -8.57, 18.22 and -3.85. Similarly components for $\underset{\sim}{S}_2$ are 20.79/20.81 of those for R_2^o , i.e. -7.28, 19.03, 4.10 . $\underset{\sim}{S}_1 + \underset{\sim}{S}_2$ add to a vector with components 15.85, 37.25, 0.25. However, $\underset{\sim}{Q}_{12}$ has components 40.47/40.55 of those for $\underset{\sim}{R}_{12}^o$, i.e. 15.85, 37.24, 0.24 . These are only very slightly different from those for $\underset{\sim}{S}_1 + \underset{\sim}{S}_2$, but serve to illustrate the point. The difference occurs because the length S of $\underset{\sim}{S}$, compared with R , depends on the κ value, and this is different from sample 1 to sample 2, and different again from the value for the pooled samples. There will be even less reason for $\underset{\sim}{Q}_{34}$ to be the same as $\underset{\sim}{S}_3 + \underset{\sim}{S}_4$, since, apart from this effect of different κ, even $\underset{\sim}{P}_{34}$ is not the same as $\underset{\sim}{R}_3^o + \underset{\sim}{R}_4^o$ to start with. This of course will be a common situation. Nevertheless, for reasonably large κ , all the differences are slight as we have seen, and one could use R^o and P^o values instead of S and Q values. Thus, returning to the test above, we would use P_{12}^o in step (a) and $R_1^o + R_2^o$ in step (b); this gives W = 0.811 and Z = 0.827. Again, Z is just significant at the 2.5% level; the critical values go up very slightly with the slight increase in W , but so also does Z . The example illustrates again that for reasonably large κ , the S and R statistics will give the same results; so also will Q and P statistics. This is especially so in the tests of a conditional nature: those for the modal axis for one sample, and those for a common modal axis for several samples. Nevertheless, since S and Q statistics are straightforward to calculate, it is suggested that they be used.

5. COMPARISON OF RESULTS.

5.1. For the axial data of Samples 1 to 4, it is instructive to compare results with those given by assuming Watson's distribution (Section 3.1). Use of this distribution, with estimation as described in Watson (1966) gives the following modal axis estimates for Samples 1 to 4 in Table 1; Sample 1, $\theta = 102.6$, $\Phi = 113.6$; Sample 2, $\theta = 78.2$, $\Phi = 111.7$; Sample 3, $\theta = 98.2$, $\Phi = 148.4$; Sample 4, $\theta = 106.2$, $\Phi = 127.9$. The results are in good agreement with those given in Table 1. Testing procedures are not yet as well developed for this distribution as for the Fisher distribution and of course it cannot be used with directed data with unequal modes.

5.2. Importance of knowledge of the type of data.

We conclude by emphasizing that it is sometimes difficult to judge the type of data from its appearance, especially when, in three dimensions, projections must be used to display the points. Axial data, for example, gathered by several different experimenters with some showing preference for recording one sense and some for another, could have the appearance of bimodal data with unequal modes. Similarly, bimodal data in which a is large, say .85, would have most observations in one hemisphere, but a few in the other, and might be thought to be Fisher distributed. In the first case, the techniques of the present paper would be used both times, and, at least for $\underset{\sim}{A}$, the same estimates would be found. In the second case, however, use of the Fisher distribution where (3) above should be used would generally produce very different κ estimates and possibly fairly different estimates of $\underset{\sim}{A}$. Thus it is important to know the type of data before deciding on the analysis.

Acknowledgements. This work was partly supported by the National Research Council of Canada, and also by the U.S. Office of Naval Research, Contract Nonr 225(52); the author expresses his thanks to both agencies for this support.

TABLE 1

Three-dimensional axial data: estimation procedures for Sample 1

Sample 1: Coordinates θ, Φ of 25 vectors: i is the vector number.

i	θ	Φ	i	θ	Φ	i	θ	Φ	i	θ	Φ
1	80	190	7	110	95	13	115	130	19	70	70
2	110	70	8	110	120	14	100	140	20	80	90
3	115	70	9	110	120	15	115	140	21	65	100
4	130	80	10	125	120	16	125	140	22	75	110
5	100	90	11	100	125	17	120	150	23	75	110
6	105	90	12	105	125	18	110	115	24	80	115
									25	50	150

Estimation of modal axis $\underset{\sim}{A}$: final estimate underlined.

$\underset{\sim}{v}$ direction cosines l	m	n	R	coordinates of R θ	Φ	Vector numbers of vectors reversed
1	0	0	12.21	84.3	332.3	1; 7 to 18; 21 to 25
0	1	0	20.20	102.0	109.7	1
0	0	1	11.68	41.5	296.9	2 to 18
-0.329	0.921	-0.208	20.54	100.8	115.2	none
-0.418	0.889	-0.188	20.54	100.8	115.2	none

Estimation of κ .

$R^o = 20.54$; $Y_o = R^o/N = 0.8218$ $\qquad \kappa_1 = 5.61$

new R.H.S. $\qquad M(\kappa_1) = 0.8201$ $\qquad \kappa_2 = 5.56$

new R.H.S. $\qquad M(\kappa_2) = 0.8200$ $\qquad \kappa_3 = 5.56$; $\underline{\hat{\kappa} = 5.56}$

true final resultant $R_1^o = 20.54$; adjusted Fisher resultant $S_1 = 20.50$

TABLE 2

Three-dimensional axial data: results for 4 samples, and pooled samples.

Sample No.	Resultant					Size	Adjusted	
	X	Y	Z	θ	Φ	R^o	S	$\hat{\kappa}$
1	-8.59	18.26	-3.86	100.8	115.2	20.54	20.50	5.56
2	-7.29	19.05	4.10	78.6	110.9	20.81	20.79	5.95
3	16.74	-8.11	2.95	99.0	154.2	18.84	18.64	3.92
4	-12.25	15.09	-6.03	107.2	129.1	20.35	20.26	5.28
1&2	-15.88	37.31	0.24	89.7	113.1	40.55	40.47	5.25
3&4	28.16	-24.98	8.63	77.1	318.4	38.62	38.15	4.21

REFERENCES

[1] Anderson, T.W. and Stephens, M.A. (1972). Tests for randomness of directions against equatorial and bimodal alternatives. Biometrika, 59, 613-622.

[2] Fisher, R.A. (1953). Dispersion on a sphere. Proc. Roy. Soc. A 217, 295-305.

[3] Pearson, E.S. and Hartley, H.O. (1972). Biometrika Tables for Statisticians, Vol. 2. Cambridge: Cambridge University Press.

[4] Stephens, M.A. (1962). Exact and approximate tests for directions. 2. Biometrika, 49, 547-552.

[5] Stephens, M.A. (1967). Tests for the dispersion and for the modal vector of a distribution on a sphere. Biometrika, 54 211-223.

[6] Stephens, M.A. (1969). Multisample tests for the Fisher distribution. Biometrika, 56, 169-182.

[7] Watson, G.S. (1956). Analysis of dispersion on a sphere. Mon. Not. Roy. Astr. Soc.: Geophys. Suppl. 7, 153-159.

[8] Watson, G.S. (1966). The statistics of orientation data. Jour. Geology, 7, 786-797.

[9] Watson, G.S. and Williams, E.M. (1956). On the construction of significance tests on the circle and the sphere. Biometrika, 43, 344-352.

M.A. Stephens
Dept. of Applied Mathematics
McMaster University
Hamilton, Ontario
Canada

Applied Statistics, R.P. Gupta, (editor)

TESTING FOR EQUALITY OF MEANS IN TWO GAMMA DISTRIBUTIONS

T.W.F. Stroud

1. INTRODUCTION

In an important paper on the asymptotic theory of likelihood ratio tests, Wald (1943) introduced another, easier-to-compute large-sample testing procedure using a quadratic form based on unrestricted maximum likelihood estimates (MLEs). This test is straightforward to apply to hypotheses of complicated form which arise in practice (Stroud, 1974a) and indeed may be computed automatically without deriving the analytic form of the test statistic (Lord, 1973).

The asymptotic (large-sample) distribution of this quadratic form, referred to hence as the "Wald statistic", is central chi-square under the null mypothesis and noncentral chi-square under a sequence of local alternatives converging to the null at the rate $n^{-1/2}$. This result has recently been derived for exponential families assuming mild regularity conditions (Stroud, 1973). A common problem the statistician faces when trying to use such asymptotic results is that of not knowing whether his sample size is large enough for the asymptotic results to be approximately correct. In this report we study this problem when the null hypothesis is the equality of means of two gamma distributions, with both shape parameters and both scale parameters unknown and possibly unequal, and when the data consists of uncensored samples of possibly unequal sizes.

The question of equality of means is one of several aspects of the gamma distribution studied in a Montecarlo simulation described elsewhere (Stroud, 1974b). Samples were generatod of sizes $n=10$, 25, 100 and 400 with values of the shape parameters ranging from $\alpha=1$ to 24. The actual (n,α) combinations used and the simulation results for frequency of rejection (under both null and alternative values) of the hypothesis of equality of means are described in Section 4, following discussion of an application in Section 2 and introduction of the test statistics in Section 3. The test statistics are the Wald statistic , a modified form of the Wald statistic, and also the normal-theory t-test with conservative and nonconservative critical points. Conclusions are summarized in Section 5.

2. APPLICATION TO WEATHER MODIFICATION EXPERIMENTS

In cloud-seeding experiments, designed to study the possibility of stimulating rain, it is a fairly common practice to use the gamma distribution as a model for the distribution of precipitation, conditional on this precipitation being nonzero. The usual assumption is that the effect of cloud seeding, if it exists, will be multiplicative, i.e., that cloud seeding may change the scale parameter but not the shape parameter of the distribution. The data presented in Schmid (1967, Table V), however, seemed to indicate an increase in the shape parameter under cloud seeding, and for this reason Schmid was reluctant to make the multiplicative assumption. If the seeded and unseeded distributions are allowed to differ in both parameters, one may still ask whether or not there is a significant change in the average precipitation on days with precipitation, i.e., whether the means of the two gamma distributions are different.

3. THE MODEL AND THE TEST STATISTICS

The gamma density is

$$f(x;\alpha,\beta) = x^{\alpha-1}e^{-x/\beta}/\Gamma(\alpha)\beta^{\alpha} , \quad x > 0 \tag{3.1}$$

where α is the shape parameter and β the scale parameter, both positive. The mean of the distribution is $\mu = \alpha\beta$ and the variance is $\alpha\beta^2 = \mu^2/\alpha$. The MLEs $\hat{\alpha}$ and $\hat{\beta}$ are the solutions of

$$\begin{aligned} \log \hat{\alpha} - \psi(\hat{\alpha}) &= \log \bar{x} - \bar{w} \\ \hat{\beta} &= \bar{x}/\hat{\alpha} \end{aligned} \tag{3.2}$$

where $\bar{x} = (\Sigma x_i)/n$, $\bar{w} = (\Sigma \log x_i)/n$ and $\psi(\alpha)$ is the digamma function, defined by $\psi(\alpha) = d \log \Gamma(\alpha)/d\alpha$, and which is tabulated (e.g., Abramowitz and Stegun, 1964). The first equation of (3.2) must be solved using tables, e.g., Greenwood and Durand (1960).

Let samples of sizes n_1 and n_2 be observed from gamma distributions with parameters (α_1,β_1) and (α_2,β_2) , respectively. We wish to test the null hypothesis of equal means:

$$H_0 : \alpha_1\beta_1 - \alpha_2\beta_2 = 0 .$$

For a hypothesis such as this one which is given by a single scalar equation, the Wald statistic is equal to the square of the MLE of the left side of H_0 divided by the MLE of the variance of the numerator. Its approximate distribution is central or noncentral χ^2 with 1 d.f. For H_0 as stated above, the expression for the Wald statistic is $(\bar{x}_1-\bar{x}_2)^2/\{(\hat{\alpha}_1\hat{\beta}_1^2;n_1) + (\hat{\alpha}_2\hat{\beta}_2^2/n_2)\}$, but to permit use against a one-sided alternative (if desired) it is more usful to compute the square root (whose distribution is approximately normal):

$$W = \frac{\bar{x}_1 - \bar{x}_2}{\sqrt{\frac{\hat{\alpha}_1\hat{\beta}_1^2}{n_1} + \frac{\hat{\alpha}_2\hat{\beta}_2^2}{n_2}}} = \frac{\bar{x}_1 - \bar{x}_2}{\sqrt{\frac{\bar{x}_1^2}{n_1\hat{\alpha}_1} + \frac{\bar{x}_2^2}{n_2\hat{\alpha}_2}}} \tag{3.3}$$

The denominator terms $\bar{x}_1^2/n_1\hat{\alpha}_1$ and $\bar{x}_2^2/n_2\hat{\alpha}_2$ are the MLEs of the variances of $\bar{x}_1$ and $\bar{x}_2$, i.e., of $\hat{\mu}_1^2/n_1\alpha_1$ and $\mu_2^2/n_2\alpha_2$, respectively. However, under the null hypothesis μ_1 and μ_2 are equal, so that it would be reasonable to substitute a common estimate $\bar{x}$ in the denominator in place of $\bar{x}_1$ and $\bar{x}_2$, yielding the modified (square root) Wald statistic

$$W^* = \frac{\bar{x}_1 - \bar{x}_2}{\bar{x}\sqrt{\frac{1}{n_1\hat{\alpha}_1} + \frac{1}{n_2\hat{\alpha}_2}}}, \tag{3.4}$$

where $\bar{x} = (n_1\bar{x}_1+n_2\bar{x}_2)/(n_1+n_2)$. W^* has the same central (but not noncentral) asymptotic distribution as W.

If instead of estimating the variance of $\bar{x}_1 - \bar{x}_2$ by MLE we use the method of moments, the corresponding result is the large-sample normal-theory t statistic:

$$t = \frac{\bar{x}_1 - \bar{x}_2}{\sqrt{\frac{s_1^2}{n_1} + \frac{s_2^2}{n_2}}} \tag{3.5}$$

where s_1^2 and s_2^2 are the sample variances. Since the sample variances are consistent estimators of the true variances (though less efficient than MLEs), this t-statistic is asymptotically normal under sampling from the gamma model. We conjecture that certain well-known facts about normal sampling might carry over into gamma sampling, i.e., that rejection regions obtained using normal distribution tables are nonconservative and that a conservative test may be obtained by using Students' t-distribution with $\min(n_1-1, n_2-1)$ degrees of freedom. In order to check these conjectures, t was calculated and the simulation and rejection frequencies for both normal and "conservative" Student critical regions were noted. In doing so, we also attempt to provide power comparisons between tests based on t and tests based on Wald and modified-Wald statistics.

The approximate degrees of freedom given by Welch (1947) were not used because they rely more heavily on normal sampling theory for their justification.

4. MONTECARLO RESULTS

For each of the selected combinations of n_1, n_2, α_1 and α_2 (see Table 1), 800 independent samples of independent gamma observations were generated, with samples of size n_1, shape parameter α_1 and scale parameter 1, alternating with samples of size n_2, shape parameter α_2 and scale parameter 1. This yielded 400 pairs of samples, each pair consisting of an (n_1, α_1) sample and an (n_2, α_2) sample. Each gamma observation was obtained as the negative logarithm of the product of α pseudo-random uniform numbers.

Table 1 shows the number of times (out of 400) the hypothesis $\mu_1 = \mu_2$ was rejected when the tests based on W, W^* and t were applied with nominal significance level .05. The portion of Table 1 labelled "Hypothesis True" was based on scale parameters $\beta_1 = 1$ and $\beta_2 = \alpha_1/\alpha_2$, implying $\mu_1 = \mu_2$. The statistics $\bar{x}_2, \hat{\alpha}_2$, and s_2^2 were easily transformed to the new β_2 value from the generated value $\beta_2 = 1$ upon multiplication by β_2, 1 and β_2^2, respectively. The column labelled "t" refers to the statistic (3.5) with critical points calculated from the Student distribution with $\min(n_1-1, n_2-1)$ degrees of freedom, whereas the "z" column refers to the same statistic with standard normal tables.

For the portions of Table 1 labelled "Hypothesis False," values of β_2 were calculated such as would yield a power (probability of rejection) equal to 0.5, according to the approximating noncentral chi-square distribution (Wald, 1943; Stroud, 1973). These values were

$$\beta_2 = \frac{1 - \frac{3.84}{n_2\alpha_2}}{1 \pm \sqrt{1 - \left(1 - \frac{3.84}{n_1\alpha_1}\right)\left(1 - \frac{3.84}{n_2\alpha_2}\right)}} ,$$

with the minus sign yielding $\mu_2 > \mu_1$ and the plus sign $\mu_2 < \mu_1$. The rows of Table 1 are arranged according to values of $\min(\alpha_1, \alpha_2)$ within values of $\min(n_1, n_2)$.

TABLE 1

Number of rejections in 400 trials of the Hypothesis $\mu_1 = \mu_2$, tested at .05 nominal level by Wald test W , modified Wald test W* , normal-theory tests t (conservative) and z (nonconservative); when $\beta_1 = 1$, $\mu_1 = \mu_2$ and when $\beta_1 = 1$, $\mu_1 \neq \mu_2$, β_2 chosen to give asymptotic power of 1/2 .

				Hypothesis True				Hypothesis False				Hypothesis False			
				$\mu_2 = \mu_1$				$\mu_2 < \mu_1$				$\mu_2 > \mu_1$			
n_1	n_2	α_1	α_1	W	W*	t	z	W	W*	t	z	W	W*	t	z
100	10	8	1	46	30	39	49	243	89	203	233	217	357	141	219
100	10	3	3	42	39	27	43	215	135	174	211	230	309	161	213
25	10	8	8	33	31	19	32	205	181	159	194	225	258	169	214
25	10	13	21	40	39	14	36	212	197	157	204	209	225	149	197
100	10	13	13	36	41	17	31	211	173	162	201	227	262	169	217
25	100	1	5	26	26	24	32	215	316	192	218	212	126	197	207
25	25	2	10	30	27	20	25	205	240	176	200	218	186	199	214
25	100	18	2	18	18	10	16	210	206	184	197	198	194	173	196
25	25	3	5	23	21	21	22	211	229	189	208	207	196	185	197
25	25	15	24	22	24	14	19	209	219	189	205	211	208	183	205
100	25	21	21	22	23	15	20	200	184	178	196	215	233	197	213
100	100	1	1	20	21	20	20	199	202	196	201	208	213	207	215
100	100	10	18	21	22	21	22	213	220	209	215	187	184	180	182
100	100	15	24	23	23	22	24	183	188	179	182	211	218	213	218
400	400	1	1	14	14	14	14	181	181	182	183	225	224	225	227
400	400	2	18	15	15	14	14	185	197	184	187	211	205	209	212
400	400	3	5	15	16	16	16	184	185	184	184	215	212	215	215
400	400	8	10	16	16	15	15	210	211	208	209	190	189	188	190
400	400	13	15	18	18	18	18	209	209	208	208	190	190	187	188
400	400	21	24	18	17	17	18	180	180	179	180	217	217	216	216

Looking at the "Hypothesis True" portion to check the true level of the tests, we note that a binomial random variable x with $n = 400$ and $p = .05$ falls in the region $13 \leq x \leq 29$ 95% of the time, so that when the number of rejections is in this range there is no evidence that the nominal level is incorrect. This holds for each of the four regions W, W^*, t and z in all cases when $\min(n_1,n_2) \geq 100$. When $\min(n_1,n_2) = 25$, t performs conservatively once and W and z each perform nonconservatively once. The average number of rejections over the 6 runs with $\min(n_1,n_2) = 25$ for each of the four critical regions was 23.5, 23.2, 17.3 and 22.3 respectively, which seems to indicate that for this sample size the t critical region is slightly conservative and the other three are slightly nonconservative. The first row of Table 1 indicates that when $\min(n_1,n_2)=10$ even the "conservative" t region may be nonconservative when one or more of the shapes parameter values is low (i.e., when one of the gamma distributions is markedly non-Gaussian).

Turning to the "Hypothesis False" portion of Table 1, and noting that a binomial random variable x with $n = 400$ and $p = .50$ falls in the region $181 \leq x \leq 219$ 95% of the time, we see that when $\min(n_1,n_2) \geq 100$ the nominal power is fairly close, though for 2 of the 9 such runs all 4 tests had rejection frequencies slightly outside these limits. If we look at the six runs with $\min(n_1,n_2) = 25$ we see that the W and z tests appear to have the right nominal power, but that the behaviour of W^* is quite asymmetric with respect to the two selected β_2 values, according to rows 6 and 7, especially 6. Clearly the use of $\bar{x}$ in the denominator of (3.4) may result in a badly estimated variance of $\bar{x}_1-\bar{x}_2$ when the null hypothesis is false.

The average number of rejections over the 12 tests with $\min(n_1,n_2) = 25$ (6 runs, 2 tests per run) for each of the four critical regions was 209, 211, 187, and 205, respectively. Comparing these figures with the average rejection rates with H_0 true, we can see no indication that either Wald test is more powerful than a normal-theory t-test with the same actual level. The situation is similar for larger sample sizes.

5. CONCLUSION

In summary, within the scope of the selected (n,α) combinations, it appears that the asymptotic theory is quite accurate when applied to the case of $\min(n_1,n_2) \geq 100$ and is fairly accurate (except for W^*, for which the correct noncentral distribution has not been evaluated) when $\min(n_1,n_2) = 25$. Discrepancies between asymptotic probabilities and relative frequencies realized in simulation are more striking when either α_1 or α_2 is low. Since the simulation did not deal with values of α less than 1, the conclusions cannot be applied here and further study is needed. There is no evidence that any of the hypothesis testing procedures studied is more efficient than any of the others.

References

[1] Abramowitz, M. and Stegun, I.A., ed. (1964). Handbook of Mathematical Functions. National Bureau of Standards, Washington, D.C.

[2] Greenwood, J.A. and Durand, D. (1960). Aids for fitting the gamma distribution by maximum likelihood. Technometrics, 2, 55-65.

[3] Lord, F.M., (1973). Automated hypothesis tests and standard errors for nonstandard problems. Research bulletin 72-42 (rev.1973), Educational Testing Service, Princeton, N.J.

[4] Schmid, P. (1967). On "Grossversuch III", a Randomized Hail Suppression Experiment in Switzerland." Proc. 5th Berkeley Symp. on Math. Statist. and Prob., Vol. 5, 141-159.

[5] Stroud, T.W.F. (1973). Noncentral convergence of Wald's large-sample test statistic in exponential families. Ann. Statist., 1, 161-165.

[6] Stroud, T W.F. (1974a). Comparing regressions when measurement error variances are known. Psychometrika, 39, 53-68.

[7] Stroud, T.W.F. (1974b). Approximate confidence intervals and tests for the gamma distribution: a Montecarlo evaluation. Research bulletin in preparation, Educational Testing Service, Princeton, N.J.

[8] Wald, A. (1943). Tests of statistical hypotheses concerning several parameters when the number of observations is large. Trans. Amer. Math. Soc., 54, 426-482.

[9] Welch, B.L. (1947). The generalization of Student's problem when several different population variances are involved. Biometrika, 34, 28-35.

T.W.F. Stroud
Department of Mathematics
Jeffery Hall
Queen's University
Kingston, Ontario
Canada K7L 3N6

Applied Statistics, R.P. Gupta, (editor)

LAGUERRE SERIES FORMS OF THE DISTRIBUTIONS OF CLASSICAL TEST-STATISTICS AND THEIR ROBUSTNESS IN NON-NORMAL SITUATIONS

M. L. Tiku

Summary.

Approximations to the power functions of Student's t (used for testing an assumed value of the mean of a normal population), Snedecor's F (used for testing the equality of the variances of two normal populations) and Fisher's F (used in analysis-of-variance), in situations of non-normality are obtained in terms of Laguerre polynomials and Gamma density functions. The numerical values of type I error and the power of Student's t and Fisher's F, and the values of type I error of Snedecor's F, in situations of non-normality are computed. Student's t and Fisher's F are shown to be insen - itive to moderate non-normality.

Introduction

Many investigations have been made to study the effect of non-normality on type I errors of the tests based on Student's t, Snedecor's F, and Fisher's F (see references [1], [3], [4], [6], [8], [10], [11], [12], [13], [17], [18], [29], [30], [35] and [36]) but very few attempts are there to investigate the effect on the power of these tests. Donaldson [8] obtained values of the power of Fisher's F for normal, exponential and log-normal distributions through Monte Carlo methods. Srivastava [25, 26] obtained approximations to the power-functions of Student's t and Fisher's F basing his derivations on the first four terms of the Edgeworth series. This series represents only some of the near-normal populations (see reference [2]). An entirely different approach is adopted here to obtain approximations to these power-functions. The distributions of the components in t^2 and F are formally expanded in terms of Laguerre polynomials and Gamma density functions. The results derived are not restricted to populations of any special form like Edgeworth series. They also provide a simple and elegant method of judging the degree of approximation of the distributions for large samples. It may however be noted that these Laguerre series expansions (equation (4.2)) might have their own limitations which are still to be investigated.

1. Numerical Values of Power of Student's t.

Let $(x_1, x_2, \ldots, x_n)$ be a random sample from $N(\mu, \sigma)$, a normal population with mean μ and variance σ^2. Then to test

$$H_0: \mu = 0 \text{ against } H_1: \mu \neq 0 \tag{1.1}$$

the test most frequently employed is based on Students $t = \bar{x}\sqrt{n}/s$; $\bar{x} = \sum_i x_i/n$ and $s^2 = \sum_i (x_i - \bar{x})^2/(n-1)$. The hypothesis H_0 is rejected at the α% level of significance if the value of $|t|$ exceeds $t_{\frac{1}{2}\alpha}$. The tail-area to the right of $t_{\frac{1}{2}\alpha}$ of Student's t distribution having $\nu = n - 1$ degrees of freedom being $\frac{1}{2}\alpha$; α is called the type I error and is determined by the equation

$$\alpha = 1 - \int_{-t_{\frac{1}{2}\alpha}}^{t_{\frac{1}{2}\alpha}} p(t, H_0)\,dt, \tag{1.2}$$

where $p(t, H_0)$ denotes the probability density function of t when H_0 is true. The power of this test is given by

$$1-\beta = \text{Prob}(|t| \geq t_{\frac{1}{2}\alpha} | H_1) = 1 - \int_{-t_{\frac{1}{2}\alpha}}^{t_{\frac{1}{2}\alpha}} p(t, H_1)\,dt. \tag{1.3}$$

Locks et.al. [15] have a table of the normal-theory power $1-\beta$. Note that for $\delta \frac{\sqrt{n}\mu}{\sigma} = 0$, $1-\beta = \alpha$. The problem is to study the effect of non-normality on α and $1-\beta$. Denote their true values by α^* and $1-\beta^*$ in a non-normal situation e.g., when the above random sample comes from a population other than normal. The problem is then of investigating the differences $\alpha^*-\alpha$ and $\beta-\beta^*$.

If a non-normal population has mean μ, variance σ^2 and standard cumulants $\lambda_r = \kappa_r/\sigma^r$ $(r=3, 4, \ldots)$, proceeding on the same lines as in the mathematical appendix it can be shown that to order $n^{-5/2}$

$$1-\beta^* \simeq (1-\beta) + \lambda_3 A\delta + \lambda_4 B - \lambda_3^2 C - \lambda_5 D\delta + \lambda_3\lambda_4 E\delta - \lambda_6 F + \lambda_4^2 G + \lambda_3\lambda_5 H \tag{1.4}$$

where $1-\beta$ is the normal-theory power and A, B, C, ... are non-normality corrective-functions. The values of these functions are given in Table 1.3 for $\nu=n-1=10$ and 60. Note that for a normal distribution $\lambda_r=0$ for $r=3, 4, \ldots$.

(1.5) Consider for example Pearson's [19, pp. 101-2] distribution having $\lambda_3=0.76$, $\lambda_4=1.23$, $\lambda_5=3.0$ and $\lambda_6=11.0$. This distribution very closely approximates to (i) the Pearson type IV curve, (ii) the non-central t distribution, and (iii) the Johnson S_u curve.

Suppose the sample comes from one of these non-normal distributions. The values of $\beta-\beta^*$, the differences between the power $1-\beta^*$ under this non-normal situation and the normal-theory power $1-\beta$, calculated from (1.4) are given in Table 1.1. Note that for $\delta=0$, $\beta-\beta^*=\alpha^*-\alpha$, the difference between the values of the non-normal and normal-theory type I error. For moderately large sample size, say $n\geq 30$, the effect of the above non-normality on the power seems to be unimportant. However, the effect of this non-normality on the type I error is negligible even for small samples.

Table 1.2 gives the values of the differences $\beta-\beta^*$ for several combinations of values of λ_3 and λ_4 for the above values of $\lambda_5=3$ and $\lambda_6=11$. It is clear from these values that the effect of λ_3, say $|\lambda_3|>0.5$, on the type I error and the power of the two-sided t-test is considerable; the effect of other standard cumulants is unimportant. It seems then that for a large number of symmetrical, or very nearly so, and moderately non-normal populations there is no reason why the two-sided normal-theory t-test should not be employed. Similar conclusions have been reached by Gayen [10] and Srivastava [25] about the suitability of the one-sided t-test.

2. Numerical values of type I error of Snedecor's F.

Let $(x_1,x_2,\ldots,x_m)$ and $(y_1,y_2,\ldots,y_n)$ be two independent random samples from $N(\mu_i,\sigma_i)$, i=1, 2. To test

$$H_0: \sigma_1 = \sigma_2 \text{ against } H_1: \sigma_1 > \sigma_2 \tag{2.1}$$

the test most frequently employed is based on the statistic $F = s_1^2/s_2^2$, the ratio of the two sample variances; H_0 is rejected if the calculated value of F is found to be greater than the tabulated value F_α for a pre-assigned type I error α.

Suppose the two samples come from two non-normal populations having means (μ_1,μ_2), variances (σ_1^2,σ_2^2) and the standard cumulants (λ_r,λ_r'), r=3,4,..., respectively. If n is the smaller of the two sample sizes, it can be shown that to order n^{-2} the non-normal type I error α^* is given by (see Tiku [29])

Table 1.1

Values of the difference β-β* for Pearson's distribution

δ	v= 10: 1-β	10: β-β*	20: 1-β	20: β-β*	30: 1-β	30: β-β*	60: 1-β	60: β-β*
				α=0.05				
0.0	0.05	.008	0.05	.005	0.05	.004	0.05	.002
+1	.146	-.081	.156	-.078	.159	-.070	.163	-.055
-1		.102		.093		.083		.061
+2	.428	-.112	.461	-.075	.472	-.056	.483	-.038
-2		.135		.089		.069		.044
+3	.767	.082	.811	.074	.824	.064	.837	.047
-3		-.065		-.070		-.062		-.047
+4	.952	.020	.972	-.002	.978	-.004	.982	-.004
-4		-.026		-.004		-.000		-.002
				α=0.01				
0.0	.01	.000	.01	.002	.01	.002	.01	.001
+1	.043	-.005	.051	-.008	.054	-.011	.057	-.014
-1		.005		.016		.019		.018
+2	.186	-.125	.228	-.130	.244	-.120	.260	-.094
-2		.146		.148		.134		.102
+3	.468	-.066	.560	-.007	.590	.013	.619	.022
-3		.096		.020		-.001		-.017
+4	.770	.122	.859	.079	.883	.053	.905	.026
-4		-.098		-.079		-.056		-.030

Table 1.2

Values of the difference $\beta-\beta^*$ for $\alpha=0.05$ and $\nu=20$; $\lambda_5=3$, $\lambda_6=11$

δ	$\lambda_4=-1.0$	0.0	1.0	2.0	3.0	δ	$\lambda_4=-1.0$	0.0	1.0	2.0	3.0
	$\lambda_3=-.5$						$\lambda_3=.5$				
0	.005	.003	.002	.003	.004	0	.005	.003	.002	.003	.004
1	.051	.057	.063	.071	.078	1	-.063	-.057	-.050	-.042	-.033
2	.039	.047	.055	.062	.071	2	-.071	-.060	-.052	-.044	-.036
3	-.060	-.060	-.060	-.059	-.058	3	.052	.051	.050	.048	.048
4	.002	.002	.002	.004	.005	4	.005	.002	-.001	-.002	-.003
	$\lambda_3=0.0$						$\lambda_3=1.0$				
0	.001	.000	-.001	-.001	-.002	0	.012	.010	.009	.010	.012
1	-.006	-.000	.007	.015	.024	1	-.120	-.114	-.107	-.099	-.090
2	-.015	-.005	.005	.015	.025	2	-.131	-.117	-.104	-.090	-.076
3	.003	.003	.003	.004	.005	3	.135	.133	.130	.128	.124
4	.003	.001	-.002	-.004	-.006	4	.006	.003	-.000	-.003	-.006

Table 1.3

Values of the non-normality corrective of functions for normal-theory type I error $\alpha=0.05$

ν	δ	$1-\beta$	10^5A	10^5B	10^5C	10^5D	10^5E	10^5F	10^5G	10^5H
10	0	.050	616	-259	-1712	-513	-1158	-19	136	-129
	1	.146	-12722	902	-97	-211	-143	53	16	97
	2	.428	-8271	1784	1185	114	487	62	27	144
	3	.767	3273	238	-358	16	2	-49	-83	-28
	4	.952	757	-556	-378	-28	-99	-18	26	-28
60	0	.050	385	-32	-384	-52	-90	0	7	0
	1	.163	-7689	259	17	-13	6	3	8	6
	2	.483	-2686	337	190	13	34	1	-0	3
	3	.837	2070	-68	-113	-2	-1	-2	-2	-2
	4	.982	-104	-76	-14	0	-1	0	1	0

$$\alpha^* \simeq \alpha + (\lambda_4 A + \lambda_4' A') - (\lambda_3^2 B + \lambda_3'^2 B') - (\lambda_6 C + \lambda_6' C') + (\lambda_4^2 D + \lambda_4'^2 D') - \lambda_4 \lambda_4' E; \qquad (2.2)$$

α being the normal-theory type I error. The numerical values of the non-normality corrective-functions A, A', etc., are given in Gayen [11] and Tiku [29]. The following Table 2.1 shows the values of the differences $\alpha^*-\alpha$ for the distribution (1.5) computed from (2.2) under the null hypothesis H_0 in (2.1). It is clear from these values that non-normality has considerable effect on the type I error of Snedecor's F test. The power of F is given by (2.2) with F_α replaced by θF_α, $\theta=\sigma_2^2/\sigma_1^2$. Calculations show that non-normality has also considerable effect on the power of the above F-test.

Table 2.1

Showing the values of $\alpha^*-\alpha$; $\lambda_4=\lambda_4'=1.23$, $\lambda_3^2=\lambda_3'^2=0.58$, $\lambda_6=\lambda_6'=11.0$, $v_1=m-1$, $v_2=n-1$; $\alpha=0.05$

v_2	$v_1=$ 4	5	6	8	12	24
4	.0081	.0087	.0088	.0090	.0089	.0087
5	.0088	.0101	.0110	.0114	.0116	.0107
6	.0106	.0115	.0125	.0130	.0135	.0126
8	.0121	.0136	.0148	.0160	.0174	.0166
12	.0129	.0155	.0172	.0200	.0222	.0230
24	.0122	.0157	.0180	.0226	.0279	.0319
40	.0101	.0136	.0171	.0221	.0329	.0363
60	.0086	.0121	.0155	.0210	.0327	.0377

3. Numerical values of the power of Fisher's F.

In one-way-classification for analysis of variance, let x_{ij} $(i=1,2,\ldots,c;\ j=1,2,\ldots,n)$ be the j-th observation in the i-th group. Assume the mathematical model

(3.1) $$x_{ij} = a + g_i + e_{ij} \quad , \quad \sum_i g_i = 0$$

where a is a constant, g_i is the i-th group effect and e_{ij} is a random error.

Write

(3.2) $$S_1 = n\sum_i (x_{i.}-x_{..})^2 \quad \text{and} \quad S_2 = \sum_{ij}\sum (x_{ij}-x_{i.})^2$$

(3.2) where

$$x_{i.} = \sum_j x_{ij}/n,\ x_{..} = \sum_{ij}\sum x_{ij}/N,\ N = nc.$$

To test the null hypothesis

(3.3) H_0: $g_1=g_2=\ldots=g_k=0$ against H_1: $g_i \neq 0$ for some $i=1,2,\ldots,c$,

under the assumption that e_{ij}'s are independently distributed as $N(0,\sigma)$ for all i and j, the statistic $F=f_2S_1/f_1S_2$ $(f_1=c-1,\ f_2=N-c)$ is used (Scheffe [23]); H_0 is rejected if the value of F is found to be greater than the tabulated value F_α

for a pre-assigned level of significance α. If $p(F,H_1)$ denotes the probability density function of F when H_1 is true, the power of F is determined by

$$1-\beta = \int_{F_\alpha}^{\infty} p(F,H_1)\,dF; \tag{3.4}$$

Tang [28] and Tiku [32] have tables of $1-\beta$. Assume that e_{ij}'s are not normally distributed and have a common non-normal distribution with mean zero, variance σ^2 and standard cumulants λ_r, $r=3,4,\ldots$. Define

$$\lambda = n\sum_i (g_i/\sigma)^2, \quad \mu = n\sum_i (g_i/\sigma)^3, \quad \delta = n\sum_i (g_i/\sigma)^4 \tag{3.5}$$

and note that if H_0 is true, λ, μ and δ are all zero. To study the effect of non-normality on the power, the statistic F in the non-normal case will be denoted by F^*. An approximation to order N^{-2} for the power $1-\beta^*$ of F^* is obtained as (see the mathematical appendix)

$$1-\beta^* \simeq (1-\beta)-\lambda_3 A+\lambda_4(B+B_1\delta)-\lambda_3^2 C+\lambda_5 D-\lambda_6 E+\lambda_4^2 F \tag{3.6}$$

where $1-\beta$ is given by (3.4), and A, B, etc., are non-normality corrective-functions. The values of these functions are given in Table 3.3 and give an idea of their relative contributions.

To calculate the values of $\beta-\beta^*$, the differences between the non-normal and normal-theory power, we note that for $\sum_i g_i=0$ (see Tang [28]) either the value of μ is zero and the value of δ is minimum and equal to λ^2/N or $c\lambda^2/N(c-1)$ according as c is even or odd, or μ has an extreme positive or negative value $\pm(c-2)\lambda^{3/2}/\sqrt{\{N(c-1)\}}$ and δ is maximum and equal to $(c^2-3c+3)\lambda^2/N(c-1)$, depending on which one of the two extreme situations, namely,

(i) the c groups form two distinct equal sets with a single group falling mid-way between for odd c, the effect in one set cancelling the effect of the other set, and

(ii) of the c groups, (c-1) are quite similar and only one is different, its effect counter-balancing the rest,

is true. In practice, the situation may be any where between these two extremes.

Suppose e_{ij}'s have a common distribution in one of the forms in

(1.5) with λ_4=1.23, λ_3^2=0.58, λ_5=3.0 and λ_6=11.0. We consider three extreme cases, namely

(a) $\mu = 0$ and δ is minimum

(b) μ is maximum positive and δ is maximum

and (c) μ is maximum negative and δ is maximum.

The values of the difference $\beta-\beta^*$ are given in Table 3.1, for various values of $\emptyset = \sqrt{\{\lambda/(f_1+1)\}}$. The values in case (a) generally lie between the values for cases (b) and (c), and the latter do not differ from each other by more than a few units in the third decimal place (see Table 3.1). Note that for $\emptyset$=0, $\beta-\beta^*=\alpha^*-\alpha$, the difference in the non-normal and normal-theory type I errors.

The values of $\beta-\beta^*$ for case (a) for various values of λ_4 and λ_3^2 and $\lambda_6 = 0$ are given in Table 3.2. Calculations show that a value of λ_6 as high as twenty does not change these values by more than $\pm$0.009.

From Tables 3.1 and 3.2 it is clear that the effect of moderate non-normality on type I error and the power of the F-test is unimportant. However, the population "kurtosis" λ_4 has a greater effect than the population "skewness" λ_3.

General Comments: It may be noted that equations (1.4), (2.2) and (3.6) include corrections only due to the first few population standard cumulants. For moderately non-normal populations it is expected that the contributions due to higher order standard cumulants will not be important (see also Tables 1.3 and 3.3) especially for large samples. This might not be true for extremely non-normal populations in which case these cumulants could be very large in magnitude and approximations which ignore these cumulants might not be useful. It is clear that the accuracy of (1.4), (2.2) and (3.6) also depends on the relative magnitude of the successive standard cumulants λ_r. If (as in the case of chi-square and certain other distributions) $\lambda_r=0(q^{1-\frac{1}{2}r})$, $q>1$ being some constant, these approximations will provide very accurate values.

Relevance of the above tests for extreme non-normality: It is possible that in situations of extreme non-normality the above tests are not at all appropriate, see Box & Tiao [3] on the subject of criterion and inference robustness. In the normal case, for example the sample mean and variance in Student's t, are sufficient statistics and the t-test is based on these, but in other cases there may be different sufficient statistics, such as Σx_i and Πx_i in a Gamma distribution, or none at all; quite another type of test may then be more appropriate than the t. It seems that in situations of extreme non-normality some thought should be given to finding alternative test procedures, although this in general may be difficult.

Table 3.1

Values of $\beta-\beta^*$, the difference between the power $1-\beta^*$ and the normal-theory power $1-\beta$ for Pearson's distribution, $\alpha = 0.05$

		$f_1=4$			$f_1=8$			$f_1=12$		
ϕ		$f_2=10$	20	60	$f_2=10$	20	60	$f_2=10$	20	60
0.0	$1-\beta$	.050	.050	.050	.050	.050	.050	.050	.050	.050
	(a)	.000	-.001	-.001	.000	-.000	-.000	.004	.001	-.001
1.0	$1-\beta$	.262	.319	.367	.286	.380	.473	.301	.422	.555
	(a)	.006	.008	.006	.007	.008	.006	.007	.007	.006
	(b)	.003	.004	.003	.004	.003	.001	.005	.002	.003
	(c)	.010	.013	.010	.010	.014	.012	.010	.014	.010
1.5	$1-\beta$	.550	.655	.730	.614	.771	.874	.650	.832	.939
	(a)	.017	.013	.002	.011	.007	-.001	.006	.003	-.003
	(b)	.013	.012	.006	.008	.009	.006	.005	.006	.003
	(c)	.021	.014	-.001	.014	.005	-.008	.009	-.001	-.012
2.0	$1-\beta$	.820	.907	.950	.881	.968	.993	.908	.986	.999
	(a)	.009	-.003	-.003	.001	-.006	-.002	-.002	-.005	-.001
	(b)	.013	.005	.002	.005	.005	.000	.002	-.001	.000
	(c)	.005	-.011	-.009	-.003	-.010	-.005	-.006	-.010	-.001
2.5	$1-\beta$	.958	.989	.997	.981	.999	1.000	.989	1.000	1.000
	(a)	-.007	-.006	-.001	-.006	-.002	-.000	.005	-.001	-.000
	(b)	-.006	-.002	.000	-.003	-.001	.000	-.002	-.000	-.000
	(c)	-.009	-.009	-.002	-.011	-.003	-.000	-.008	-.001	-.000

Table 3.2

Values of the difference $\beta-\beta^*$ for $\alpha=0.05$ and 0.01; $f_1=8$, $f_2=20$; $\lambda_6=0$

		$\lambda_3^2=0$				$\lambda_3^2=1.5$				$\lambda_3^2=3$			
ϕ	$1-\beta$	$\lambda_4=-1.5$	0	1.5	4	$\lambda_4=-1.5$	0	1.5	4	$\lambda_4=-1.5$	0	1.5	4
0.0	.050	.003	0	-.004	-.010	.007	.004	-.000	-.006	.011	.008	.004	-.007
	.010	.001	0	-.002	-.005	.005	.004	.002	-.001	.008	.007	.006	-.001
1.0	.380	-.015	0	.017	.049	-.010	-.005	.012	.044	-.014	-.011	.006	.038
	.153	-.009	0	.012	.041	-.007	-.002	.010	.039	-.005	-.004	.008	.037
1.5	.771	-.007	0	.005	.008	-.006	.001	.006	.009	-.004	.003	.008	.011
	.489	-.023	0	.021	.050	-.026	-.003	.018	.047	-.030	-.007	.014	.043
2.0	.968	.009	0	-.008	-.021	.010	.001	-.007	-.020	.010	.001	-.007	-.020
	.844	.005	0	-.006	-.019	.008	.003	-.003	-.016	.011	.006	-.000	-.013
2.5	.997	.001	0	-.002	-.004	.001	.000	-.002	-.004	.000	-.001	-.003	-.005
	.981	.011	0	-.009	-.021	.011	-.000	-.009	-.021	.011	-.000	-.099	-.021
3.0	1.000	.000	0	-.000	-.000	.000	-.000	-.000	-.000	.000	-.000	-.000	-.000
	.999	.001	0	-.001	-.004	.001	-.000	-.000	-.000	.002	.001	.000	-.003

Table 3.3

Values of the non-normality corrective terms in equation (3.6) for α=0.05

$f_1=8,\ f_2$	$10^7A/\mu$	10^5B	10^7B_1	10^5C	$10^7D/\mu$	10^5E	10^5F
				$\phi=0.5$			
10	3449	-96	-15	-512	566	-14	-26
20	7695	20	-67	-20	859	-2	-3
30	10672	56	-58	76	861	1	6
60	15361	65	29	82	656	2	5
				$\phi=1.0$			
10	2970	1004	123	297	111	46	83
20	5876	1047	358	355	-77	35	45
30	7064	885	507	201	-178	20	19
60	7939	551	694	48	-212	6	2
				$\phi=1.5$			
10	253	1326	40	450	-142	20	-66
20	-1268	388	12	-89	-166	-22	-50
30	-2345	52	-32	-106	-108	-18	-22
60	-3620	-121	-101	-39	-31	-6	-2
				$\phi=2.0$			
10	-560	-292	-9	-128	-29	-44	-34
20	-818	-538	-27	-41	29	-5	22
30	-704	-344	-27	11	30	2	10
60	-476	-128	-20	10	16	2	1
				$\phi=2.5$			
10	-178	-595	-5	-71	13	-2	42
20	-52	-101	-2	20	6	5	-2
30	-19	-26	-1	7	2	1	-2
60	-4	-3	-0	1	0	0	-0

4. Mathematical Appendix

Distribution of Fisher's F: To obtain the distribution of

$$F^* = f_2S_1/f_1S_2 \tag{4.1}$$

write

$$x = \frac{1}{2\rho}S_1/\sigma^2 \quad\text{and}\quad y = \tfrac{1}{2}S_2/\sigma^2,\ \rho=(f_1+2\lambda)/(f_1+\lambda) \tag{4.2}$$

and consider the formal expansion (see also Tiku [29 to 34]

$$p(x,y) \simeq \{\sum_{r=0}^{4}\sum_{s=0}^{4} \beta_{rs}\, L_r^{(m)}(x)L_s^{(k)}(y)\}p_m(x)p_k(y) \quad . \tag{4.3}$$

Here

$$(4.4)\quad \begin{cases} m=\frac{1}{2}p=\frac{1}{2}(f_1+\lambda)^2/(f_1+2\lambda),\ k=\frac{1}{2}f_2,\ p_m(x)=\frac{1}{\Gamma(m)}\ e^{-x}\ x^{m-1}\ ,\ \text{and} \\ L_r^{(m)}(x)=\frac{1}{r!}\ \sum_{j=0}^{r}(-1)^r\ \binom{r}{j}\ \frac{\Gamma(m+r)}{\Gamma(m+j)}x^j \qquad (j=0,1,2,\ldots) \end{cases}$$

is the r-th Laguerre polynomial associated with $p_m(x)$.

Note that under normality assumption of the 'error' distribution

$$(4.5)\qquad p(x,y)\simeq p_m(x)p_k(y)$$

i.e., x and y are independently distributed; $p_m(x)$ being Patnik's [16] approximation to a non-central chi-square distribution.

Writing $X = x-m$ and $Y = y-x$ we have the relation

$$(4.6)\quad E(X^rY^s) = (1/2\sigma^2)^{r+s}(1/\rho)^r E\{(S_1-E(S_1))^r(S_2-E(S_2))^s\}\ .$$

Using the tables of the product moments of S_1 and S_2 (David & Johnson [6]) and proceeding on the same lines as Tiku [29] we obtain (4.7); see also reference [5], [7], [9], [14], [29] and [30]. For $\lambda = 0$, (4.7) agree with the corresponding expressions in Tiku [29, p. 85]. Note that β_{rs}, $r+s\leq 4$, are all the coefficients in terms of the population standard cumulants of order N^{-2} (see reference [6]).

$$\beta_{00} = 1,\ \beta_{10} = \beta_{01} = 0$$

$$\beta_{20} = \frac{1}{N}\ \frac{f_1^2}{(p+2)(f_1+2\lambda)}\ \lambda_4,\ \beta_{11} = \frac{1}{N}\ \frac{f_1}{(f_1+\lambda)}\ \lambda_4,\ \beta_{02} = \frac{1}{N}\ \frac{f_2}{(f_2+2)}\ \lambda_4$$

$$(4.7)\quad \beta_{30} = -\ \frac{(f_1+\lambda)}{(f_1+2\lambda)^2(p+2)(p+4)}\left\{\frac{f_1^3}{N^2}\ \lambda_6+\frac{4}{N}f_1(f_1-1)\lambda_3^2+8\lambda_3\mu+12\frac{f_1}{N}\ \frac{\lambda^2}{(f_1+\lambda)}\ \lambda_4 -8\frac{\lambda^2}{(f_1+\lambda}\right\}$$

$$\beta_{21} = -\ \frac{f_1}{N(f_1+2\lambda(p+2)}\ \left\{(f_1/N)^2\lambda_6+4\frac{\lambda^2}{(f_1+\lambda)}\ \lambda_4\right\}$$

$$\beta_{12} = -\ \frac{1}{N(f_1+\lambda)(f_2+2)}\ \left\{(f_2/N)\lambda_6+4\lambda_3^2\right\}$$

$$\beta_{03} = -\ \frac{1}{N(f_2+2)(f_2+4)}\ \left\{(f_2^2/N)\lambda_6+4(f_2-f_1-1)\lambda_3^2\right\}\ ,\ \text{and so on.}$$

From (4.3) to (4.7) the probability density function of

$$w = kx/my = F^*/(1+\lambda/f_1)$$

is obtained as

$$p(w) \simeq p_{oo}(w) + \sum_{2\le r+s\le 4}\sum \beta_{rs}\, p_{rs}(w) \tag{4.8}$$

where

$$p_{rs}(w) = \frac{\Gamma(\frac{1}{2}p+r)\,\Gamma(\frac{1}{2}f_2+s)}{r!\,s!\,\Gamma(\frac{1}{2}p)\,\Gamma(\frac{1}{2}f_2)} \sum_{i=0}^{r} (-1)^i \binom{r}{i} \left\{ \sum_{j=0}^{s} (-1)^j \binom{s}{j} p(w;\tfrac{1}{2}p+i,\tfrac{1}{2}f_2+j) \right\}, \text{ and}$$

(4.9)

$$p(w;\tfrac{1}{2}p+i,\tfrac{1}{2}f_2+j) = \frac{(p/f_2)^{\frac{1}{2}p+i}}{\beta(\frac{1}{2}p+i,\frac{1}{2}f_2+j)} \quad \frac{w^{\frac{1}{2}p+i-1}}{(1+pw/f_2)^{\frac{1}{2}(p+f_2)+i+j}}$$

<u>Moments of F^*</u>: From (4.8) and (4.9) we obtain the following expression for the mean and the variance of F^*:

$$\mu_1' \simeq (1+\lambda/f_1)(1+2/N) + (1/f_1N)\lambda_4\ \lambda + 0(1/N^2) \tag{4.10}$$

and

$$\mu_2 \simeq (2/p)\left[\frac{(p+2)}{(f_1+2)}\left(1+\frac{f_1+6}{N}\right)\left(1+\frac{\lambda}{f_1}\right)^2 + \frac{1}{2}\,\frac{p+2}{N}\left\{\frac{p}{p+2} - \left(1+\frac{\lambda}{f_1}\right)\right\}\lambda_4 \right.$$
$$\left. -\frac{1}{N}\left(\frac{p+2}{f_1+2}\right)\left(1+\frac{\lambda}{f_1}\right)\lambda_4\lambda - \frac{2}{N(f_1+2)(f_1+2\lambda)}\left(1+\frac{\lambda}{f_1}\right)\lambda_4\lambda^2 + 0(1/N^2)\right] \tag{4.11}$$

For $\lambda = 0$, (4.10) and (4.11) agree with the corresponding expressions in Gayen [11] and Tiku [29]. It may be noted that to $0(N^{-2})$ the above expressions do not contain terms in the population standard cumulants other than λ_4 .

<u>Power-function of F^*</u>: The power

$$1-\beta^* = \text{Prob}(F^* \ge F_\alpha) = \text{Prob}(w \ge F_\alpha/(1+\lambda/f_1)) \tag{4.12}$$

is obtained from (4.8) as

$$1-\beta^* \simeq (1-\beta) + \sum_{2\le r+s\le 4}\sum \beta_{rs}\, \frac{\Gamma(\frac{1}{2}p+r)\,\Gamma(\frac{1}{2}f_2+s)}{r!\,s!\,\Gamma(\frac{1}{2}p)\,\Gamma(\frac{1}{2}f_2)}\, I_{r,s}(u) \tag{4.13}$$

where $u = 1/\{1+f_1(f_1+\lambda)F_\alpha/f_2(f_1+2\lambda)\}$, and

$$(4.14) \qquad I_{r,s}(u) = \sum_{i=0}^{r} (-1)^i \binom{r}{i} \left\{ \sum_{j=0}^{s} (-1)^j \binom{s}{j} I_u(\tfrac{1}{2}f_2+j, \tfrac{1}{2}p+i) \right\};$$

$I_u(a,b)$ being Karl Pearson's [21] imcomplete beta-function. From Soper's [24] equations explicit algebraic expressions for (4.14) can be obtained (see also Tiku [29]). Wise [39] gives useful formulae for computing incomplete beta-functions. It can be proved that (4.14) is uniformally convergent (see also Whittaker & Watson [38]).

Retaining only the terms of $0(N^{-2})$ in (4.7) we obtain the explicit equation (3.6). The expressions for the terms on the right hand side of (3.6) can easily be obtained from (4.7) and (4.13). For example

$$(4.15) \qquad 1-\beta = I_u(\tfrac{1}{2}f_2, \tfrac{1}{2}p) + \frac{1}{24} \frac{\lambda^2}{(f_1+2\lambda)\rho^3} \left\{ 4I_{3,0} + 3\frac{(f_1+4\lambda)}{(f_1+2\lambda)} I_{4,0} \right\}$$

$$(4.16) \qquad A = \frac{1}{6\rho^3} \left\{ I_{3,0} + 3\frac{\lambda}{(f_1+\lambda)\rho} I_{4,0} \right\}$$

and so on. Incidentally, note that (4.15) is Tiku's [31] approximation to the power-function of normal-theory F.

5. General comment. The motivation for the above derivations of the expressions of the power functions in terms of Laguerre polynomials and Gamma density functions came from the fact that the Pearson coefficients $\beta_1 = \mu_3^2/\mu_2^3$ and $\beta_2 = \mu_4/\mu_2^2$ of the two components in Student's t^2 , Snedecor's F and Fisher's F lie close to the Tyep III lines $\beta_2 = 3+1.5\beta_1$. For example the (β_1, β_2) points of S_1/σ^2 are as follows:

Table 5.1

The (β_1,β_2) points of S_1/σ^2; $\lambda_3=0.76$, $\lambda_5=3.0$, $\lambda_6=11.0$;

$\mu=0$, $\delta=\lambda^2/N$, $\phi=\sqrt{\{\lambda/(f_1+1)\}}$

$\lambda_4 = 2.23$						$\lambda_4 = 4.23$					
f_1	f_2	ϕ	β_1	β_2	Diff*	f_1	f_2	ϕ	β_1	β_2	Diff*
4	10	0.0	3.841	9.619	0.85	4	10	0.0	4.240	10.233	0.87
		1.0	1.745	5.903	.29			1.0	2.262	6.658	.26
		2.0	0.642	4.050	.09			2.0	0.895	4.410	.07
		3.0	0.312	3.508	.04			3.0	0.444	3.694	.03
	60	0.0	2.457	6.947	.26		60	0.0	2.751	7.469	.34
		1.0	1.205	4.769	-.01			1.0	1.333	5.037	.04
		2.0	0.439	3.636	-.02			2.0	0.489	3.728	-.01
		3.0	0.212	3.306	-.01			3.0	0.237	3.351	-.01
12	10	0.0	1.789	6.121	.44	12	10	0.0	1.667	5.992	.49
		1.0	.893	4.561	.22			1.0	1.164	4.872	.13
		2.0	.351	3.612	.09			2.0	.552	3.835	-.01
		3.0	.175	3.304	.04			3.0	.291	3.431	-.01
	60	0.0	1.070	4.820	.22		60	0.0	1.229	5.072	.23
		1.0	.532	3.857	.06			1.0	.650	4.050	.08
		2.0	.202	3.318	.02			2.0	.256	3.405	.02
		3.0	.099	3.155	.01			3.0	.127	3.199	.01

Diff: $= \beta_2-(3+1.5\beta_1)$, i.e., the difference between the true β_2 and the β_2-value on Type III line.

It is clear from these values that the (β_1,β_2) points of S_1/σ^2 lie close to the Type III or chi-square lines. In such situations expansions in terms of Laguerre polynomials and Gamma density functions seem to provide accurate approximations (see Pearson & Tiku [20], Roy & Tiku [22] and Tiku [30 to 37]). The Gamma density function in the expansion is chosen such that its first two moments are identical with the first two moments of $\frac{1}{2}S_1/\rho\sigma^2$. Equating the first two moments on both sides of $x \approx \frac{1}{2}S_1/\rho\sigma^2$ (x being a Gamma variate with mean m), i.e.,

$$m=\tfrac{1}{2}(f_1+\lambda)/\rho \quad \text{and} \quad m = (\tfrac{1}{2}\frac{f_1^2}{N}\lambda_4+\tfrac{1}{2}f_1+\lambda)/\rho^2 \approx \tfrac{1}{2}(f_1+2\lambda)/\rho^2 \tag{5.1}$$

we obtain

$$m = \tfrac{1}{2}(f_1+\lambda)^2/(f_1+2\lambda) \quad \text{and} \quad \rho = (f_1+2\lambda)/(f_1+\lambda) . \tag{5.2}$$

The Gamma density function $p_m(x)$ used in equation (4.3) has therefore its first two moments common with $\frac{1}{2}S_1/\rho\sigma^2$, and similarly for the distribution of $\frac{1}{2}S_2/\sigma^2$.

Acknowledgements.

Thanks are due to the referee for helpful comments. Thanks are also due to the National Research Council for a research grant,and to Mr. Sushil Kumra for assistance in computation. I am grateful to Miss Linda Westfall for typing this manuscript.

References

[1] Bartlett, M.S. (1955). "The effect of non-normality on the t-distribution. Proc.Camb.Phil.Soc. 31, 223-6.

[2] Barton, D.E. & Dennis, K.E. (1952). The conditions under which Gram-Charlier and Edgeworth curves are positive and unimodal. Biometrika 39, 425-7.

[3] Box, G.E.P. & Tiao, C. (1964). "A note on criterion robustness and inference robustness". Biometrika 51, 169-73.

[4] Box, G.E.P. & Watson, G.S. (1962). "Robustness to non-normality of regression tests". Biometrika 49, 93-106.

[5] David, F.N. (1959). "The z-test and symmetrically distributed random variables". Biometrika 46, 123-9.

[6] David, F.N. & Johnson, N.L. (1951). "The effect of non-normality on the power function of the F-test in the analysis of variance". Biometrika 38, 43-57.

[7] David, F.N. (1969. "Note on the application of Fisher's k statistics". Biometrika 36, 383-93.

[8] Donaldson, T.S. (1968). "Robustness of the F-test to errors of both kinds of the correlation between the numerator and denominator of the F-ratio". J. America. Stat. Assn. 63, 660-7.

[9] Fisher, R.A. (1930). "Moments and product moments of sampling distributions". London Math.Soc. 30, 199-238.

[10] Gayen, A.K. (1949). "The distribution of Student's t in random samples of any size from non-normal universes". Biometrika 36, 353-69.

[11] Gayen, A.K. (1950). "The distribution of the variance ratio in random samples of any size from non-normal universes". Biometrika 37, 236-55.

[12] Geary, R.C. (1947). "Testing for non-normality". Biometrika 34, 209-42.

[13] Hack, H.R.B. (1958). "An empirical investigation into the distribution of the F-ratio in samples from two non-normal populations". Biometrika 45, 260-5.

[14] Kendall, M.G. (1940). "The derivation of multivariate sampling formula from univariate formula by symbolic operation". Annals of Eugenics 12, 392-402.

[15] Locks, M.O., Alexander, M.J. & Byars, B.J. New Tables of the non-central t distribution. Aeronatuical Research Laboratories, 1963.

[16] Patnaik, P.B. (1949). "The non-central chi-square and F distributions and their applications". Biometrika 36, 202-32.

[17] Pearson, E.S. & Adyanthaya, N.K. (1929). "The distribution of frequency constants in small samples from non-normal symmetrical and skew populations". Biometrika 21, 259-86.

[18] Pearson, E.S. (1932). "The analysis of variance in cases of non-normal variation". Biometrika 23, 114-33.

[19] Pearson, E.S. (1963). "Some problems arising in approximating to probability distributions using moments". Biometrika 50, 95-112.

[20] Pearson, E.S. & Tiku, M.L. (1970). "Some notes on the relationship between the distributions of central and non-central F". Biometrika 57, 175-9.

[21] Pearson, Karl. (1934). Tables of the incomplete beta-function. Cambridge University Press.

[22] Roy, J. & Tiku, M.L. (1962). "A Laguerre series approximation to the sampling distribution of the variance". Sankhya 24, 181-4.

[23] Scheffe, H. (1967). The analysis of Variance. John Wiley & Sons, Inc.

[24] Soper, H.E. (1919). "The numerical evaluation of the incomplete beta-function". Tracts of computers, No. VI. Cambridge University Press.

[25] Srivastava, A.B.L. (1958). "Effect of non-normality on the power-function of t-test". Biometrika 45, 421-29.

[26] Stivastava, A.B.L. (1959). "Effect of non-normality on the power of the analysis of variance test". Biometrika 46, 114-22.

[27] Szego, G. (1939). Orthogonal Polynomials. Amer. Math.Soc. Colloquim Publications, XXIII.

[28] Tang, P.C. (1938). "Power of the F-test in terms of non-central F-distribution". Statistics Research Mem:2, 126-49.

[29] Tiku, M.L. (1964). "Approximating the general non-normal variance-ratio sampling distributions. Biometrika 51, 83-95.

[30] Tiku, M.L. (1964). "A Laguerre product series expansion of the distribution of variance-ratios in two-way-classification". J. Indian Soc. Agric. Stat. 16, 301-16.

[31] Tiku, M.L. (1965). "Laguerre series forms of non-central chi-square and F-distributions". Biometrika 52, 415-27.

[32] Tiku, M.L. (1967). "Tables of the power of the F-test". J. Amer. Stat. Assn. 62, 525-39.

[33] Tiku, M.L. (1966). "A note on approximating to the non-central F distribution". Biometrika 53, 606-10.

[34] Tiku, M.L. (1965). "Chi-square approximations for the distributions of goodness-of-fit statistics U_n^2 and W_n^2". Biometrika 52, 630-3.

[35] Tiku, M.L. (1971). "Power function of the F-test under non-normal situations". J. Amer. Stat. Assn. 66, 913-6.

[36] Tiku, M.L. (1971). "Student's t distribution under non-normal situations". Aust. Jour. Statist., 14, 103-8.

[37] Tiku, M.L. (1971). "A note on the distribution of Hotelling's generalized T_0^2". Biometrika 58, 237-41.

[38] Whittaker, E.T. & Watson, G.N. Modern Analysis. Cambridge University Press.

[39] Wise, M.E. (1950). "The incomplete beta-function as a contour integral and a quickly converging series for its inverse". Biometrika 37, 208-18.

M. L. Tiku
Dept. of Applied Mathematics
McMaster University
Hamilton, Ontario
Canada L8S 4K1

Applied Statistics, R.P. Gupta, (editor)

INSTEAD OF GAUSS-MARKOV LEAST SQUARES, WHAT?

John W. Tukey*

ABSTRACT

The method of least squares is by now very classical and well embedded in consequences and related concepts. It is high time that it be used as a springboard to more useful techniques, techniques that are clearly related to their takeoff point, techniques that diverge neither too little nor too much. (This is the proper treatment for any well developed statistical or scientific idol.) We enquire briefly into (a) the meaning of the coefficients, (b) what "if not" replaces each of 7 "ifs" in a statement of the Gauss-Markov theorem, (c) how we ought to think about discontinuities in variance structure.

1. Introduction

Our topic today starts from least squares and involves many subtopics, as we shall see. What we shall say also belongs to a supertopic - the management of idols.

Neither unquestioning acceptance or iconoclasm is a proper way to manage a scientific idol, wherever we may be in the historical development of its role. At the earliest stages we need only suspend our questions for a time while we develop its consequences. Soon thereafter we ought to follow the Scots proverb, that a man may have "a gey fine hoose" but he must "sit loose to it". Later yet, when the idol stands surrounded by well-developed concepts and details, we ought to use it only as a take-off point, as a basis from which to move to more realistic situations.

Chemists will recognize these stages fairly clearly in the history of the classical model of the structure of a gas as a collection of hard, round spheres. Early on, it led to an understanding of the simplest properties of gases. Later, it was the takeoff point for more and more detailed models of gas-phase molecular interaction. And it is still used as a stage in the education of young chemists.

Statisticians can recognize something similar in the history of the Gaussian shape of distribution - the shape often miscalled "normal" - where there has been both more blind acceptance and more iconoclasm than was appropriate. It is four decades since one form of iconoclasm - distribution-free techniques (often called non-

*Prepared in part in connection with research at Princeton University sponsored by the U.S. Atomic Energy Commission.

parametric methods) - began to be popular. Yet even today we see much of seemingly unquestioning acceptance of Gaussian distribution as an underlying hypothesis.

Today's approach toward "robust methods" - whether fully formalize as in Huber's important work, or rather richly empirical, as in the Princeton robustness study - is more appropriate to our current state of knowledge and understanding. In it, we start from the Gaussian and move away a more or less measured distance. For Huber this distance may be [9] amount of contamination, Prohorov distance, or total variation. For the Princeton robustness study [1] this distance is shown by the nonconstancy of "denominator" in situations where the distribution of the observations correspond to the distribution of

$$\frac{\text{unit Gaussian}}{\text{an independent denominator}}.$$

In each case there is more than just a tie to the Gaussian model - there is a real use of that model as a starting point. Notice, however, that, in each case, we do not stay at the starting point - rather we move away from it more than a little and less than a lot.

How can we treat the famous idol of "least squares" in a similar way? We ought not destroy it utterly, neither ought we accept it without question. Rather we should begin with it and then take a realistic step away. Such a step may well be one of many, and it ought be neither too small nor too large.

Some think of "least squares" as providing a fitted function--an interpolation formula if you like--to be judged only by the quality of its fit. Their lot is easier than that of those who think of "least squares" as a way to estimate coefficients, blithely thought of as easily made meaningful. This paper treats the woes of those who first seek coefficients and then seek to interpret them. (The others avoid most of the difficulties described in sections 3 and 4 below.)

Comments by my colleagues J. M. Chambers, R. W. Hamming, R. L. Obenchain and especially C. L. Mallows have led to the elimination of errors and confusions. Those that remain are the writer's responsibility.

2. The Gauss-Markov theorem

The classical statement, classically taken as the reason why we should use least squares, bears the name of Gauss and the name of Markov. Its essentials are simple:

[1] if the central value of y for given x's is $\Sigma\beta_i x_i$, where the β_i are wholly unknown.

[2] if we know the x's without error.

[3] if the variance of the jth y is σ^2/w_j , where w_j is known but σ^2 is not.

[4] if the covariances of distinct y's are zero. (We can, of course, replace zero by $h_{jk}\sigma^2$ with h_{jk} known by complicating the answer slightly.)

[5] if we are willing to confine ourselves to estimates of the β_i that are linear functions of the given y's .

[6] if we are willing to take the central values in [1] as means and confine ourselves to estimates that are unbiased in mean. (For more general nonlinear problems we will have to considerably restate both this "if" and the next.)

[7] if we are willing to judge the quality of our estimates by their variances (for the given set of x's).

[*] then the best estimate of either any individual β or any linear combination of β's is to be had by "least squares".

We have put down seven "ifs" and one conclusion. At our present state of knowledge, we ought to question each of the seven "ifs" - we ought to use each to point a direction in which to move a suitable distance away from our idol.

Our basic maxim, on meeting an "if" that has been in place long enough for consequences and related concepts to have been well developed, should be to ask "AND WHAT IF NOT?".

We will do just this,

- in this section about the 1st if.
- then about the meaning of the results.
- then about ifs 2 to 7.
- then about historical continuity or discontinuity of the structure of variability.

(For more on computational techniques, and such concepts as "catchers" and "biorthogonality" see Beaton and Tukey 1974.)

nonlinear problems

The first "if" called for the parameters to enter linearly into the fitted expression. Most of you know that we cannot stop there, that the problems of the real world involve many parameters that do not enter linearly. What of them?

The working answer is classical:

- start somewhere sensible (if possible).
- differentiate w.r.t. the parameters.
- fit the resulting linear expression.
- iterate as often as seems helpful.

Such a program

- can have computational difficulties.
- could be bothered by curvature.
- requires restatement of unbiasedness.

We know something about ways to clear computational difficulties ([7], also [3], [4] and [13]), more about the effects of curvature ([2], [6], [8], [10], [11] than one might think, and what is probably enough about replacing unbiasedness (see section 7 below).

Undoubtedly there is much to be usefully said in this direction, but little of it will be said here. For the present, we will draw from the nonlinear case only a clear guide toward using CHANGES in possible fits, toward using vectors defined over data sets that take the form

change in fit = one potential fit MINUS another

(Going to the limit will give us the derivatives on which fitting is often based in nonlinear problems.)

The first "if" also calls for some $\Sigma\ \beta_i x_i$ to fit exactly, something that cannot be passed over lightly, either. We have thought far too little about (i) the consequences when the closest $\Sigma\ \beta_i x_i$ is only an approximation, (ii) how many more x's is it worth adding to make the fit a given amount closer (if we know how much closer), (iii) what we ought to do with observations associated with unusual x-vectors (sometimes we should cherish them and sometimes we should almost forget them). These aspects of the first "if" are not being adequately considered here. (For a part of these questions see Box (1966).)

* the general normal equations *

When y stands for the observed value, and $\hat{y}$ for that fitted - in our case

$$\hat{y} = \Sigma\hat{\beta}_i x_i$$

then the equations

$$\Sigma(\text{weight})(\text{any contemplated change in fit})(y-\hat{y}) = 0$$

where the summation is over data points and the parallelism is over any set of contemplated changes in fit (any set of $Y-\hat{y}$, where the Y's are each of the same form as $\hat{y}$) are equivalent to

$$\Sigma(\text{weight})(\text{any change in fit})y = \Sigma(\text{weight})(\text{the same change})\hat{y}$$

Either can well be called "the normal equations".

The formulation just given generalizes the usual statement, which takes the derivatives of the fit with respect to each of a chosen set of parameters as the only set of changes to be considered.

To ask that certain equalities hold is to ask that any linear combinations of them hold and vice versa. The consequences of the more general normal equations are identically those of any restricted set. The implications of our understanding of the more general formulation seem well worth while.

* carriers and matchers *

We use the word "carrier" to refer to an available change in fit. In the nonlinear case it will be the derivative of the fit in some direction of parameter space, taken at the present fit. In the linear case it will be a possible change in fit - something that could occur from any fit as a starting point.

We emphasize and reemphasize "change in fit" - some will feel that "fit" alone should be enough - at least in the linear case. Were we to stick to the form of the second "if" exactly, it would. But we need not and, in practice, will not. If the central value of y is

$$q(x_1, x_2, \ldots, x_k) + \Sigma\beta_i x_i$$

all that we have said about the linear case goes through unchanged, although changes in fit involve only $\Sigma(\delta\beta_i)x_i$, while every fit also includes $q(x)$.

It is convenient to have a word for the things that appear in the normal equations. We suggest

$$\text{matcher} = (\text{weight})\cdot(\text{change in fit})$$

With this choice, we can write the second form of the normal equations as

$$\Sigma(\text{matcher})(\text{data}) = \Sigma(\text{same matcher})(\text{fit})$$
FOR ANY AND ALL MATCHERS.

* "least squares" *

The name is "least squares" - yet so far we have been wise enough not to square anything - and not to explicitly minimize anything. It is easy to see that minimizing

$$\Sigma(\text{weight})(y-\hat{y})^2$$

requires the derivative w.r.t. h of

$$\Sigma(\text{weight})(y-\hat{y}-h(\text{change}))^2$$

to vanish at $h=0$, i.e. that the normal equation

$$\Sigma(\text{weight})(\text{change})(y-\hat{y}) = 0$$

hold.

The "minimize potential" formulation of least squares seems of use only to show uniqueness -- absent in many nonlinear problems -- and the like. The "balance of forces" formulation of our general normal equations seems both simpler and a better guide to insight.

3. Meanings, anyone?

We have done our best to get away from any particular coordinate system in parameter space. We have, so far, emphasized a "fit", rather than a "set of parameter values" -- we have stuck even closer, perhaps, to the essence of the matter than to even our formulation of the Gauss-Markov theorem, which mentions estimates of parameters and their linear combinations. Many, once it is called to their attention, will not like this. They want estimated values of specific parameters to which they want to ascribe specific meanings.

In particular, they think that:

- the value of b_3 tells them of "the effect of x_3"
- the value of b_7 tells them of "the effect of x_7"
- and so on. Is there any truth in this? If any, how much?

It is a matter of simple algebra (e.g. see Beaton and Tukey, 1974) to see that if we replace x_3 by $x_3 + 12x_7$ we will NOT change b_3 but we WILL change b_7 . (There is nothing about this that is restricted to least squares. So long as we can identify our fit by the properties of its residuals $\{y_i - \hat{y}_i\}$ the result must hold.) We did nothing to x_7, yet b_7 changed. Both values cannot represent something described by only the words "the effect of x_7" .

Chemists are used to having partial derivatives depend on what is "held constant". A heat capacity is a derivative of heat input w.r.t. temperature. All physical chemists know that

$$C_p = \text{heat capacity at constant pressure}$$

and

$$C_v = \text{heat capacity at constant volume}$$

are different. Indeed, for a gas, their difference has a simple interpretation.

Mathematicians and statisticians have rarely bothered to get such matters clearly in their heads. Less often have they got them in the heads of their students and readers. The strength of the implied words

"ceteris paribus" = other things equal

and the importance of which "ceteris" is meant have been greatly underemphasized.

Let us call the collection of all possible changes in fit

$$\{\sum_i \beta_i x_i, \text{all}\{\beta_i\}\}$$

our POSSE. Consider the SUBPOSSE

$$\{\sum_{i \neq 7} \beta_i x_i ; \text{all}\{\beta_i\}\}$$

of all CHANGES in which β_7 is UNALTERED, so that x_7 does not appear. Beaton and Tukey [3] show that $\hat{\beta}_7$ is determined by the COMBINATION of

$$x_7$$

AND

the subposse with $\beta_7 = 0$

and with nothing simpler. Just as a heat capacity involves both temperature (thought of as changing) and another variable or variables (thought of as fixed), so each regression coefficient involves both "its variable" thought of as changing) and a SUBPOSSE of possible changes (thought of as "taken care of" elsewhere).

The most we can say is that b_7 tells us about the apparent effect of x_7 IN THE PRESENCE of a specific subposse. Indeed, it can be seen that the choice of the subposse is often more important than the choice of x_7 .

In that part -- which is large -- of the use of regression where we try to attach meaning to our coefficients, there is nothing more important than to think -- and think hard -- about what subposse is appropriate when assessing a particular β_i for a particular purpose.

Indeed, there seems no reason to suppose, if we want to find both an estimate b_3 of a coefficient of x_3 and an estimate b_7 of a coefficient of x_7, both for specific purposes, that it will always be desirable to find them both from the same fit -- least squares or otherwise.

4. Imprecise x's and the consequences of imposed change

Let us begin with one x, asking about the linear dependence of y upon x. The words "errors in both variables" signpost a field that books on statistics enter much too rarely. For this there are at least two reasons:

- the terrain is indeed swampy.
- we have all been taught to shun "error".

In physical science we hope and usually trust that we are measuring both what we say we are measuring and what we want to measure. The degree to which this is true is always somewhat limited, in part by the conceptual questions that gave rise to Bridgeman's operationalism, in part by limited frequency responses of our instruments to changing inputs, in part by real complications -- such as, for example, interferences by other substances in classical chemical analysis.

As we move into either engineering or the life sciences, what we can afford to measure drifts away from what we would like to measure, and, even when this is admitted, drifts -- perhaps somewhat less far -- from what we say we are measuring. By the time we come to concepts of great interest to all of us -- for example: "poverty", "health", and "educational attainment" -- what is measured is far from being close to what we would like to measure -- what we would like to measure provided we could both describe it precisely and then measure it.

The more generally important the topic, the more likely it is that we deal with "both variables measured indirectly and imperfectly". (Had we always used these words instead of "both variables measured with error", a few more texts and lecture courses might have considered these important problems.)

* altered circumstances *

Why should indirect and imperfect measurement matter so much to us? Not for passive forecasting, something we might like to do when x is available before y. For the relation between x and y in the past may well continue into the future. But rather for a reason that seems unnatural to the classically trained statistician or data analyst: the future may be different. Not just different because of unspecified things extraneous to variables with which we are involved (we return to this question far below) but different because of what people -- perhaps guided by our results -- are going to do -- do both deliberately and with an intent to change one or more y's

IF X, which we cannot or do not measure, is the quantity that really influences y (whether directly or through some Y we cannot or do not measure does not matter), and IF the relation can be taken, well enough for present purposes, as

$$y = \beta^* X + \alpha + \text{hash}$$

IF x is an indirect and imperfect measure of X, and IF that relationship can be taken, also well enough for present purposes, as

$$x = X + (\text{hash of variance } \sigma^2)$$

where the hash is uncorrelated with X (we are optimistic to believe in a regression coefficient of exactly unity, but we may sometimes get away with this optimism) then what do we expect when we fit

$$y = \beta x + \alpha' + (\text{hash})$$

-- what, indeed?

The classical result is

$$\beta = \text{ave}\{\text{est}\beta\} = \frac{\text{var}\{X\}}{\sigma^2 + \text{var}\{X\}} \beta^*$$

a simple appearing result. If we could only assess σ^2? Clearly if the size of β^* matters, a value of β may not be good enough.

It is not easy to assess σ^2 -- or to do anything else to help us from β to β^*. One large reason for this is that our

$$\text{hash of variance } \sigma^2$$

includes the systematic ways in which x fails to reflect X -- as by being also influenced by other variables U, V, and W -- as well as the irregularly fluctuating ones. Thus σ^2 is not confined to errors of measurement, whose size is itself not easy to assess.

There was a time when it was hoped that all could be saved by finding an

instrumental variate

z and estimating β^* as the ratio

$$\frac{\text{slope of regression of } y \text{ on } z}{\text{slope of regression of } x \text{ on } z}$$

This seemed easy -- not really easy, but perhaps only bearably difficult -- so long as we spoke of "errors", since it seemed enough to make the measurement errors in z uncorrelated with those of y and those of x. In the more precise terms of "indirectly and imperfectly" we recognize ourselves as deeply challenged, for we must now understand our indirections and imperfections well enough -- and have enough potential z's to select from -- to find a z whose indirections and imperfections are uncorrelated both with those of x and with those of y.

We should not be surprised that "instrumental variates" were not a quick and easy solution to all our problems, but we should not give up hope that we will find important instances where they will help us in the future -- after much hard thought in each case.

* nonzero causation? *

The case of one x is surely not the only case where "indirection and imperfection" will trouble us. As we include more x's, each almost certain to be troubled with indirection and imperfection, we are led to using more and more instrumental variables -- something that calls for more and more hard thought. But this is not all.

A simple -- and often terribly misleading -- argument, an argument that purports to show that x_1 -- or what x_1 is closely related to -- contributes causally to y, is to fit a regression of y on both x_1 and on a collection of other x's -- x's that are plausibly closely related to the other plausible causes -- and note that b_1 is significantly $\neq 0$. Here "contributes causally" has to mean "IF we could leave fixed the other plausible causes, but, by taking certain actions, alter X_1 -- incidentally altering the x_1 to which it is closely related -- in a definite direction then y would change in a corresponding way.

So long as both (1) the other x's are only approximately equal to their associated plausible causes and (2) the plausible cause to which x_1 is related is correlated with one or more of the other plausible causes, the appearance of a β_1 different from zero does

NOT imply that x_1 -- or its associated quantity, X_1 -- contributes causally to y .

This seeming paradox is easy to explain. Let X_1 and X_2 , neither directly measurable, be two plausible causes. Let x_1 , which is measured, point toward X_1 , and let x_2 do the same as X_2 . Let y depend on X_2 alone. Let X_1 and X_2 be correlated. (Were they not, the question of X_1's relationship to y would be much more easily resolved and we would not be trying such an analysis.) Now both X_1 -- and consequently x_1 -- and x_2 tell us about the value of X_2 . The best we can learn about X_2 from what we have will thus combine x_1 and x_2 . Thus regression of y on these two will involve both.

There are few if any arguments from empirical observation that establish causality with certainty. Most arguments will be often right (by accident) and from sometimes to often wrong (by nature). We can ask to do as good a job of approaching causality as we can. This will often lead to looking at several alternate arguments. But -- alone or combined -- these arguments are rather unlikely to be completely leakproof.

All this difficulty does not mean that we should not use such arguments. Rather we should give much attention to testing and diversifying them -- and we should "sit loose" to their conclusions. The problem of indirection and imperfection in both -- or all -- variables is real -- it cannot be papered over by benign neglect -- we need to face up to it in a very serious way. Thus the second "if" is a real can of worms. (For another class of problems see Box (1966).)

5. Weighting and misweighting -- the 3rd and 4th ifs

Let us turn, briefly, from the bad news to the good. In the third "if" we assumed that the variance of the jth y was σ^2/w_j with the w_j KNOWN. What if our knowledge were wrong?

The question is easy to answer for the trivial case where least squares leads to a weighted arithmetic mean (Tukey 1948) and answerable with more difficulty in general (Tukey, unpublished). The answer is the same in both cases. If the ratios

$$\frac{\text{actual weight}}{\text{proper weight}}$$

are nearly the same for all data points, the increase in variance is small. Specifically, if

$$A \le \frac{\text{actual weight}}{\text{proper weight}} \le rA$$

then the variance of the least squares estimate of any linear combination of the parameters is

$$\le \frac{(r+1)^2}{4r}$$

times as large as if the proper weights were used. The values of this bound look as follows:

r:	1	2	3	4	5
bound:	1	$1\frac{1}{8}$	$1\frac{1}{3}$	$1\frac{9}{16}$	1.8

Ordinarily the variance of estimate will not increase by a factor nearly so much larger than 1 as the bound suggests. Thus we do quite well in practice with $r = 2$ (or conceivably $r = 3$). Of course large r can ruin us. (Conventional estimates of the variances and covariances of our estimates can be affected much more seriously.)

If we deal with the fourth "if" as suggested when it appeared, and introduce covariances $h_{jk}\sigma^2$, we would need to face being wrong about the h_{jk}. The extreme ratios of variances, actual to proper, are now replaced by the ratio of the extreme determinantal roots for actual-vs-proper covariances, so that

$$r = \frac{\lambda_{max}}{\lambda_{min}}$$

where the λ's are the roots of

$$|H_{jk} - \lambda h_{jk}| = 0$$

where h_{jk} are actual and H_{jk} are proper.

The overall picture is that geodesists and astronomers, with their usually simple weighting systems, have acted very reasonably. Precise weighting is neither needed nor very helpful. The only thing we need to fear greatly is using a large weight when a small one was needed. (The converse is not nearly as troublesome.)

6. Nonlinear estimate -- the 5th if

We come now to the 5th "if". In the days when even the simplest arithmetic was a trial, we shuddered at the calculations of least squares -- I, for one, can recall how proud Carl Bennett was of a girl who could solve about a dozen linear equations in a dozen unknowns in about a day -- this didn't include collecting sums of squares and products. Today we would put even very simple least squares calculation on the computer, so a calculation that takes no more than a few times as much arithmetic as linear least squares is likely to be thought of as easy.

I am told that Morven Gentleman, for example, disposes of greater efficiency for least squares calculation with the remark that "key-punching the input costs one or two orders of magnitude more than doing the calculation", while John Chambers likes to point to quite large linear least-squares computations where the cost of printing and plotting the answers is now 10 times the cost of computation.

To understand this 5th "if", we need to be aware of a second classical result -- one that complements the Gauss-Markov theorem -- one that states that when all the "errors" are Gaussian (and the parameters enter linearly) the optimum estimates will be linear in the observations. The lessons from these two results are clear:

- if we can do better than least squares (when all but the 5th "if" hold) it will be by using estimates that are not linear in the observations.
- such estimates are likely to be most useful when we are far from the Gaussian situation.

The commonly considered estremes of non-Gaussianity are two, one uncommon in practice, the other all too common. At the uncommon extreme the "error" distribution is short-tailed -- more like the rectangular. It is in such unlikely cases that "least 4th powers" and the like may occasionally outdo least squares.

At the all too common extreme we have to face an apparently long-tailed distribution -- it matters little whether the appearance comes from medium tails and varying spread, from long tails and constant spread, or from an intermediate situation. Here we know how to improve on least squares by a considerable margin, although we do not yet know how to "fine tune" our procedures really close to their highest quality.

The simplest approach is to introduce an additional weighting factor into the normal equations and accepting an iterative calculation. This requires replacing, for a suitable set of matchers

$$\Sigma(\text{matcher})\cdot(y-\hat{y}) = 0$$

by

$$\Sigma w\left\{\frac{y-\hat{y}_0}{cS_0}\right\}\cdot(\text{matcher})\cdot(y-\hat{y}) = 0$$

where

$$w(u) = \begin{cases} 0 & u << 0 \\ \text{intermediate} & u < 0 \\ 1 & u \text{ near } 0 \\ \text{intermediate} & u > 0 \\ 0 & u >> 0 \end{cases}$$

where $\hat{y}_0$ refers to the previous iteration, where c is a constant and where S_0 is a measure of scale based on the collected residuals $\{y-\hat{y}_0\}$ of the previous iteration (perhaps the median absolute deviation).

The choices

$$w(u) = \begin{cases} (1-u^2)^2 & \text{for } |u| \leq 1 \\ 0 & \text{else} \end{cases}$$

$$S_0 = \begin{cases} \text{median absolute deviation, or} \\ \frac{1}{2}(\text{interquartile range or H-spread}) \end{cases}$$

c somewhere from 4 to 9

are often useful. (For exemplification on both real data and artificially messed-up data -- and further discussion -- see Beaton and Tukey, 1974.)

Beaton, Tukey and Van Hassel [4] have now prepared a program implementing these choices which all interested may try. Those who wish to modify their own programs can clearly to do easily.

Given any likelihood of long-tailed or long-tail-like behavior, such a modification is much safer than least squares, and will often teach us much more about what is really going on. Today we feel moderately happy about the 5th "if":

- we know it is often very important.
- we know how to deal with it quite well.

7. Unbiasedness and the 6th if

What is the strength of unbiasedness in classical least squares? Just to say that

$$\text{ave}\{b_i\} = \beta_i$$

is to conceal the essentials just as far as we know how to do this. All of us with experience with real data know that

- there is something of value in some concept or concepts resembling unbiasedness.
- there is no sacredness in having the bias exactly zero -- so long as the bias is a small fraction of the standard deviation of estimate, its existence as "bias" will not matter.

How can we reconcile these seemingly disparate facts?

A useful first step is to divide the unbiasedness condition into three parts, namely:

$$\frac{\partial}{\partial \beta_j}\text{ave}\{b_i\} = 0 \qquad \text{for all } i \neq j$$

$$\frac{\partial}{\partial \beta_i}\text{ave}\{b_i\} = 1 \qquad \text{for all } i$$

$$\text{ave}\{b_i\} = \beta_{i0} \text{ for one choice } \{\beta_{10}, \beta_{20}, \ldots, \beta_{k0}\}$$

Of these, when we consider them carefully, we are most interested in the first, moderately interested in the second, and least interested in the last.

* insulation *

We can strengthen the first of these considerably by considering, instead, a quite strong result -- one that holds in linear least squares -- namely

$$\frac{\partial b_i}{\partial \beta_j} = 0 \quad \text{for} \quad i \neq j$$

This says the following:

- if we modify y to $y + hx_j$ then the change in b_i, for $i \neq j$, vanishes faster than h.

In the linear case, where b_i is a linear combination of the y's, the change in b_i has to be linear in h and hence has to vanish identically. (In a nonlinear case, we dare only ask for a zero derivative as $h \to 0$.) This means that a change which can be perfectly absorbed by changing one b_j will be so absorbed -- no other b being altered. This is not a statistical property, not something that happens only on average. It is a specific property, about what happens to a few (here two at a time suffice) alternative sets of numbers. Thus it is both strong and clear.

So long as there are no linear dependencies among the x's, this is equivalent to:

- when y is changed to $y + hx_j$, the residuals $y - \hat{y}$ remain the same.

And this will happen so long as the fit can be recognized (identified, even defined) by the properties of its residuals. For the residuals of any fit to y will be the same as those of some fit to $y + hx_j$. With the two collections of possible sets of residuals the same, any procedure selecting fits on the basis of residuals will select the same fitted set of residuals.

This property, which I have called elsewhere

"insulation of b_i from the value of β_j "

is thus a reasonably common property.

Our only problem is dealing with criteria for fitting that have multiple solutions. For such a criterion, $y \to y + hx_j$ may move us from one solution to another. Eventually we may have to give some thought to avoiding such possibilities. If we use the sort of estimate mentioned in the last section, beginning with "S_0 large"-- really with "$w(u) = 1$ for all data points" -- we can have no computational problem, since the first iteration provides all needed insulation.

*transparency *

The next part of "unbiasedness" has to do with what happens to b_i , itself, when $y \to y + hx_i$. If it costs us little, we would like to have $b_i \to b_i + h$. This we will certainly have -- for the linearly-appearing parameter case -- for all the estimates with which we are concerned here.

When what is fitted is non-linear in the parameters, again we can only ask about derivatives, requiring

$$\frac{\partial b_i}{\partial \beta_i} \approxeq 1 \quad ,$$

to have what we may call "transparency" of our estimating processes to changes in the parameters. (We cannot require = 1, because we can only EXACTLY get rid of biases in linear situations. We have to be realistic and satisfy ourselves with $\approxeq 1$, perhaps somewhere in [.95, 1.05] or even in [.99, 1.01].)

* quality and the 6th if *

We need also to recognize that there is reason for the generalized form of the Cramer-Rao inequality, that one should usually "descale" one's quality measure by looking at

$$\frac{\mathrm{var}\{b_i\}}{\left(\dfrac{\partial b_i}{\partial \beta_i}\right)^2}$$

rather than at the variance alone, so that the qualities of $.99b_i$, b_i, and $1.01b_i$ are all the same. (We have already planned to keep

$$\frac{\partial b_i}{\partial \beta_i}$$

as close to unity as we reasonably can.)

So much, at least today, for the 6th "if".

8. Variance of estimate as a criterion -- the 7th if

As we move to high-quality resistant, nonlinear estimates, we make the shapes of our estimate distributions more nearly Gaussian. As a consequence, variance of estimate is a better criterion. Judging by the experience for the simplest case -- location -- where it does well enough in moderate and large-sized samples, we may confidently expect variance to do well when there are, effectively, rather more than ten observations per parameter. (We can hope that it does well as soon as there are rather more than twenty well-distributed degrees of freedom for error.)

If we need something better, we would do quite well to adopt the 4.2% pseudovariance of estimate, defined as the variance of a Gaussian distribution whose upper and lower 4.2% points (at $\pm\ \sigma\sqrt{3}$) match the upper and lower 4.2% points of the distribution of estimates.

9. Forecasting

One great use of least squares -- and now, perhaps, of techniques related to, but distinct from , least squares -- is to supply forecasting formulas with which to peer into the future. We are accustomed to look only at the statistics of the past in choosing such formulas. IF the future is to be like the past -- and IF we know nothing else than the statistics of the past -- we may be wise to do this, but these are large IFs and our standard policy is to ask "And WHAT IF NOT?".

* genetic selection *

Some insight comes from the practices of plant breeding for grain yield, where the quantity it is desired to improve can be easily taken apart, often as

$$\begin{aligned}\text{log wt of grain} \equiv\ & \text{log number of rows of kernels}\\ &+ \text{log number of kernels per row}\\ &+ \text{log ave wt of a kernel}\end{aligned}$$

which we may write, for the observed quantities, as

$$y = x_1 + x_2 + x_3$$

If we measured "without error" we should clearly make our selection decisions on the basis of y, and could gain exactly NOTHING by calculating -- or, horrors, using -- the individual x_i. But this is not the case. The quantities we measure -- the phenotype values -- are not the quantities, satisfying a similar relation

$$Y = X_1 + X_2 + X_3$$

that really concern us. When we allow for both the variances and covariances of the X_i and the variances (and covariances) of the $x_i - X_i$, it is perfectly possible that

$$1.07x_i + .94x_2 + 1.02x_3$$

is a better guide for our selection decisions than y itself.

This is only likely to be true so long as the variances and covariances concerned do not change. In the genetic situation, such consistancy is a reasonable hope. A change in climate -- or a year of catastrophic weather -- clearly changes the breeding problem. (A massive infusion of new genetic variability does so less obviously.)

* economic forecasting *

In the economic world, such constancy is not always present, as the oil crisis has recently and vigorously reminded both consumers and econometric modellers. Almost all the latter had, presumably because history showed nothing but cheap oil, either never considered oil prices as important variates or had let the statistics of recent history sweep oil prices out of their equations. (Much as a plant breeder who had seen only 8-rowed maize might find row numbers disappearing from his equations.)

What can we do to prepare for changing variance structures? Is not the ability to do well in changing times one of the greatest advantages that an economic forecast can possess?

These are broad questions -- we can hardly deal with any large part of them today. But we can deal with one simple example.

* parallel carriers with a consistent scale *

Suppose y, whatever it may be, is to be forecast linearly from $x_1, x_2, \ldots, x_k$ each of which is a measure of general national economic health, expressed in similar units, so that every pair is at least moderately highly correlated. (I would like to thank R.L. Obenchain and H.D. Vinod for calling my attention to the importance of such problems and the need for new techniques to deal with them. Notice that we are making an assumption that depends very much on a coordinate system. We can no longer change coordinates freely in parameter space.) Especially if there have been no discontinuities of behavior over the years from which a regression is fitted, the individual coefficients will not be at all well determined. There will be various ways to divide up the needed dependence among $b_1, b_2, \ldots, b_k$ each of which gives nearly the same values for

$$\hat{y} = \Sigma b_i x_i$$

The precise choice of the b_i will seem to depend on the precise historical variance-covariance structure of the x's. We dare not be shocked by

$$1.4x_1 - .3x_2 - .1x_3$$

giving the apparently best performance, especially when any of

$$x_1$$

$$.5x_1 - .5x_2 + x_3$$

$$\frac{1}{3}x_1 + \frac{1}{3}x_2 + \frac{1}{3}x_3$$

all appear to give very nearly equal performance.

It seems likely that economic judgement would suggest that, of those mentioned,

$$\frac{1}{3}x_1 + \frac{1}{3}x_2 + \frac{1}{3}x_3$$

would be most likely to withstand drastically changed conditions well. How can we arrange to move, more or less automatically, in such a direction? To make coefficients nearly alike when this does not cost* very much? Notice that we are taking a view quite different from the accustomed desire to be parsimonious about the number of

* Cost, so long as historical variances and covariances continue to apply.

x's we use. In this section we are thinking, instead, that using more x's is something like increasing the sample size, something that offers us protection against catastrophe.

* a z-trace *

The practice of "ridge regression" seems rampant, though plotting the modified estimates of coefficients against the log of the ridge parameter (rather than against its raw value) leaves many of us confused as to where to stop -- or even whether to start. The sort of idea that leads to ridge traces suggests, however, that we might do well to minimize, not

$$\Sigma\ (\text{weight})(y - \hat{y})^2$$

but rather

$$c_{12}(b_1 - b_2)^2 + c_{13}(b_1 - b_3)^2 + \ldots + \Sigma(\text{weight})(y - \hat{y})^2$$

for suitable c_{ij} .

One suggestion, generating what might be called a z trace, -- would be to put $c_{ij} = \lambda z_{ij}^2$ where z_{ij} is Fisher's z expression of the correlation between x_i and x_j and λ is the trace parameter. (We can also do $c_{ij} = \lambda z_{ij}$ by a simple modification of what follows.)

If we are to fit by least squares, this is equivalent to adding $k(k-1)/2$ dummy data points -- one for each (unordered) pair ij -- for which

- all weights are λ
- all y's are zero
- for the dummy corresponding to $i < j$, all x's but two are zero, these being $x_i = + z_{ij}$, and $x_j = - z_{ij}$.

(We could even go over to nonlinear estimation with $w(u)$ forced to =1 for the dummy points, though we might want also to think about changing the form of our fit, perhaps combining the $b_i x_i$ in a nonlinear way.)

Doing this -- with λ as large as can be used without increasing the historically-estimated variance of forecast more than, say, 5% -- would be a way to bring the $\{b_i\}$ more nearly in line with one another without paying too much of an insurance premium (against changes) for doing this. Who will try it?

10. Summary

We have looked hard at linear least squares, and the seven "ifs" that are part of the Gauss-Markov theorem. We have seen that:

- the meaning of a regression coefficient is more subtle than we might at first think, but we can say what it is.
- nonlinear least squares offers us guidance about how to discuss -- and think about -- the linear case, and brings in a few problems of its own.
- the fact that our x's usually "measure" what we would really like to use in only a rather approximate sense is more important, and has more consequences, than we might have thought.
- weighting need only be approximate, making an "if" or two much less worrisome.
- using nonlinear estimates in either linear or nonlinear least squares is now easy and effective.
- (approximate) unbiasedness can be easily reformulated in terms of insulation and approximate transparency.
- "variance of estimate" is not a bad criterion, and how to improve it if necessary.
- forecasting often needs to be considered in terms of possible changes of variance-covariance structure, something that can lead to rather different numerical answers.

It will become relatively rare, as we absorb these points, that we will want to do "plain least squares", but we will want to do something that is one or more moderate-sized steps from least squares very very often. We will think harder about meanings of coefficients -- and about the great impact of indirect and imperfect measurement, but we will continue to do the best we can -- which will usually be based upon -- but not equal to -- least squares. It is to be hoped that we will "sit loose" to those aspects of our results that now appear less than certain.

And it is to be hoped that we will adopt broadly similar attitudes and considerations in dealing with other statistical idols.

REFERENCES

[1] D.F. Andrews, P.J. Bickel, F.R. Hampel, P.J. Huber, W.H. Rogers, and J.W. Tukey, (1972). Robust Estimates of Location: Survey and Advances, Princeton University Press, Princeton, New Jersey.

[2] E.M.L. Beale, (1960). Confidence regions in nonlinear estimation. J. Roy. Stat. Soc. 22: 41-76

[3] A.E. Beaton and J.W. Tukey (1974). The fitting of power series, meaning polynomials, illustrated on band-spectroscopic data. Technometrics 16: 147-185.

[4] A.E. Beaton, J.W. Tukey, and W. Van Hassel (1974). w-estimators and biweight. To appear as an Educational Testing Service RB.

[5] G.E.P. Box (1966). Use and abuse of regression. Technometrics 8: 625-629.

[6] M.J. Box (1971). Bias in nonlinear estimation. J. Roy. Stat. Soc. B33: 171-190.

[7] John M. Chambers (1973). Fitting nonlinear models: numerical techniques. Biometrika 60: 1-13.

[8] Irwin Guttman and Duane A. Meeter (1965). On Beale's measures of nonlinearity. Technometrics 7: 623-637.

[9] Peter J. Huber (1972). The 1972 Wald Lecture. Robust statistics, A review. Ann. Math. Stat. 43: 1041-1067.

[10] A. L. Nagar (1959). The bias and moment-matrix of the general k-class estimators in structural equations. Econometrica 27: 575-595.

[11] J.D. Sorgan and W.M. Mikhail (1971). A general approximation to the distribution of instrumental variables estimates. Econometrica 39: 131-169.

[12] J.W. Tukey (1948). Approximate weights. Annals of Math. Statist. 19: 91-92.

[13] D.G. Watts (1974). Improving convergence in nonlinear least squares by partitioning. (Proceedings of this Conference)

John W. Tukey
Department of Statistics
Princeton University
Princeton, New Jersey
U.S.A.

or
Bell Laboratories
Princteon, New Jersey
U.S.A.

Applied Statistics, R.P. Gupta, (editor)

TIME SERIES ANALYSIS AND ECONOMETRIC MODEL CONSTRUCTION

by Arnold Zellner*

I. Introduction

Since the 1920's, econometricians have produced a large number of econometric models. In fact it is probably accurate to state that an econometric model-building industry has emerged. Workers have built models of consuming units, firms, banks, markets, urban economies, transport systems, educational systems, regional economies, national economies, etc.[1] These modeling efforts have as one objective the production of results that will be useful in understanding the operation of the particular entity being modeled. In the process of achieving understanding, it is of course the case that propositions derived from economic theory are often utilized in models. A second objective of most modeling efforts is that of prediction. For example, several models of the U.S. economy are in operation that yield quarterly forecasts of such variables as gross national product, employment, unemployment, the price level, etc. Third, models are often employed to analyze the probable effects of alternative policies and/or to provide a basis for implementing good control policies. In the case of a model of a national economy, it is important to be able to predict the effects of a proposed tax cut or an increase in government spending. In some cases, experiments with optimal control policies are being studied. Finally, models can be employed in attempts to assess the probable effects of changes in parameter values, lag structures, etc., an area that is called "structural analysis." In summary, econometric models can contribute to (a) achieving the scientific objective of understanding economic behavior, (b) obtaining better predictions and forecasts and (c) improving policy-makers' attempts to control economic systems so as to achieve better performance.

Whether, econometric models will in the future make more substantial contributions in the areas mentioned above, obviously depends critically on the quality of econometric models. The quality of econometric models depends not only on the state of our

* Research financed by National Science Foundation Grant GS-40033 and by income from the H.G.B. Alexander Endowment Fund, Graduate School of Business, Univ. of Chicago.

[1] This and other footnotes are presented at the end of the paper.

economic knowledge and the quality of our data but also on the methodology of econometric model construction. While much progress has been made in mathematical, statistical and computer methodology, it is the main point of this paper that much more has to be done in order to produce econometric models that are of high quality, that is that predict reasonably well and that can be used with a reasonable degree of confidence in analyzing policy problems and in obtaining good control policies.

The schematic diagram in Figure 1 illustrates some of the activities involved in econometric model construction. With respect to any project, including an econometric modeling project, it is important to develop an initial statement of the project's objectives. A clear-cut statement of objectives will be helpful in planning and executing a project. The next step is a feasibility study to determine whether the project is "doable" given the state of our knowledge, data, methodology and time, personnel and budgetary constraints. Also, the feasibility study should appraise alternative approaches for realizing the objectives of a project. If an econometric modeling approach is selected, it will usually be necessary to review past studies relating to the project's objectives and to engage in some preliminary data gathering and analysis. The purpose of these activities is to produce an initial model, denoted M_o in Fig. 1. Generally speaking, it is thought that the initial formulation of a model should be as simple as possible but potentially adequate to realize the objectives of the project. Once M_o has been formulated, it can be subjected to mathematical and computer simulation analysis to determine that it is logically consistent and to establish its properties. In this work, that resembles what an engineer does with a model of a new airplane in wind-tunnel experiments, a model-builder will often use crude estimates of parameters and experiment with alternative functional forms for relationships. In this way, he builds up a familiarity with the properties of M_o. Further, data can be brought to bear on the initial variant M_o to check that the implications of the model are in agreement with the information in the data. The results of these mathematical analyses, computer simulation experiments and statistical analyses may suggest that M_o is deficient in certain respects. On discovering the nature of the deficiencies of M_o, it will be necessary to reformulate M_o to provide a new variant, denoted M_1 in Fig. 1. Then M_1 is subjected to mathematical, computer and statistical analyses to determine if it is an adequate formulation. Often additional

Fig. 1

Model Building Activities: A Schematic Representation*

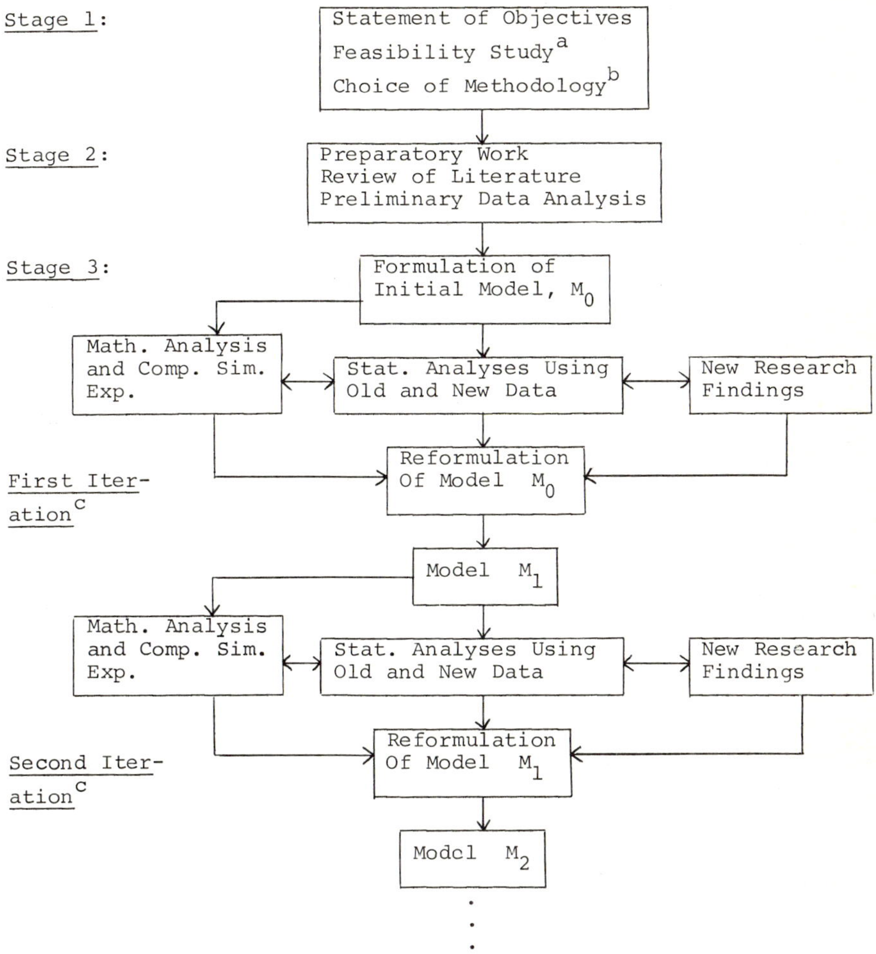

*For further discussion of this approach to modeling, see H.R. Hamilton, S.E. Goldstone, J.W. Milliman, A.L. Pugh III, E.R. Roberts and A. Zellner, Systems Simulation for Regional Analysis: An Application to River-Basin Planning, MIT Press, 1969, and A. Zellner, "The Care and Feeding of Econometric Models," U. of Chicago, Graduate School of Business Selected Paper No. 35.

[a] It is assumed that this study shows the project to be feasible.

[b] It is assumed that a "modeling" approach is selected.

[c] The iterative procedure may disclose problems in the original formulation of goals, feasibility, and methodology so that refining and reformulation of the effort may not be confined solely to the model itself.

"iterations" will be necessary in order to obtain a variant that is satisfactory. In fact, model improvements usually continue to be made even after a model is in operation just as with yearly model improvements in the automobile industry.

In the present paper, attention will be directed toward one part of the modeling problem, namely that of using data to "check-out" the implications of linear dynamic econometric models. The approach, described and applied earlier, involves drawing out the logical implications of a linear dynamic econometric model for the forms and parameter values of processes on individual variables and for the transfer functions associated with the model.[2] Given that these implications are clearly set forth, data and statistical methods can be employed to assess whether the implications are consistent with the information in the data. If a model's implications are consistent with the information in the data, then our confidence in the model is enhanced. If, however, a model's implications are not consistent with information in the data, then reformulation is necessary.

In what follows, we shall specify a multiple time series process (MTSP) in Section II. Processes for individual variables, that we call the final equations (FEs), implied by the MTSP will then be derived. Further, it will be shown how a MTSP can be specialized to assume the form of a linear dynamic simultaneous equation model (SEM). The transfer functions (TFs) associated with the SEM will be derived and their relation to the SEM's FEs will be established. Testable implications of the SEM's TFs and FEs will be indicated. In Section III, some illustrative examples will be provided while in Section IV some issues and problems in statistical methodology will be considered. Last, in Section V, some concluding remarks are presented.

II. Analysis of Linear Dynamic Econometric Models.

In approaching the problem of constructing a dynamic econometric model, it is possible to incorporate little or much prior information in an initial formulation. In what follows, we shall begin by considering an approach that initially uses relatively little prior information and then proceed to variants that incorporate more prior information. If our data are extensive, of good quality and contain much information, then it is possible to rely more on the information in the data and less on prior information in formulating a model. However, when the information in our data is limited, there will generally be a greater dependence on prior information gleaned from economic theory and other sources in formulating a model.

If we are able to specify the observable variables $\underline{z}_t' = (z_{1t}, z_{2t}, \ldots, z_{pt})$ that probably appear in a model and are willing to assume that a linear multiple time series model or multivariate autoregressive-moving average (ARMA) process is an adequate approximation, we can follow Quenouille (1957) and Zellner and Palm (1973) by writing

$$(2.1) \qquad \underset{p\times p}{H(L)}\ \underset{p\times 1}{\underline{z}_t} = \underset{p\times p}{\underline{\bar{\theta}}} + \underset{p\times p}{F(L)}\ \underset{p\times 1}{\underline{e}_t} \qquad t = 1,2,\ldots,T$$

as the process that may be generating the $p\times 1$ vector $\underline{z}_t$.[3] In (2.1), L denotes a lag operator such that $L^n \underline{z}_t \equiv \underline{z}_{t-n}$, $\underline{\bar{\theta}}' = (\bar{\theta}_1, \bar{\theta}_2, \ldots, \bar{\theta}_p)$ is a vector of constants, $\underline{e}_t$ is a $p\times 1$ vector of unobservable random errors, and $H(L)$ and $F(L)$ are each $p\times p$ matrix lag operators with elements that are finite polynomials in the lag operator L. That is, $H(L) = \{h_{ij}(L)\}$ and $F(L) = \{f_{ij}(L)\}$ where the typical elements are given by $h_{ij}(L) = \sum_{\ell=0}^{r_{ij}} h_{ij}^{(\ell)} L^{\ell}$ and $f_{ij}(L) = \sum_{\ell=0}^{q_{ij}} f_{ij}^{(\ell)} L^{\ell}$, where r_{ij} and q_{ij} denote the degrees of $h_{ij}(L)$ and $f_{ij}(L)$, respectively. With respect to the $p\times 1$ error vector, $\underline{e}_t$, it is assumed that

$$(2.2) \qquad \begin{aligned} E\underline{e}_t &= \underline{0} \\ E\underline{e}_t \underline{e}_{t'}' &= \delta_{tt'} I_p \end{aligned} \qquad \text{for all } t, t'$$

where $\delta_{tt'}$ is the Kronecker delta and I_p is a $p\times p$ unit matrix. Possible contemporaneous and serial correlations and differing variances of the elements of the error process can be introduced via appropriate specification of the matrix $F(L)$.

If the matrix $F(L)$ has elements of degree zero in L, (2.1) is a pure AR process. If $H(L)$ has all elements of degree zero in L, the process in (2.1) is a pure MA process. When both $H(L)$ and $F(L)$ contain polynomials not all of degree zero, (2.1) represents a mixed ARMA process. In general, if $H(L)$ is invertible, that is $H(L)$ has roots with values outside the unit circle, we can multiply both sides of (2.1) by the adjoint matrix $H^*(L)$ associated with $H(L)$ to obtain:

(2.3a) $$|H(L)|\underline{z}_t = \underline{\theta} + H^*(L)F(L)\underline{e}_t$$

where $|H(L)|$ is the determinant of $H(L)$, a polynomial of finite degree in L, and $\underline{\theta}' = (\theta_1,\theta_2,\ldots,\theta_p)$ is a vector of constant parameters. Explicitly, the equations of the system in (2.3a), that we call the final equations (FEs), are given by:

(2.3b) $$|H(L)|z_{it} = \theta_i + \underline{\alpha}_i'\underline{e}_t, \quad i = 1,2,\ldots,p$$

where $\underline{\alpha}_i'$ denotes the i'th row of $H^*(L)F(L)$. It is seen from (2.3b) that the processes for individual elements of $\underline{z}_t$ are in the ARMA form. If there are no common factors in the terms on both sides of (2.3a), then the order and parameters of the AR parts of the processes in (2.3b) should be identical, a point that can be checked in empirical analyses. If empirical analyses yield the result that the order and parameters of the AR parts of the FEs are identical, then the vector $\underline{z}_t$ may have been generated by the process in (2.1) without any additional restrictions on $H(L)$. If however, the order and parameters of the AR parts of the FEs are found to be different for different elements of $\underline{z}_t$, then it is possible that $H(L)$ has a special form, e.g. $H(L)$ might be a triangular or block diagonal matrix. By study of the empirical results for the FEs it may be possible to infer some possible properties of the matrix $H(L)$. In particular, for large systems if $H(L)$ were completely unrestricted, the processes in (2.3a) would have high order AR parts, an implication that is not in agreement with empirical findings that point to relatively low order AR parts of FEs for many economic variables.[4] Whether or not it is possible to infer properties of $H(L)$ from empirical findings relating to the FEs, it is important to note that when FEs' forms have been determined from the data and their parameters have been estimated, these equations can be used to generate predictions.

One assumption, usually employed in econometric work, that specializes the forms of the matrices $H(L)$ and $F(L)$, is that some of the variables in $\underline{z}_t$ are exogenously determined, that is generated by processes that are independent of the processes generating the remaining variables. If we partition $\underline{z}_t$ as follows, $\underline{z}_t' = (\underline{y}_t'\underline{x}_t')$ where $\underline{y}_t$ is a $p_1 \times 1$ vector of endogenous variables and $\underline{x}_t$ is a $p_2 \times 1$ vector of exogenous variables, with $p = p_1 + p_2$, then we can write the system in (2.1) as:

$$\begin{bmatrix} H_{11}(L) & H_{12}(L) \\ H_{21}(L) & H_{22}(L) \end{bmatrix} \begin{bmatrix} \underline{y}_t \\ \underline{x}_t \end{bmatrix} = \begin{bmatrix} \underline{\bar{\theta}}_1 \\ \underline{\bar{\theta}}_2 \end{bmatrix} + \begin{bmatrix} F_{11}(L) & F_{12}(L) \\ F_{21}(L) & F_{22}(L) \end{bmatrix} \begin{bmatrix} \underline{e}_{1t} \\ \underline{e}_{2t} \end{bmatrix} \tag{2.4}$$

where $H(L)$, $F(L)$, $\underline{\bar{\theta}}$ and $\underline{e}_t$ have been partitioned to correspond to the partitioning of $\underline{z}_t$. The assumption that $\underline{x}_t$ is exogenous gives rise to the following restrictions:

$$H_{21}(L) \equiv 0 \ ; \ F_{12}(L) \equiv 0 \ ; \text{ and } \ F_{21}(L) \equiv 0 \ . \tag{2.5}$$

With these restrictions imposed on (2.4), we have:

$$\underset{p_1\times p_1}{H_{11}(L)} \ \underset{p_1\times 1}{\underline{y}_t} + \underset{p_1\times p_2}{H_{12}(L)} \ \underset{p_2\times 1}{\underline{x}_t} = \underset{p_1\times 1}{\underline{\bar{\theta}}_1} + \underset{p_1\times p_1}{F_{11}(L)} \ \underset{p_1\times 1}{\underline{e}_{1t}} \tag{2.5a}$$

and

$$\underset{p_2\times p_2}{H_{22}(L)} \ \underset{p_2\times 1}{\underline{x}_t} = \underset{p_2\times 1}{\underline{\bar{\theta}}_2} + \underset{p_2\times p_2}{F_{22}(L)} \ \underset{p_2\times 1}{\underline{e}_{2t}} \ . \tag{2.5b}$$

The equations in (2.5a) are referred to as the <u>structural equations</u> while those in (2.5b) are processes generating the random exogenous vector of variables, $\underline{x}_t$.[5]

The assumption that $\underline{x}_t$ is exogenous, leading to the restrictions in (2.5), has important implications for the forms of the FEs. Given that $H_{22}(L)$ in (2.5b) is invertible, the FEs for the elements of $\underline{x}_t$ are:

$$|H_{22}(L)|\underline{x}_t = \underline{\bar{\bar{\theta}}}_2 + H^*_{22}(L)F_{22}(L)\underline{e}_{2t} \ , \tag{2.6}$$

where $|H_{22}(L)|$ and $H^*_{22}(L)$ are the determinant and adjoint matrix of $H_{22}(L)$, respectively. On substituting for $\underline{x}_t$ in (2.5a) from (2.6) and multiplying both sides by the adjoint matrix of $H_{11}(L)$, namely $H^*_{11}(L)$, the FEs for the elements of $\underline{y}_t$ are:

$$\begin{aligned} |H_{11}(L)||H_{22}(L)|\underline{y}_t = \underline{\bar{\bar{\theta}}}_1 &- H^*_{11}(L)H_{12}(L)H^*_{22}(L)F_{22}(L)\underline{e}_{2t} \\ &+ |H_{22}(L)|H^*_{11}(L)F_{11}(L)\underline{e}_{1t} \ , \end{aligned} \tag{2.7}$$

where $\underline{\bar{\bar{\theta}}}_1$ is a vector of constants. On comparing (2.6) and (2.7), it is seen that the FEs for the elements of $\underline{y}_t$, the vector of endogenous variables, have AR parts that have orders equal to or greater than those of the FEs for the elements of $\underline{x}_t$ given in (2.6). If $|H_{22}(L)|$ is a polynomial of degree zero in L , the AR parts of (2.6) and (2.7) are of the same order. Usually, the order of the AR part of (2.7) will exceed that of (2.6), a point that can be checked empirically. Further the order of the MA errors in (2.7) will usually be larger than those of (2.6), another point that can be checked empirically. Last, if as is usually the case, more is assumed about the elements of $H(L)$ and $F(L)$, then the implications of such assumptions for the forms of the FEs can be deduced and checked empirically.

Another set of equations that is important in analyzing econometric and other models is the set of <u>transfer functions</u> (TFs).[6] The TFs can be obtained from the structural equations in (2.5a) by multiplying both sides by $H_{11}^*(L)$, the adjoint matrix associated with $H_{11}(L)$. This operation yields the following set of TFs:

$$(2.8)\qquad |H_{11}(L)|\underline{y}_t = \underline{\theta}_1^* - H_{11}^*(L)H_{12}(L)\underline{x}_t + H_{11}^*(L)F_{11}(L)\underline{e}_{1t} \ .$$

As Pierce and Mason have noted, the AR parts of the TFs, $|H_{11}(L)|y_{it}$, $i = 1,2,\ldots,p_1$, are of the same order and have identical parameters if no cancelling occurs in (2.8), a point that can be checked empirically. On the other hand, should $H_{11}(L)$ be given a particular structure, for example it might be assumed block-diagonal then the AR parts of (2.8) could differ. By using data to investigate the forms of the TFs, it is possible to check the empirical implications of specific assumptions regarding the form of $H_{11}(L)$. In addition, assumptions regarding $H_{11}(L)$, $H_{12}(L)$, and $F_{11}(L)$ generally imply testable propositions about the lags and parameter values in the terms of (2.8) involving current and lagged elements of $\underline{x}_t$ and $\underline{e}_{1t}$.

Given that data have been employed to determine the forms of the FEs for the elements of $\underline{x}_t$ in (2.6) and the TFs in (2.8), it is possible to substitute the estimated FEs for $\underline{x}_t$ in (2.8) to obtain the implied FEs for the elements of $\underline{y}_t$. These can be compared with the FEs for the elements of $\underline{y}_t$ determined directly from the data. If the two sets of results are in agreement, this provides a further consistency check on results. If, however, some of the elements of $\underline{x}_t$ are not exogenous or some other specification errors are present, it is probable that the two sets of results for the FEs for elements of y_t will not be compatible.

Given that the results of FE and TF analyses are compatible with what has been assumed about the matrices H(L) and F(L) in (2.1) and other assumptions and if sufficient identifying information is present, the parameters of the structural equations in (2.5a) can be estimated.[7] Given that the structural equations are available, they can be used for analyzing the possible effects of changes in structural parameters a problem that can not generally be analyzed with just FEs and/or TFs. However, these latter equation systems are extremely valuable in the process of checking the implications of given structural equations against the information in the data. That is, since the forms of FEs and TFs can often be determined from information in the data, these empirical results can play an important role in ruling out many variants of the structural equations, an important consideration in efforts to formulate structural equations that are consistent with the information in a set of data. Last, while the role of the FEs and TFs in model-building has been emphasized, their uses in approaching the problems of prediction and control should not be overlooked.

III. Examples Illustrating General Approach.

In this section we present some examples illustrating the general points made in the preceding section. In this connection, it should be remarked that very few econometricians emphasize or are aware of the implications of linear dynamic structural equations for TFs and FEs. In fact many econometricians view the ARMA FEs as being naive, alternative, mechanical models.[8] Thus it is thought useful to analyze some special cases to illustrate the points made in Section II.

The first example that we analyze is a supply and demand model constructed to explain the variation in the daily price (P_t) and quantity (Q_t) of hogs at eight large terminal markets in the U.S.[9] The two relations in the model, a demand equation and a supply equation are shown below:

Demand Equation:

$$(3.1)\quad \Delta\log P_t = \beta_{10t} + \beta_{11}\Delta\log Q_t + \beta_{12}\Delta\log Q_{t-1} + \beta_{13}\Delta\log Q_{t-2} + \beta_{14}\Delta\log Wh_{t-1} + u_{1t} \quad t = 1,2,\ldots,T$$

Supply Equation:

(3.2) $\Delta\log Q_t = \beta_{20t} + \beta_{21}\Delta\log P_{t-1} + u_{2t} \qquad t = 1,2,\ldots,T$

where

P_t = average price in \$/hundred wt. of all barrows and gilts sold.

Q_t = total number of head of barrows and gilts sold on day t.

Wh_{t-1} = price of output on day $t-1$ measured as a weighted average of the prices of major wholesale pork cuts in \$/hundred wt. at Chicago.

u_{1t}, u_{2t} = error terms.

β_{10t}, β_{20t} = intercept terms plus dummy variables for different days of the week in both relations and a seasonal dummy in the supply relation.

The two endogenous variables in (3.1)-(3.2) are $\Delta\log P_t$ and $\Delta\log Q_t$. If we write $\underline{y}_t' = (y_{1t}, y_{2t}) = (\Delta\log P_t, \Delta\log Q_t)$ we can express (3.1)-(3.2) as

$$\begin{bmatrix} 1 & -\beta_1(L) \\ -\beta_2(L) & 1 \end{bmatrix}\begin{bmatrix} y_{1t} \\ y_{2t} \end{bmatrix} = \begin{bmatrix} \beta_{10t} \\ \beta_{20t} \end{bmatrix} + \begin{bmatrix} 0 \\ \beta_{14} \end{bmatrix} x_t + \begin{bmatrix} u_{1t} \\ u_{2t} \end{bmatrix} \tag{3.3}$$

where $x_t = \Delta\log Wh_{t-1}$ is assumed exogenous, $\beta_1(L) = \beta_{11}+\beta_{12}LP\beta_{13}L^2$ and $\beta_2(L) = \beta_{21}L$. The determinant and adjoint matrix of the martix lag operator on the l.h.s. of (3.3) are $1-\beta_1(L)\beta_2(L)$, a third degree polynomial in L, and

$$\text{Adj}\begin{bmatrix} 1 & -\beta_1(L) \\ -\beta_2(L) & 1 \end{bmatrix} = \begin{bmatrix} 1 & \beta_1(L) \\ \beta_2(L) & 1 \end{bmatrix},$$

respectively. Thus the TFs associated with (3.3) are:

$$[1-\beta_1(L)\beta_2(L)]y_{1t} = \beta_{10t} + \beta_1(L)\beta_{20t} + \beta_{14}\beta_1(L)x_t + u_{1t} + \beta_1(L)u_{2t} \tag{3.4a}$$

$$[1-\beta_1(L)\beta_2(L)]y_{2t} = \beta_{20t} + \beta_2(L)\beta_{10t} + \beta_{14}x_t + u_{2t} + \beta_2(L)u_{1t}. \tag{3.4b}$$

It is seen that the AR parts of the TFs are identical, each of order 3. Further in (3.4a), the lag polynomial hitting $x_t = \Delta\log Wh_{t-1}$ is of degree 3 while in (3.4b) x_t appears with no lag. If u_{1t} and u_{2t} are assumed non-autocorrelated, the error in (3.4a) is in the form of a second order MA process while that in (3.4b) is in the form of a first order MA process. These are some of the testable implications of the structural equations in (3.3) that can be checked by analyzing the TF system in (3.4). In addition, if the process generating x_t is assumed to be

$$\phi_a(L)x_t = \theta_o + \theta_b(L)\epsilon_t \qquad t = 1,2,\ldots,T \tag{3.5}$$

where $\phi_a(L)$ and $\theta_b(L)$ denote finite lag polynomials of degrees a and b, respectively, ϵ_t is a non-autocorrelated error with zero mean and constant finite variance, and θ_o is a constant and if $\phi_a(L)$ is invertible, we have $x_t = \theta_o' + \phi_a^{-1}(L)\theta_b(L)\epsilon_t$. If we substitute this last expression into (3.4), we obtain the following FEs for y_{1t} and y_{2t}:

$$\begin{aligned}\phi_a(L)[1-\beta_1(L)\beta_2(L)]y_{1t} &= \theta_o'' + \phi_a(L)[\beta_{10t} + \beta_1(L)\beta_{20t}] \\ &\quad + \beta_1(L)\theta_b(L)\epsilon_t + \phi_a(L)u_{1t} + \phi_a(L)\beta_1(L)u_{2t}\end{aligned} \tag{3.6a}$$

$$\begin{aligned}\phi_a(L)[1-\beta_1(L)\beta_2(L)]y_{2t} &= \theta_o''' + \phi_a(L)[\beta_{20t} + \beta_2(L)\beta_{10t}] \\ &\quad + \beta_{14}\theta_b(L)\epsilon_t + \phi_a(L)u_{2t} + \phi_a(L)\beta_2(L)u_{1t}\end{aligned} \tag{3.6b}$$

In Table 3.1 the degrees of the lag polynomials appearing in (3.6a-b) are indicated. From the information in Table 3.1, we see that the AR parts of both (3.6a) and (3.6b) are of order $3+a$, with $a \geq 0$. The error terms of (3.6a) are in a MA form with order equal to $\max\{2+b, 2+a\}$ while the order of the MA errors in (3.6b) is equal to $\max\{b, 1+a\}$. Thus, if the econometric model in (3.1)-(3.2) is correctly specified, the final equations in (3.6) should exhibit AR parts of order $3+a$, $a \geq 0$, and autocorrelated MA error terms.

It is interesting to note that the workers who formulated and estimated the model in (3.1) - (3.2) analyzed the data for daily prices using Box-Jenkins' techniques to identify the underlying process. Their analysis yielded the result that daily price in the eight markets studied follows a random walk, $P_t = P_{t-1} + \eta_t$, a result which they failed to note is apparently

Table 3.1
Degrees of Lag Polynomials in (3.5)-(3.6)+

Equation	AR Polynomial	Polynomial for:		
		ϵ_t	u_{1t}	u_{2t}
(3.5)	a	b	-	-
(3.6a)	3+a	2+b	a	2+a
(3.6b)	3+a	b	1+a	a

+It is assumed that u_{1t}, u_{2t} and ϵ_t are serially uncorrelated. Thus the degrees of the error term polynomials shown in the table are lower bounds given no cancelling.

in conflict with the implications of their econometric model, in particular equation (3.6a) above. Also the model that they determined for quantity, Q_t, using Box-Jenkins' techniques is:

$$(3.7) \qquad (1-0.90L^5)(1-1.44L + 0.47L^2)Q_t = (1-0.70L^5)(1-1.52L + .66L^2)a_t$$

where a_t is a non-autocorrelated error term with zero mean and constant variance. The finding in (3.7) coupled with the finding of a random walk for price seems at variance with the implications of the econometric model in (3.1)-(3.2), particularly those shown in Table 3.1. It is possible that reformulation of (3.1)-(3.2) could result in FEs and TFs in accord with the information in the data.

In concluding discussion of this published example, it is important to note that the econometric analysis and the time series analysis were presented as alternative approaches for analyzing the data, a view shared by many econometricians at present. However, from what has been presented above, it is clear that it is possible to synthesize the econometric and time series approaches, a synthesis that provides a number of new ways to check the implications of structural econometric models, to discover errors in model formulation, and to suggest ways in which a model's formulation can be improved.

The second example that we take up is the analysis of a simple monetary model formulated along lines suggested Friedman.[10] The equations of the model are:

(3.8) Money Demand: $Y_t = Ae^{\gamma_1 i_t} M_t^D \qquad A, \gamma_1 > 0$

(3.9) Money Supply: $M_t^S = BE^{\gamma_2 i_t} H_t \qquad B, \gamma_2 > 0$

(3.10) Market Clearing: $M_t^D = M_t^S = M_t$

(3.11) Fisher Equation: $i_t = c_t + r_{Yt}^*$

(3.12) Anticipations Equation: $r_{Yt}^* - r_{Yt-1}^* = \beta(r_{Yt-1} - r_{Yt-1}^*) \qquad \beta > 0$

The money demand equation, (3.8), links nominal money demanded M_t^D tc nominal income Y_t and the nominal interest rate i_t. The money supply equation (3.9) relates money supply M_t^S to the nominal interest rate i_t and "high-powered" money, H_t. Equation (3.10) states that the variables adjust to equate M_t^D and M_t^S to the observed money supply M_t. The Fisher equation relates the nominal interest rate i_t to the anticipated proportionate rate of change of nominal income, $r_{Yt}^* = \log Y_t^* - \log Y_{t-1}^*$, where Y_t^* is the anticipated value of nominal income for period t as of period $t-1$. The quantity $c_t = \rho_t - q_t^*$, the real rate of interest, ρ_t, minus the anticipated proportionate rate of growth of real income. Initially it is assumed that $c_t = c$, a constant. Last, equation (3.12) is a widely used assumption about how anticipations are formed wherein r_{Yt-1} is the actual proportionate rate of growth of nominal income in period $t-1$.

In (3.8)-(3.12), M_t^D, M_t^S, r_{Yt}^* and r_{Yt-1}^* are not measured in our data. Thus we substitute from (3.10) in (3.8) and (3.9), take the logs of both sides, and first difference the resulting equations and (3.11) to obtain:

(3.13) $$r_{Yt} = \gamma_1 \Delta i_t + r_{Mt}$$

(3.14) $$r_{Mt} = \gamma_2 \Delta i_t + r_{Ht}$$

(3.15) $$\Delta i_t = \Delta r_{Yt}^*$$

(3.16) $$\Delta r_{Yt}^* = \beta(r_{Yt-1} - r_{Yt-1}^*)$$

where $\Delta \equiv 1-L$, $r_{Yt} = \Delta \log Y_t$, $r_{Mt} = \Delta \log M_t$ and $r_{Ht} = \Delta \log H_t$. Since (3.16) can be expressed as $[1-(1-\beta)L]r_{Yt}^* = \beta L r_{Yt}$, we have $[1-(1-\beta)L]\Delta r_{Yt}^* = \beta L \Delta r_{Yt}$. Then using $\Delta i_t = \Delta r_{Yt}^*$, from (3.15), we can replace (3.15) and (3.16) by

$$[1-(1-\beta)L]\Delta i_t = \beta L \Delta r_{Yt} \quad . \tag{3.17}$$

Thus the three equations of the model involving the three observable endogenous variables r_{Yt}, Δi_t and r_{Mt} and the observable exogenous variable r_{Ht} are given by (3.13), (3.14) and (3.17). We can express these three equations, with error terms added, as follows:

$$\begin{bmatrix} 1 & -1 & -\gamma_1 \\ 0 & 1 & -\gamma_2 \\ -\beta L(1-L) & 0 & 1-(1-\beta)L \end{bmatrix} \begin{bmatrix} r_{Yt} \\ r_{Mt} \\ \Delta i_t \end{bmatrix} = \begin{bmatrix} 0 \\ 1 \\ 0 \end{bmatrix} r_{Ht} + \begin{bmatrix} u_{1t} \\ u_{2t} \\ u_{3t} \end{bmatrix} , \tag{3.18a}$$

or in more general form,

$$H_{11}\underline{y}_t = \underline{\alpha} r_{Ht} + \underline{u}_t \tag{3.18b}$$

where H_{11} denotes the matrix on the l.h.s. of (3.18a), $\underline{y}_t' = (r_{Yt}, r_{Mt}, \Delta i_t)$, $\underline{\alpha}' = (0,1,0)$ and $\underline{u}_t' = (u_{1t}, u_{2t}, u_{3t})$, a vector of error terms.

The TFs associated with (3.18) are obtained by premultiplying (3.18) by the adjoint matrix H_{11}^* associated with H_{11}, namely,

$$|H_{11}|\underline{y}_t = H_{11}^*\underline{\alpha} r_{Ht} + H_{11}^*\underline{u}_t \quad , \tag{3.19}$$

where $|H_{11}| = 1-(1-\beta)L - \beta(\gamma_1+\gamma_2)L(1-L)$, a second degree polynomial in L, and

$$H_{11}^* = \begin{bmatrix} 1-(1-\beta)L & 1-(1-\beta)L & \gamma_1+\gamma_2 \\ \gamma_2\beta L(1-L) & 1-(1-\beta)L-\gamma_1\beta L(1-L) & \gamma_2 \\ \beta L(1-L) & \beta L(1-L) & 1 \end{bmatrix} \tag{3.20}$$

Explicitly, the TF system in (3.19) is:

$$[1-(1-\beta)L-\beta(\gamma_1+\gamma_2)L(1-L)] \begin{bmatrix} r_{Yt} \\ r_{Mt} \\ \Delta i_t \end{bmatrix} = \begin{bmatrix} 1-(1-\beta)L \\ 1-(1-\beta)L-\gamma_1\beta L(1-L) \\ \beta L(1-L) \end{bmatrix} r_{Ht} + \begin{bmatrix} v_{1t} \\ v_{2t} \\ v_{3t} \end{bmatrix} \tag{3.21a}$$

with $\underline{v}_t = H_{11}^*\underline{u}_t$, where $\underline{v}_t' = (v_{1t}, v_{2t}, v_{3t})$.

On examination of the l.h.s. of (3.21), it is seen that the structural equations of the monetary model imply that the three TFs have AR parts of order two with identical parameters. In addition, the implied lags on r_{Ht}, the rate of growth of high-powered money, in the TFs are shown below:

TF for:	Order of Lag on r_{Ht}
r_{Yt}	1
r_{Mt}	2
Δi_t	2

It should also be noted that (3.21a) involves some additional strong restrictions on the parameters of the TFs that can be appreciated by rewriting (3.21a) as

$$(3.21b) \qquad [1-(1-\beta+\eta)L+\eta L^2]\begin{bmatrix} r_{Yt} \\ r_{Mt} \\ \Delta i_t \end{bmatrix} = \begin{bmatrix} 1-(1-\beta)L \\ 1-(1-\beta+\eta_1)L+\eta_1 L^2 \\ \beta L-\beta L^2 \end{bmatrix} r_{Ht} + \begin{bmatrix} v_{1t} \\ v_{2t} \\ v_{3t} \end{bmatrix}$$

where $\eta = (\gamma_1+\gamma_2)\beta$ and $\eta_1 = \gamma_1\beta$. The following are the restrictions on the parameters of (3.21b):

(i) The sum of the coefficients of r_{Yt-1} and r_{Yt-2} equals the coefficient of r_{Ht-1} in the TF for r_{Yt} .

(ii) The coefficient of r_{Ht} should equal one in the TF for r_{Yt} and r_{Mt} and be equal to zero in the TF for Δi_t .

(iii) In each of the three TFs, the sum of the AR parameters is equal to β , the coefficient of $r_{Ht-1} - r_{Ht-2}$ in the TF for Δi_t .

(iv) In the TF for Δi_t , the sum of the coefficients of r_{Ht-1} and r_{Ht-2} is zero.

(v) The sum of the coefficients of r_{Ht-1} and r_{Ht-2} in the TF for r_{Mt} equals the coefficient of r_{Ht-1} in the TF for r_{Yt} .

Given data on the variables and assumptions regarding the error terms, the above restrictions can be tested.

With respect to the error vector $\underline{v}_t$ in (3.21), given by $\underline{v}_t = H_{11}^{*}\underline{u}_t$, it is the case that assumptions about the properties of $\underline{u}_t$ imply testable implications about the properties of $\underline{v}_t$. For example, if the $\underline{u}_t$'s are assumed to be serially uncorrelated, then the elements of $\underline{v}_t$, given by

$$(3.22)\qquad \underline{v}_t = \begin{bmatrix} v_{1t} \\ v_{2t} \\ v_{3t} \end{bmatrix} = \begin{pmatrix} [1-(1-\beta)L]u_{1t} + [1-(1-\beta)Lu_{2t} + (\gamma_1+\gamma_2)u_{3t} \\ \gamma_2\beta L(1-L)u_{1t} + [1-(1-\beta)L-\gamma_1\beta L(1-L)u_{2t}+\gamma_2 u_{3t} \\ \beta L(1-L)u_{1t} + \beta L(1-L)u_{2t} + u_{3t} \end{pmatrix}$$

will be autocorrelated. In fact, under the assumption that the u_{it}'s are serially uncorrelated, the MA processes on the v_{it}'s will have the following properties given that $\gamma_1, \gamma_2 \neq 0$ and $0 < |\beta| < 1$:

Error Term	Order of MA Process
v_{1t}	1
v_{2t}	2
v_{3t}	2

In empirical analyses, these implications regarding the orders of the MA error processes, that are lower bounds, can be tested empirically.

If the process for r_{Ht} is given by

$$(3.23)\qquad \emptyset_p(L)r_{Ht} = \theta_q(L)u_{4t}$$

where $\emptyset_p(L)$ and $\theta_q(L)$ are polynomials in L of finite degrees, p and q, respectively, and u_{4t} is a non-autocorrelated error term with zero mean, and constant variance, that is distributed independently of the structural error terms, we can determine properties of (3.23), including the values of p and q, from analysis of data on r_{Ht}. Further, if we premultiply the TF system in (3.21a) by $\emptyset_p(L)$ and substitute $\emptyset_p(L)r_{Ht} = \theta_q(L)u_{4t}$ from (3.23) we obtain the following FEs:

$$(3.24)\qquad |H_{11}(L)|\emptyset_p(L)\begin{bmatrix} r_{Yt} \\ r_{Mt} \\ \Delta i_t \end{bmatrix} = \begin{bmatrix} 1-(1-\beta)L \\ 1-(1-\beta)L-\gamma_1\beta L(1-L) \\ \beta L(1-L) \end{bmatrix}\theta_q(L)u_{4t} + \emptyset_p(L)\begin{bmatrix} v_{1t} \\ v_{2t} \\ 4_{3t} \end{bmatrix},$$

where $|H_{11}(L)| = 1-(1-\beta)L-\beta(\gamma_1+\gamma_2)L(1-L)$. It is seen that the AR parts of (3.24) are identical. Further properties of the FEs are indicated in Table 3.2.

Table 3.2

Selected Properties of FEs in (3.23)-(3.24)

Variable	Order of AR	Order of MA Error[+]
r_{Yt}	2+p	max(1+q, p+n_1)
r_{Mt}	2+p	max(2+q, p+n_2)
Δi_t	2+p	max(2+q, p+n_3)
r_{Ht}	p	q

[+] n_i, i=1,2,3, denotes the order of the MA process for v_{it}, i=1,2,3, respectively. Since, as shown above, the $n_i > 0$, i=1,2,3, the orders of the MA errors in the FEs are bounded from below.

Given that we determine the forms of the FEs and estimate them, it is possible to check whether the information in a sample of data is consistent with the implications of the model. If inconsistencies are encountered, the structural equations can be reformulated as shown in Zellner and Palm (1974). For example, the assumption of unitary income elasticity of real balances can be relaxed. Anticipations can be modeled in accord with Muth's rational expectations hypothesis, etc.

Empirical analyses of monthly data for the U.S., 1953-72, led to the following findings using large sample likelihood ratio tests and Box-Jenkins' techniques to discriminate among alternative ARMA processes.

Table 3.3

Empirical Findings for Final Equations[+]

Variable[++]	Period of Analysis		
	1953-72	1953-62	1963-72
i_t: (3MTB)	(2,1,3)or(3,1,3)	(2,1,3)or(3,1,3)	(2,1,3)or(3,1,3)
i_t: (4-6MPCP)	(2,1,3)or(3,1,2)	(2,1,3)or(3,1,2)	(2,1,3)or(3,1,2)
$\ell n\ M_{1t}$	(3,1,3)	(2,1,2)or(3,1,3)	(2,1,3)or(3,1,3)
$\ell n\ Y_t$	(0,1,4)	(0,1,4)	(0,1,4)
$\ell n\ H_t$	(0,1,0)	(0,1,0)or(2,1,0)	(0,1,0)

[+]Entries in the table in the form (p,d,q) denote a p'th order AR and q'th order MA process for the d'th difference of a variable, notation employed by Box and Jenkins.

[++]3MTB $\equiv$ 3 month Treasury Bill interest rate

4-6MPCP $\equiv$ 4-6 month Prime Commercial Paper interest rate

$M_{1t} \equiv$ currency plus demand deposits

$Y_t \equiv$ personal income

$H_t \equiv$ high-powered money

It is seen from Table 3.3 that the processes identified from 240 monthly observations, 1953-72, are all relatively low order schemes. In fact, that for $\ell n\ H_t$ is a random walk. Also, the identification of processes for the sub-periods, 1953-62 and 1963-72 produced results not far different from those obtained for the full period 1953-72.

Upon comparing the empirical results in Table 3.3 for the 1953-72 period with the model's implications set forth in Table 3.2, the empirical finding for $\ell n\ H_t$, a (0,1,0) process indicates that $p=0$ and $q=0$ in Table 3.2. Given $p=0$, the model implies that the AR parts of the FEs for i_t , $\ell n\ M_t$ and $\ell n\ Y_t$ should all be second order --- see Table 3.2. The empirical findings in Table 3.3 contradict this implication in that the AR part of the FE for $\Delta \ell n\ Y_t = r_{Yt}$ is of zero'th order, that for $\Delta \ell n\ M_{1t} = r_{M_1 t}$ is of order 3 and that for Δi_t is of order 2 or 3. Since these empirical results are not in accord with the model's implications, other variants of the model were formulated and checked against the empirical findings for FEs. A reasonable variant of the structural model that is consistent with the properties of the empirically determined FEs has been formulated.[11] Its TFs are currently being analyzed.

IV. Issues and Problems in Statistical Methodology.

The issues and problems of statistical methodology that arise in the present context are similar to those arising in many other contexts although here time series considerations are central. With respect to formulation of an initial model, it is the case that subject matter knowledge, data availability, and the objectives of a model-builder will be important. Also, the availability of mathematical, computer and statistical techniques often conditions the formulation of an initial variant of a model. Given that an initial variant of a structural model has been formulated, its FEs and TFs can be derived. Statistical problems arise in checking whether the implied FEs, TFs and other assumed properties of the structural equations are consistent with the information in the data. This range of issues has been called the problem of model identification by Box and Jenkins. Given that a model has been identified, problems of parameter identification, estimation and prediction arise. Below a brief review of some aspects of these problems in relation to the procedures described in Section II is provided.

4.1 Identification of Models.

As mentioned above, the problem of identifying models from the data is an important part of model construction. In connection with FEs, we have employed estimated autocorrelation and partial autocorrelation functions in the well-known manner suggested by Box and Jenkins (1970). These procedures have been very helpful in determining forms of individual FEs. While useful, the Box-Jenkins' procedures are somewhat informal. To be more formal, we have employed large-sample likelihood ratio tests to discriminate among nested models. That is, given alternative values of p, d and q, one can formulate associated likelihood functions, evaluate them at maximum likelihood estimates, and form $2 \ell n\lambda$, where λ is the ratio of maximized likelihood functions, that will have a large sample χ^2 distribution. As shown earlier, this procedure is quite operational.[12] However, it is a large sample procedure that works just for nested pairs of FEs. To handle the problem of both nested and non-nested FEs, we have suggested the use of large sample approximate Bayesian posterior odds ratios using an approach essentially similar to that put forward by Jeffreys and Lindley for approximating posterior odds ratios.[13] Further, it may be that spectral techniques can also be useful in identifying individual FEs.

Since error terms in different FEs are generally not independent, it would be desirable to have multivariate identification procedures for the set of FEs. This is somewhat complicated since the range of alternative models is often broad. However work is in progress to get large-sample likelihood ratio tests and approximate Bayesian posterior odds ratios relating to sets of FEs.

As regards identification of individual TFs, the Box-Jenkins' suggested procedures have not been too helpful in our work. We have employed large sample likelihood ratio tests and indicated how to obtain large sample approximate posterior odds ratios. As with the FEs, multivariate generalizations of these procedures would be extremely valuable.

With respect to the structural equations, it may be that some of the identification procedures suggested by Quenouille (1957) and Tse (1973) that utilize estimated serial correlation matrices will be found helpful. In addition, multivariate spectral techniques may be useful. To date, we have little information about the application and fruitfulness of these methods for identifying structural equation systems or classes of structural equation systems that are consistent with the information in a set of data.[14]

4.2 Estimation of Equation Systems.

In the process of identifying FEs and TFs using large sample likelihood ratio tests or Bayesian posterior odds ratios, parameter approximate estimates are needed. Then too, investigators are usually interested in inferring the values of parameters appearing in FEs, TFs and structural equations. Thus there is a need for good procedures for estimating the values of parameters appearing in these equation systems. To date work has been concentrated on obtaining estimates that have good large sample properties, either by sampling theory or Bayesian procedures. In large samples, it should be appreciated that maximum likelihood estimates are generally good approximations to Bayesian posterior means for any non-dogmatic prior distribution for the parameters. Thus, as Jeffreys has pointed out, maximum likelihood estimates have a large sample Bayesian justification relative to quadratic loss functions. Some work on estimating parameters of time series models employing "diffuse" and "informative" prior distributions for the parameters has appeared in the literature.[15] At present, we shall briefly review some operational large sample methods that we have employed in our past work and that we are developing.

With respect to individual FEs and individual TFs, we have been employing the Box-Jenkins' (1970) non-linear least squares approach to estimate parameters' values. In their approach, which they have given a Bayesian interpretation, the non-linear least squares problem is solved using Marquardt's algorithm.[16] In using this approach, we have found that it is necessary to be parsimonious with respect to introducing parameters. When "too many" parameters are employed, the likelihood function is often found to be rather flat and rather imprecise parameter estimates are obtained. Regarding TFs, use of "too many" input variables can cause similar problems.

Just as with sets of "seemingly unrelated" regression equations with correlated error terms, it is possible to get improvements in the efficiency of estimation by joint estimation of a set of FEs or a set of TFs.[17] The approach that we are developing for estimating sets of FEs involves the following steps:[18] (1) Estimate each equation using the Box-Jenkins' procedure and obtain estimates of the error terms, the residuals; (2) use the residuals so obtained to get a consistent estimate of the contemporaneous variance covariance matrix of the error terms, denoted $\hat{\Sigma}$; (3) insert lagged residuals in place of lagged error terms; and (4) apply generalized least squares, based on $\Sigma = \hat{\Sigma}$, to obtain estimates of the AR and MA parameters in the same way that "seemingly unrelated" regression coefficient estimates are obtained. Step (4) is a linear operation that can be performed readily using existing computer programs. A similar approach is being developed for application to estimation of sets of TFs. Since part (4) of the procedure involves work with a linear system, estimation with linear restrictions cutting across equations can be readily performed. Also, large sample tests of linear hypotheses can be carried through as with regression systems. That it is possible to test linear hypotheses cutting across equations and to impose linear restrictions across equations is important since as shown above FEs and TFs associated with structural models will often have identical AR parameters.

Estimation of parameters in the structural equations is possible if the structural parameters are identified. Hannan (1971) has provided identification conditions for parameters of AR structural equations with MA error terms. Given that these conditions are met, parameters of individual structural equations can be estimated consistently by an approach described and applied in recent work.[19] This approach is currently being extended to provide an operational method for jointly estimating parameters in a set of structural equations that resembles in broad outline the three-stage least-squares estimation technique.

4.3 Problems of Prediction.

How one obtains predictions depends on which equation systems have been satisfactorily identified. First, if just the FEs have been identified, they can be used to generate predictions of the future values of variables. Second, if TFs have been identified, they too can be used to obtain predictions given values of the input variables (the elements of $\underline{x}_t$) in future periods. When input variables' values in future periods are not known, it will be necessary to predict them using the FEs for the input variables or outside information. Third, if the structural equations' forms have been determined and estimated, it is possible to derive the associated FEs and TFs from the estimated structural equations that we refer to as the restricted FEs (RFEs) and restricted TFs (RTFs). The RFEs and RTFs can be employed to generate predictions. Further work is needed to appraise the relative properties of predictions yielded by the procedures indicated above.[20]

V. Concluding Remarks.

Elements of an approach to econometric model-building have been described above. It is recognized that many problems that arise in model-building have not been considered, for example missing and poor data, non-linearities, seasonality, possible changes in parameters' values, aggregation problems, etc. While much could be said about each of these problems, it was thought worthwhile to abstract from them in this paper. In terms of what has been presented, further work on methods of model identification, estimation and prediction would be valuable. Experience in applying the modeling approach described above to actual problems and assessing the quality of the resulting models in various dimensions is beginning to accumulate. That this is the case is most important since in the last analysis of the proof of the pudding is in the eating.

Footnotes

[1]See Nerlove (1966) for a description of early models of national economies. Hickman (1972) provides information regarding properties of several U.S. national models that are currently in operation. In Hamilton et al. (1969), a review of several regional models is provided. Econometric models of various agricultural markets have been developed and reported in technical bulletins of the Agricultural Marketing Service, U.S. Dept. of Agriculture. Other models have been reported in economic journals including Econometrica, Journal of Econometrics, International Economic Review, Review of Economics and Statistics, American Economic Review, Journal of Political Economy, etc. The Journal of Economic Literature provides abstracts and listings of papers appearing in a wide range of economic journals.

[2]See Zellner and Palm (1973, 1974).

[3]In many applications, it may be that the elements of $\underline{z}_t$ will be differences of original variables. As Box and Jenkins point out, differencing will often be useful for inducing stationarity.

[4]See Nelson (1972) and Zellner and Palm (1973, 1974) for examples of FEs that have been determined and estimated from quarterly and monthly economic data.

[5]If some of the exogenous variables are non-random, (2.5b) should be interpreted as relating just to the random exogenous variables.

[6]For a valuable discussion of transfer functions, see Box and Jenkins (1970) and Pierce and Mason (1971).

[7]Hannan (1971) has provided results relating to the identification of parameters of linear dynamic structural equations with moving-average error terms.

[8]See e.g. the discussion of Cooper's paper in Hickman (1972) and Nelson (1972).

[9]The model considered was formulated and estimated by Leuthold, MacCormick, Schmitz and Watts (1970).

[10]See Friedman (1970, 1971). This model and several variants of it have been analyzed using monthly data for the U.S. in Zellner and Palm (1974).

[11]See Zellner and Palm (1974).

[12]Large sample likelihood ratio tests have been employed in Zellner and Palm (1973, 1974) to aid in identifying FEs and TFs.

[13]See Jeffreys (1939, 1948, 1961, 1967), Lindley (1961), Zellner and Palm (1973) and Palm (1974).

[14] As Quenouille (1957) pointed out, different structural equation systems can have identical associated serial correlation matrices.

[15] See e.g. Box and Jenkins (1970), Newbold (1973), Shiller (1973) and Zellner (1971).

[16] We have employed a computer program developed by C.R. Nelson and S. Beveridge, Graduate School of Business, U. of Chicago.

[17] "Seemingly unrelated" regression systems have been considered in Zellner (1962, 1963), Zellner and Huang (1962), Telser (1964) and Parks (1967) from the sampling theory point of view and in Tiao and Zellner (1965), Zellner (1971) and Box and Tiao (1973) from the Bayesian point of view.

[18] This work is to be reported in a forthcoming paper by F. Palm and A. Zellner.

[19] See Zellner and Palm (1973).

[20] See Dhrymes (1973) for an interesting analysis of large sample properties of predictions yielded by "restricted" and "unrestricted" equation systems. McCarthy (1972) presents some finite sample results relating to "restricted" predictors for a static simultaneous equation model.

References

[1] Box, G.E.P., and G. M. Jenkins (1970): Time Series Analysis, Forecasting and Control, San Francisco: Holden-Day, Inc.

[2] ______ and G. C. Tiao (1973): Bayesian Inference in Statistical Analysis, Reading, Mass.: Addison-Wesley Publishing Company.

[3] Dhrymes, P. J. (1973): "Restricted and Unrestricted Reduced Forms: Asymptotic Distribution and Relative Efficiency," Econometrica, 41, 119-134.

[4] Friedman, M. (1970): "A Theoretical Framework for Monetary Analysis," Journal of Political Economy, 78, 193-238.

[5] ______ (1971): "A Monetary Theory of Nominal Income," Journal of Political Economy, 79, 323-337.

[6] Hamilton, R. J., S. E. Goldstone, J. W. Milliman, A. L. Pugh III, E. R. Roberts and A. Zellner (1969): Systems Simulation for Regional Analysis: An Application to River-Basin Planning, Cambridge: MIT Press.

[7] Hannan, E. J. (1971): "The Identification Problem for Multiple Equation Systems with Moving Average Error Terms," Econometrica, 39, 715-765.

[8] Hickman, G. B., editor (1972): Econometric Models of Cyclical Behavior (two volumes), New York: National Bureau of Economic Research, Inc.

[9] Jeffreys, H. (1939, 1948, 1961, 1967): Theory of Probability, Oxford: Clarendon Press.

[10] Leuthold, R. M., A. J. A. MacCormick, A. Schmitz and D. G. Watts (1970): "Forecasting Daily Hog Prices and Quantities: A Study of Alternative Forecasting Techniques," Journal of the American Statistical Association, 65, 90-107.

[11] Lindley, D. V. (1961): "The Use of Prior Probability Distributions in Statistical Inference and Decisions," in J. Neyman (editor), Proc. Fourth Berkeley Symposium on Mathematical Statistics and Probability, Vol. I, 453-468.

[12] McCarthy, M. D. (1972): "A Note on the Forecasting Properties of Two Stage Least Squares Restricted Reduced Forms -- The Finite Sample Case," International Economic Review, 13, 757-761.

[13] Nelson, C. R. (1972): "The Predictive Performance of the FRB-MIT-PENN Model of the U.S. Economy," American Economic Review, 62, 902-917.

[14] Nerlove, M. (1966): "A Tabular Survey of Macro-Econometric Models," International Economic Review, 7, 127-175.

[15] Newbold, P. (1973): "Bayesian Estimation of Box-Jenkins Transfer Function-noise Models," Journal of the Royal Statistical Society, Series B, 35, 323-336.

[16] Palm, F. (1973): "On the Bayesian Approach to Comparing and Testing Hypotheses When "Knowing Little"," manuscript, H.G.B. Alexander Research Foundation, Graduate School of Business, U. of Chicago.

[17] Parks, R. W. (1967): "Efficient Estimation of a System of Regression Equations When Disturbances are Both Serially and Contemporaneously Correlated," Journal of the American Statistical Association, 62, 500-509.

[18] Pierce, D. A. and J. M. Mason (1971): "On Estimating the Fundamental Dynamic Equations of Structural Econometric Models," paper presented to the Econometric Society Meeting in New Orleans.

[19] Quenouille, M. H. (1957): The Analysis of Multiple Time Series, London: C. Griffin and Company.

[20] Shiller, R. J. (1973): "A Distributed Lag Estimator Derived From Smoothness Priors," Econometrica, 41, 775-788.

[21] Telser, L. G. (1964): "Iterative Estimation of a Set of Linear Regression Equations," Journal of the American Statistical Association, 59, 845-862.

[22] Tiao, G. C. and A. Zellner (1965): "On the Bayesian Estimation of Multivariate Regression," Journal of the Royal Statistical Society, Series B, 26, 277 285.

[23] Tse, E. (1973): "Report of Work in Progress," presented to a special seminar, Graduate School of Business, U. of Chicago.

[24] Zellner, A. (1962): "An Efficient Method of Estimating Seemingly Unrelated Regressions and Tests for Aggregation Bias," Journal of the American Statistical Association, 57, 348-368.

[25] ______ (1963): "Estimators for Seemingly Unrelated Regression Equations: Some Exact Finite Sample Results," Journal of the American Statistical Association, 58, 977-992.

[26] ______ (1971): An Introduction to Bayesian Inference in Econometrics, New York: John Wiley and Sons, Inc.

[27] ______ and D. S. Huang (1962): "Further Properties of Efficient Estimators for Seemingly Unrelated Regression Equations," International Economic Review, 3, 300-313.

[28] ______ and F. Palm (1973): "Time Series Analysis and Simultaneous Equation Econometric Models," presented to the Oslo Meeting of the Econometric Society, August, 1973, to appear in the Journal of Econometrics.

[29] ______ and ______ (1974): "Time Series and Structural Analysis of Monetary Models of the U.S. Economy," manuscript, H.G.B. Alexander Research Foundation, Graduate School of Business, U. of Chicago.

Arnold Zellner
Graduate School of Business
University of Chicago
Chicago, Illinois 60637
U.S.A.

Applied Statistics, R.P. Gupta, (editor)

THE UNIVERSITIES' RESPONSE TO THE INCREASING DEMAND FROM GOVERNMENT AND INDUSTRY FOR APPLIED STATISTICIANS

A Panel Discussion

K. L. Weldon, Dalhousie University (Chairman): Five panelists from Universities, Government and Industry have agreed to present their views on this topic. I am sure you will find their views stimulating, and possibly even provocative. Following their presentations, the discussion will be opened to the floor.

Perhaps I should make a brief comment on the topic selected for discussion. It includes a presumption that the demand for applied statisticians from government and industry is actually increasing. Unfortunately, my evidence for this is merely anecdotal, and I would like to say in advance that if it is found thay my presumption is incorrect, then hopefully the proposals for responding to the presumed increase in demand for applied statisticians could be reversed! Meanwhile, we may assume the demand is in fact increasing.

Regrettably, the comments from the floor from Drs. H.A. David, W.T. Federer, A.G. Laurent, John W. Tukey and others were not recorded.

A.W. Douglas, Department of the Environment: Before making a few points about the use of applied statistics in the federal government and the training needs of applied statisticians, I should define the type of statistical activity I have had the most experience with and about which I will talk. My remarks will be most relevant to the activities and needs of general statistical consulting groups - that is, groups which provide statistical advice, and sometimes computing assistance, to clients on a wide variety of problems in, usually, several subject matter areas. Groups such as this exist in the Departments of Agriculture, National Helath and Welfare, and Environment, the Department of which the Branch I represent is a part. Other Departments have applied statisticians in groups or singly which work in more restricted areas - the most obvious of these being the sample survey group in Statistics Canada.

To put my remarks in their proper context a few remarks about the organization I represent might be useful. Our Branch, which is part of the Computing and Applied Statistics Directorate, was first established in the Department of Forestry in 1961 as a central statistical and scientific computing group for the whole Department which had research establishments right across the country.

This was perhaps a necessary arrangement in those days because of the shortage of statisticians, but it quickly became evident to us that good statistical consulting, and to an even greater extent good data analysis, could not be carried on at a distance. Because of this we, as the supply of statisticians increased, have been promoting the acquisition, for each regional Departmental establishment that can justify it, a statistical and scientific computing capability. I mention both statistics and scientific computing since in the world of real data one cannot these days exist efficiently without the other. For our part we act as a consulting group for most of the National Capital area establishments and undertake activities which promote communication and co-operation among field statisticians and, in general, promote the cause of statistics in the Department.

Thus, ideally, every establishment which needs a statistician would have one in-house. This ideal, naturally, has not been realized.

What kind of people are we and the regional directors looking for to fill these positions? Ideally, once more, statisticians need a solid background in theoretical statistics, training or experience in the application of statistical theory and techniques where standard statistical assumptions may not be fulfilled, a good general knowledge of the subject matter area(s) involved, and an appreciation of the constraints imposed on experimentation in this field. Now that doesn't seem like too much to ask, does it?

The question is, in the training of statisticians to fill positions like these, how close can we come to the ideal? It may not be my place as a non-academic to suggest ways in which the ideal can be approximated, but, anyway, for discussion purposes, I'll toss out a few ideas, most of which are not new.

It has been our experience that the theoretical training of most honest-to-god statisticians emerging from graduate school has been adequate, but very few have had experience or training in applying this knowledge to real life situations. How can students acquire this training? Or, indeed, is it the function of the university to provide it?

Assuming the university wishes to accept this function it could
(1) require that, say, graduate students as part of their training participate in a consulting service to fellow graduate students or staff under the guidance of a statistician staff member.
(2) introduce more comprehensive longer term statistical problems based on actual subject-matter situations into statistics courses.
(3) Where class sizes permit, use the tutorial system to advantage in this regard.

(4) introduce statistics courses which bridge the gap between theoretical courses and methods courses with more emphasis on the development of techniques under constraints, robustness properties, etc.

Students of mathematics and statistics should also be encouraged to take courses in subject-matter areas, and as summer employment, or part-time employment during the academic year, spend some time on site working with subject-matter specialists in order to gain an appreciation of their problems and to learn their vocabulary, as well as teaching them the vocabulary of statistics.

Each of these suggestions implicitly assumes that the university professors involved have some practical experience in order to carry out these programs successfully.

For a number of reasons this experience exists in only a small number of cases. One of the principal reasons, I believe, is that the reward system does not encourage the academic to gain practical experience. The "publish or perish" system makes a professor reluctant to take on practical problems which may not lead to publication in a statistical journal. Co-authorship of a paper in a subject-matter journal doesn't seem to be worth anything in terms of tenure or promotion.

Therefore, the first step, it seems to me, is to create an environment which encourages _interested_ teachers of statistics to become involved with practical problems. Once that is done (1) University statistical consulting groups could be established to provide assistance to subject-matter people in other university departments, and to local private or government institutions, _or_ at least, (2) individual professors would be more inclined to work on contract during the summer or academic year with organizations such as those mentioned.

Perhaps a word of caution is needed here. Consulting services such as those mentioned will be successful only if the consultant is willing to take second place to the subject-matter specialist: the statistician must remember that first priority is the solution of the client's problem, not that the problem if altered might provide a good topic for one of his graduate students or might result in a statistical paper.

The most common complaints we hear regarding university consultants, in addition to the one just mentioned, are that they don't take time to really understand the researcher's point of view and problem constraints, and they can't explain concepts and methods in language the client can understand.

The statistician, of course, has his own list of complaints about his clients.

But, I'm sure, none of these problems is insurmountable if the interest on both sides is genuine, and both parties are willing to spend the time to break down the barriers of communication and prejudice that now exist.

I hope my remarks haven't harped on the obvious too much. If so, you have at least identified another individual who agrees with you, and if not, then perhaps my remarks will stimulate some discussion.

G.D.J. Edwards, Science Council of Canada: ABOUT THE MATHEMATICS STUDY. The Science Council of Canada has contracted out, to a consortium of six Canadian mathematical organizations, the task of assessing the changing role of the mathematical sciences in Canadian education, science, industry and government. In a nutshell, our job is to look at the universe of mathematical activity as an ecosystem and to suggest possible linkages and feedback mechanisms which might help people at various levels of responsibility to coordinate and integrate their activities.

As you know, the influence of the mathematical sciences has expanded enormously in the last few decades. Mathematical methods have become increasingly important in such diverse areas as biology, economics, medicine, sociology, government, and industry. The modern science of statistics and the electronic computer have, of course, played a central role in these developments.

Unfortunately, this expansion in mathematical activity has been accompanied by a regrettable fragmentation among those who are professionally involved with mathematics. There are so many pockets of mathematical activity - high school teachers of mathematics; university professors; reseachers in the physical sciences, in the biological sciences, in the social sciences; operations research men and applied statisticians in government and industry - and there is so little communication between these different groups!

In particular, there is not nearly enough feedback from those who are actually using mathematics to those who are providing advanced mathematical training. During the period of rapid university expansion, most university graduates in the mathematical sciences were absorbed back into the educational system at one level or another. Now that the demand for mathematics teachers is levelling off, this pattern is changing - the majority of graduates in the mathematical sciences are morelikely to become users of mathematics or mathematical consultants.

Thus the feedback problem is of curcial importance; How can you train people for a job that you yourself know nothing about?

The Mathematics Study is proceeding on three fronts. Firstly, a questionnaire has been sent out to all graduates in the Mathematics Sciences from Canadian universities in certain selected years, in order to find out where they are now, what use they are making of their mathematical training (if any), and what they think of their mathematical training in retrospect. Secondly, we have held a series of eight one-day seminars involving academic mathematicians and users of mathematics in government and industry. These were usually focussed on specific themes - e.g. Mathematics and the Transportation Industry, Mathematics in Financial Institutions, Mathematics and the Resource Industries. On March 4, 5 and 6, we had three consecutive seminars on the role of mathematics in the fedreal government, under the following headings: Mathematics and Policy Planning, Mathematics, Statistics, and the Environment, and Mathematics and Technology. Thirdly, we have called for briefs from all and sundry, and have received more than two-hundred submissions from individuals and groups across Canada dealing with all conceivable aspects of the Mathematical Sciences in Canada.

PERSONAL EXPERIENCE. I should make it clear at the outset that I am not a statistician. My function here this morning is just to pass on to you some observations that have been made by others in the course of interviews, during the one-day seminars, or in written briefs.

Before I begin, however, I would like to say something about my own experience. As a student of applied mathematics at the University of Toronto, I took one course in Statistics in my third year. It included such topics as hypothesis testing and analysis of variance. I must confess that I never really understood these concepts, and I am still a little mystified by the whole subject. Perhaps a bit of history as to how and why these techniques developed, or at least some well-chosen motivating examples, would have helped to eliminate the obscurity. A consideration of possible uses, common abuses, and fundamental limitations of these techniques would have helped even more. As it was, I cam away from statistics without feeling that I really understood what it was all about. I imagine that many less mathematically-inclined students have come away from the subject with far worse feelings. I urqe all of you who are teachers of statistics to bear this in mind: that most of your students have no idea of what you are trying to teach them, or why. The subject has to be put into context for them before it begins to make sense to them.

THE NEED FOR APPLIED STATISTICIANS. At the present time, there seems to be a genuine shortage of applied statisticians here in Canada. Statistics Canada is forced to look outside the country for suitably trained applied statisticians, in spite of a policy of hiring Canadian. Even the universities are having difficulties in finding enough applied statisticians to fill vacancies in their departments. There is also an evident need for statistical consultants of various kinds. In a Panel Report on Biomathematics which prepared a few years ago (in connection with a Science Council Report on Basic Biology), the difficulties that many biologists face in obtaining statistical assistance is graphically documented. The same is true in the social sciences. Most large government departments have statistical consulting services, but ther are too often understaffed and over-worked. Research teams and model-building projects should involve statisticians from beginning to end to ensure the soundness of the statistical methodology; in many cases, however, the statistician is either not consulted at all, or he is asked to salvage someone's thinking after the fact.

So there certainly is a need for applied statisticians. However, this remark has to be qualified in two ways. First of all, "need" is not the same as "demand". As Patrick McTaggart-Cowan is fond of saying, "Demand is that portion of the need which is funded". Or, to paraphrase one of the government delegates to our March 5 Seminar, "What we need and what we get are two different things." The second point is that the need for statisticians depends very much upon the attitude of the person who calls himself a statistician. It isn't just a simple numbers game of "supply and demand", because there are significant additional requirements beyond mere competency which must be met in order for a statistician to be genuinely useful in applied work.

QUALITIES NEEDED. First of all, an applied statistician has to know how to cope with unstructured problems. Employers of mathematicians (not just statisticians!) have frequently remarked that most university graduates in the mathematical sciences seem unable or unwilling to cope with unstructured problems. These graduates expect to get their problems served up on a platter in a nice, neat, mathematical formulation. When they get into a job in government or industry, they often undergo a kind of culture shock; all of a sudden they find out that they are expected to formulate the problems as well as to solve them! Apparently, it never occurred to them that anyone capable of formulating a problem in mathematical terms could probably solve it as well - for only a person with considerable mathematical insight is able to cast a problem into an appropriate mathematical context.

Communication skills are also vitally important and frequently lacking. Statistical consulting work requires people who have the patience to listen to a non-mathematician describe his problem and the imagination to grasp what is relevant and what is not. It also requires people who can interpret results in simple, non-technical language to managers and decision makers who are often mathematically illiterate. As one government statistician recently observed, "A manager would rather live with a problem that he can't solve than with a solution that he can't understand". Similar considerations arise in scientific consulting owrk, where it is often necessary to take a scientist who is very competent in his own field but who has very little understanding of mathematics, and teach him enough statistics so that he can understand the implications of what he is doing. On a still more fundamental level, the applied statistician has to be prepared to "sell" his ideas by actively attempting to persuade others of the usefulness and the necessity of proper statistical methodology.

A third requirement for successful applied work is the ability to work as a peer in large teams. This requires patience, perseverance, and a willingness to cooperate with others. On the verbal level, it means being able to choose an appropriate level of discussion in order to maintain rapport with colleagues. In terms of writing, it means being able to express ideas clearly and concisely, to document conclusions at an appropriate level of detail, and to summarize. Some understanding of human psychology, the struture of bureaucracies, the mechanics of projects, and group dynamics is also extremely valuable.

UNIVERSITY TRAINING. These three observations have a number of implications for university training in the mathematical sciences. As indicated above, the major complaint from employers of mathematicians and statisticians is not about the content of the courses which are given in university, but about the students' basic attitude and orientation. Graduates in the Mathematical Sciences are not prepared to work on unstructured problems; they are not adept at communicating with non-mathematicians; they often underestimate the non-mathematical difficulties which are encountered in practical problem-solving. In short, most graduates in the mathematical sciences have unrealistic expectations with regard to non-academic employment

- expectations which cause frustration and disappointment all around.

None of the people I spoke with wants the university to become just a training school for industry. They are the first to insist that the primary function of a university should be to teach students how to think and to give them a broad, flexible education which can be adapted to a multitude of purposes. And that's the heart of the problem, because at present, university mathematical training is <u>too narrowly oriented towards an academic career of teaching and research</u>. Education should be much broader than that. It should include material that may be relevant to other types of occupations as well. Even those who become academic mathematicians would benefit from a froader education, since they will eventually be teaching non-mathematical students in various service courses. It would be helpful if they, as teachers of students in other disciplines, had a more realistic understanding of the way in which mathematics is applied in other fields of endeavour.

<u>SOME RECOMMENDATIONS</u>. A number of things can be done in order to provide students with some experience in coping with unstructured problems and in developing communication skills. Visitors from other university departments, from government, and from industry, can be invited to give seminars, lectures, or workshops. Group projects can be assigned involving real life problems, which might require some communication with subject-matter people in other disciplines. A university consulting service can be established through which senior level students can obtain practical experience under the supervision of faculty members.

Relevant work experience can be provided through summer employment or through cooperative programs such as those offered by the University of Waterloo. Joint curricula and joint seminars can be designed in conjunction with other departments. And at the post-graduate level, theses which are related to problems arising in other fields can be encouraged.

Of course, all of these suggestions have to be taken with a grain of salt. The object is not to "turn the tables", but to strike a balance between theoretical investigation and practical application. Obviously, some students will be more theoretically inclined and others more practically inclined - but this should be a matter of emphasis rather than a difference of kind. In fact the division between "pure" and "applied" is largely fictitious, for in many practical applications, the basic assumptions underlying the standard statistical techniques simply do not hold. In such cases, a good knowledge of statistical theory is essential. However, a statistician has to have the right attitude - he has to be ready to get his hands dirty before the will be able to bring his theoretical knowledge to bear on the problem. And in the final analysis, attitudes are the same as expectations, and expectations can be affected by educational experiences.

INCENTIVES FOR UNIVERSITY FACULTY. If university mathematical education is to become more diversified in order to meet the diverse needs of today's students, mechanisms must be found which will encourage professors of mathematics and statistics to gain some practical experience in other fields. They can hardly be expected to communicate attitudes or experiences to their students which they themselves have never had. In order to make this diversification possible, there sill have to be a serious reassessment of the reward system for promotion, tenure, and research grants. It may have to be accepted as a fundamental principle that people should get rewarded for doing work that needs to be done - including such things as consulting work, exposition, problem-solving, and interdisciplinary research. Mechanisms for evaluating this kind of activity will have to be developed over a period of time as experience is gained in these areas.

One of the most frequent suggestions that we have encountered is for professors in the Mathematical Sciences to spend sabbaticals or leaves of absence in other university departments, in government offices, or in industry. The NRC program of Senior Industrial Fellowship should make this a relatively easy thing to do, provided the reward system can accommodate this kind of activity and deem it as useful and necessary work. It has also been suggested that publications of a mathematical nature appearing in applied journals should be given more weight in awarding research grants. One participant at the March 5th seminar recommended the publication of a new journal which would carry "middle ground" articles offering constructive criticism of current statistical methodology in various fields, as well as articles of a purely expository nature. Still another participant suggested the establishment of an informal body to act as a clearing house or brokerage. Its function would be to put users of mathematics in touch with academics who are willing to do consulting work in the appropriate field of interest.

But whatever mechanisms are proposed or adopted, the fundamental issue is that any activity which is considered to be beneficial to the statistical community or to the mathematical community as a whole, should carry its own rewards in the form of advancement and prestige. Otherwise, such activities will suffer from neglect, and the entire profession will be the worse for it.

J. R. Rutherford, Dupont of Canada: Let me tell you about my background a little bit to show you how the views I will present today have developed. For the past six years I have been a consultant statistician in a science and technology environment with Dupont of Canada. The previous eight years I was getting Ph.D. and was teaching statistics at University. With this experience I think I have come to understand the viewpoint of the academic statistician and of the applied statistician.

Three years ago at Statistics Canada 71 in Montreal I was on a panel similar to this one. At that time I had perceived that there were differences in the two worlds - academic versus applications - but I was then unsure of what might be the trouble. This unsureness caused me to be frustrated and the resulting hostility to the academic world showed so much that, after a particularly snide comment by me, the panel chairman said that I "must have been bitten by a professor". Today the hostility should not be so apparent because I am now sure what the trouble is, what the differences between the two worlds are. I am very sure that I know the cause of the problem. I do not claim to be right as to my definition of the problem, only sure there is a difference.

As a basic premise for this discussion I state that the role of the educational system is to train people to function effectively in society and I believe that universities are a part of the educational system. I will qualify the work "effectively" later.

Academics live in a world in which certain types of intellectual activity are highly valued but effective functioning in society requires only a small amount of the academic-type intellectual skills and a very large amount of other skills, both intellectual and non-intellectual. On the non-intellectual side, the ability to deal with people, and to make love are perhaps the most important. For academics at university to be effective trainers they must find out what intellectual skills are needed and they must learn how to teach these skills. In the rest of this talk I will try to make it clear what I think these skills are in the context of applied statistics.

I define a good applied statistician as being a person who is an astute observer, who is problem oriented and who can think statistically. I will discuss the key phrases in this definition in succession but first I will describe a dichotomy of intellectual activity, a dichotomy which is extremely important and one whose existence must be explicitly recognized by all teachers.

The dichotomy is between creativity and I.Q. Everybody knows that IQ is measured by an IQ test but there is a thing called creativity which is measured by creativity tests. In the same sense that IQ tests do not measure all that is important about a person's intelligence, a creativity test does not measure all that is important about a person's creativity. However, what a creativity test does measure is a very important constituent of a person's intellectual make-up.

It is easiest to show what these tests measure by giving examples of the type of question. An IQ test will have a question posed like: "which of these five words best completes the sentence?" A creativity test is the converse "Here is a standard fable, invent three new endings to show a moral, happiness and sadness." How many definitions of the work "bolt" can you think of?" "How many uses can you suggest for a brick?" On all of these questions, the originality and unusualness of the answer contributes to the test score. A very norrow thinking conformist would say, for the last question: "you can build a brick wall, a brick floor, a brick garage, a high brick wall and a low brick wall, etc." A nonconforming creative thinker might say "You can tie a note to a brick and throw it through a window. You can hit somebody on the head with it to get their attention", and so on. People who would give these two types of answers use different intellectual process in response to the posed questions.

Of great practical importance is the fact that IQ and creativity are independent, are uncorrelated. You all know somebody who can learn a vast array of facts from a book and who can recall them instantaneously but this same person has never put these facts together in a new way in his life - briks are for building walls and that's that. On the other hand there are people who give every appearance of being quite stupid but who regularly (but with large time intervals) come up with valid solutions for hitherto unrecognized problems. Problem recognition is a creative activity and is one characteristic of creative people.

I think that the educational system is biased too strongly towards the development of IQ and does not place enough emphasis upon the development of the creative thinking processes. One approach to teaching which I think rather close to the usual university type teaching is described by Thomas Gradgrind, a school principal in a Dickens' novel. "Now what I want is, FACTS. Teach these boys and girls nothing but FACTS. FACTS alone are wanted in life". I guess I have been unable to contain all my holtility.

Contrasted with the Gradgrind approach is that in which "there is cmphasis upon the transfer of training from one subject to another, the searching for common principles in terms of which facts from quite different domains of knowledge can be related, the stressing of analogies, and similies, and metaphors, a seeking for symbolic equivalents of experience in the widest possible number of sensory and imaginal modalities, exercises in imaginative play, training in retreating from the facts in order to see them in larger perspective and in relation to more aspects of the large context thus achieved - these and still other emphases in learning would, I believe, strengthen the disposition to intuitive perception as well as intuitive thinking."

This is a quote from one item in a vast literature on the subject of creativity. The best review and summary I know of is an article I wrote "Problem Recognition, Creativity and Statistics". Anyone who wants a copy can write to me a Dupont of Canada, Kingston. I think that university teachers should organize their teaching so that the requirements for the development of creative thinking are explicitly considered and are provided in as pervasive a way as is possible.

In the context of the good applied statistician as I have defined it, creative thinking is essential. Creative people are natural observers, they can see the forest in spite of the trees. They have the ability to make value judgements, they can decide what is important and what unimportant. You all know of some students who do an analysis of a course before they do any studying. They first decide what the professor thinks is important then a proportion of this is learned for the exam. You may on occasion consider that this type of student is getting less than full value for his education dollar. I would argue otherwise - he is learning how to distinguish the important from the unimportant. Creative people can toy with concepts - Einstein called it combinatory play. A creative person is naturally inventive.

Now finally considering the key phrases of my definition of an applied statistician:

<u>Astute Observer</u>. It is essential for most people to be trained in some science or a field like economics if they are going to be able to see things. Most people can profit as observers by training in observation.

Problem Oriented. There are several points here. Problem orientation is the antithesis of disciplinarity. A student who has just completed, say, three years of study for a Ph.D. in statistics will naturally believe that statistics is as important to the rest of the world as it has been to him, he will consequently look at all problems as being opportunities to prove the importance of statistics. Disciplines exist for the convenience of academics, but real world problems have an existence of their own and their solutions do not recognize discipline boundaries.

The lack of problem orientation is partially a function of the value system of the graduate. If the graduate looks upon a newly presented practical problem primarily as being a likely source of a scholarly publication, the scientist's problem necessarily takes second place. Good scientists will not tolerate such attitudes for long.

Statistics as an academic discipline has rather narrow and well-defined limits. A practicing statistician has to be willing to accept much wider responsibility than he might be lead to expect by his academic mentors. In a typical year I might be involved with perhaps forty experiments. An engineer in a year might be involved in one or two. Thus in a yera I would be involved in the planning and design of experiments more times than would be an engineer in a career. I should become a relative expert in problem definition, objective selection, choice of factors etc. These are aspects of planning which are defined by academic statisticians to be the responsibility of the scientist and not of the statistician. But it is these scientific aspects which are most difficult to do properly. The statistics is often trivial when these are done properly and the statistics is impossible when they are not. The graduating statistician must leave university with the expectation that he will accept these wider responsibilities and it is only resonable to expect that the university would provide some training to make him intellectually capable of doing so.

In the paper I presented yesterday "The Design of Experiments and Data Analysis: A Scientific Approach" I have presented several methods and techniques which I think other statisticians could well learn to use to help them discharge these wider responsibilities.

Statistical Thinking. Finally I have come to what this conference is about. I define statistical techniques as being those which are designed to remove or prevent obscurities caused by error, bias, incompleteness, inconsistencies, confounding, etc. from sets of data so that ordinary deterministic type thinking is made possible. A creative statistician will use a statistical technique which is appropriate for the problem being considered.

In summary then, academic statisticians must learn to teach statistics in such a way that creative, statistical-thinking people are graduated. In relation to the role of the educational system in training people to be effective in society, the word "effective" can be equated with "creative".

A. J. Tingley, Dalhousie University: I will begin by recalling a few historical facts which have a bearing on the topic under discussion. The first statisticians were tax collectors -- employees of the state. The intellectual development of statistics resulted from the application of mathematics and mathematical thinking to the problems of data analysis. Probability, as distinct from statistics, received stimuli from gamblers, and later, commerce, e.g. insurance companies. The military needs for mathematical work during the Second World War gave stimulus to many aspects of applied mathematics including applied statistics and probability. Much of the applied mathematics could not possibly involve the analysis of real data: for example, in modelling the frequency of contact between Allied ships and German submarines in North Atlantic, before such contacts occurred, required probabilistic thinking with pencil and paper, to deduce the implications of parameters taking certain values. Certainly many useful results can be deduced on the basis of probability calculations alone, and not only from the analysis of real data. This being the case, applied statisticians will inevitably meet problems which do not involve statistics, but rather probability, and should have the mathematical training sufficient to be applied probabilists on occasion.

If the university graduates a rather pure breed of theoretical mathematician, this may not be all bad. Some of these will develop an interest in applied work, and will be admirably equipped by their background to become applied mathematicians.

The previous speaker (Dr. Watts) said that statisticians needed to have their own department in a university in order to allow the dependent development of statistics as a discipline distinct from mathematics. I do not think this is an important consideration. Perhaps the statisticians at Queen's feel too insecure to promote the development of statistics from within the Mathematics Department. Local conditions and personalities are no doubt the determining factors as to the best organization. The statisticians at Dalhousie have not run into any resistance from other mathematicians in achieving their goals, as far as I know. Indeed, cooperation has been frequent and effective, and resources have been allocated to statisticians on what I regard as a favourable basis. (Of course, as the recent Chairman, largely involved, I undoubtedly have a bias!) For example, several interdepartmental seminars have been organized on statistical topics, a new course in Statistical Consulting has recently been started, plans are being made to form a faculty-wide statistical consulting centre based right in the Mathematics Department. Moreover this conference has been partially supported by the Mathematics Department.

To many mathematicians, the new emphasis on applications in mathematical training is a compromise of intellectual rigour and a threat to the production and appreciation of elegant results. I personally regret to some extent that the days of the "ivory tower" have slipped into the past. Twenty years ago, the university was financed to a large extent by donations from wealthy families, and it is not surprising that its curriculum was designed to satisfy the intellectual needs of the sons and daughters of these wealthy families. These needs were for an "education", when "education" was considered to mean intellectual discipline and a familiarity with the great ideas of history. This is not the same as training for a specific profession, a concept which receives more emphasis today. When the university was privately supported, it supported the needs of the supporters, and this meant, in mathematics, a training in pure mathematics.

But this is no longer the case. The public at large supports the university through our governments, and so the university must serve a broader segment of society. Government and Industry are encouraging universities to undertake activities related more to applications, and an appropriate development in line with this shift in emphasis is the provision of programs to train applied statisticians. While I believe we can, and must, retain all the opportunities for pure mathematics that we have had in the past -- for this is the basis of all applied work -- we must recognize the new demands. If the new rules of the game call for a shift of emphasis to applications, then we must be prepared to adapt to these new rules. This situation reminds me of the following verse:

There was a young lady from Kent
Who said that she knew what it meant
When invited to dine
Given food, lots of wine,
She knew what it meant -- but she went!

We must accept the consequences of having gone, so why not enjoy it? Applied mathematics can be satisfying and enjoyable.

D. G. Watts, Queen's University: I have been asked on short notice to fill a vacancy on this panel on "The response of Canadian Universities to the demands for applied mathematicians". I have not been aware of any significant response by Canadian Universities and so cannot really address myself to the topic. However, I think I can contribute to the discussion by suggesting one important way in which Canadian Universities *should* respond, and more importantly how we *statisticians* should respond.

It is evident from what previous speakers have said, in particular Gordon Edwards, that there is a considerable demand for 'the complete problem solver' - the kind of person who can listen to a scientist, formulate the scientist's problem in a abstract sense, solve the abstract problem, interpret the solution in the original terminology, and finally communicate that solution so that the scientist fully understands and appreciates the solution. Now that description very accurately defines a competent well-rounded statistician, in my view, but I don't see many (if any) places in Canada where such a statistician is being produced. Why?

The main reason, it seems to me, is that statistics in Canadian Universities is virtually *invisible*! For example, almost *all* of the statisticians in Canadian universities are housed in Mathematics departments. And virtually one-hundred percent of all statistics courses are given by departments other than statistics. Who is responsible?

Well, it seems to me that Canadian professors of statistics are mostly to blame. We have allowed ourselves to be bought out by Mathematics departments without giving any thought to what we get and the price we have to pay. On the positive side, we statisticians have accepted positions in Mathematics departments because it is easy to become a resident in an established department, we don't have so many administrative duties, and (as remarked to me at this Conference!) we don't have to fight for a budget. But what is the price?

We professional statisticians appear to have all the advantages and at no cost. The invidious aspect of this transaction, however, is that all the cost is borne by the profession. And this cost manifests itself in many debilitating ways. First, I believe the name of a department is important in conditioning the type of personnel it attracts. Thus the type of statistician who will generally be attracted to or recruited by a Mathematics department is quite different from that attracted to a Statistics department. The same is even more true for the types of student especially if one considers the very real barriers imposed by the "prerequisites". We thus force a subtle but strong influence on statistics as a profession.

Second, peer influence is extremely important on conditioning the personnel. It takes a strong secure individual to eschew peer values and "do his or her own statistical thing" in a Mathematics department. Not only that, but in my experience a very large majority of scientists believe that most mathematicians are of little value to them in solving their real problems. Consequently a statistician in a Mathematics department will be consulted less than if the statistician were in the Statistics department. I believe that fact exerts a subtle but strong influence on statistics as a profession.

Third, we exert strong influences on students who take such courses as "Mathematics XXX, An introduction to statistics" because there are never enough good statisticians to teach all the statistical service courses. At Queen's for example, approximately 40% of the Mathematics department student throughput in second year is in statistics courses and over 30% each in third and fourth years. The statistics faculty should therefore comprise about 35% of the Mathematics department, whereas in fact statisticians comprise about 10%! Statistics as a profession is therefore exploited by the Mathematics department, because it offers statistics courses which it is not able to staff. But even more cruel is the fact that the service courses are manned by non-statisticians who have no real understanding of statistics and, further, no experience or interest in data analysis! Now in my opinion, an introductory statistics course for a general audience is one of the most difficult courses to give, and should be given only by experienced data analysts. And consider what the statistics course enrolments would be if introductory courses were taught by interested and interesting, motivated and motivating statisticians!

An additional high cost which the profession of statistics pays when there is no statistics department, is that statistics has no input to the main propaganda organs of the academic system - the university calendars. Statistics neatly becomes a non-discipline, which cannot attract students, and which has no academic status.

In summary, by neglecting our <u>profession</u> we Canadian statisticians pay dearly. Without a statistics department, members of the profession at a university:

1) are subject to subtle but strong influences which affect the development of the profession of statistics,
2) have little control over statistics programs and the admission requirements for statistics programs,
3) have no power to monitor the quality of statistics courses,
4) have no budget wtih which to control the hiring of statisticians,
5) have no procedure (in the current academic scheme) to advertize statistics courses to attract students

These are vital factors which affect the whole development of statistics - and if we Canadian statisticians don't do anything to further the profession in Canada, <u>no</u> <u>one</u> <u>else</u> <u>will</u>!

<u>K.L. Weldon</u>: I would like to thank the panelists for their presentations this morning. I record my thanks to Dr. D.G. Watts who was persuaded to contribute to the Panel after being asked just moments before this session began. My special thanks goes to Dr. R. P. Gupta, organizer of this conference for asking me to chiar the panel discussion.